Molecular Genetics of Bacteria

4th Edition

Molecular Genetics of Bacteria

4th Edition

Jeremy W. Dale
Simon F. Park
University of Surrey, UK

John Wiley & Sons, Ltd

The Atrium, Southern Gate, Chichester,
West Sussex PO19 8SQ, England

Telephone (+44) 1243 779777

Email (for orders and customer service enquiries):
cs-books@wiley.co.uk
Visit our Home Page on www.wileyeurope.com
or www.wiley.com

Other Wiley Editorial Offices

John Wiley & Sons Inc.,
111 River Street, Hoboken, NJ 07030, USA
Jossey-Bass, 989 Market Street, San Francisco, CA 94103-1741, USA

Wiley-VCH Verlag GmbH, Boschstr. 12, D-69469 Weinheim, Germany

John Wiley & Sons Australia Ltd,
33 Park Road, Milton, Queensland 4064, Australia

John Wiley & Sons (Asia) Pte Ltd,
2 Clementi Loop #02-01, Jin Xing Distripark, Singapore 129809

John Wiley & Sons Canada Ltd,
22 Worcester Road, Etobicoke, Ontario, Canada M9W 1L1

Library of Congress Cataloging-in-Publication Data

Dale, Jeremy.
Molecular genetics of bacteria / Jeremy W. Dale, Simon Park. — 4th ed.
p.; cm.
Includes bibliographical references and index.
ISBN 0 470 85084 1 (Cloth) — ISBN 0 470 85085 X (Paper)
1. Bacterial genetics. 2. Microbial genetics. 3. Molecular genetics. 4. Genetic engineering.
[DNLM: 1. Bacteria—genetics. 2. Genetic Engineering. 3. Molecular Biology. QW 51 D139m 2003] I. Park, Simon. II. Title.
QH434 .D35 2003
572.8′293—dc22

2003023103

British Library Cataloguing in Publication Data

A catalogue record for this book is available from the British Library

ISBN 0 470 85084 1 hardback
0 470 85085 X paperback

Typeset in 10.5/13pt Times by Kolam Information Services Pvt. Ltd, Pondicherry, India
Printed and bound in Great Britain by TJ International Ltd, Padstow, Cornwall
This book is printed on acid-free paper responsibly manufactured from sustainable forestry in which at least two trees are planted for each one used for paper production.

Contents

Molecular Genetics of Bacteria, 4th Edition by Jeremy Dale and Simon F. Park
 ISBN 0 470 85084 1 (cased) ISBN 0 470 85085 X (pbk)

Preface

When the first edition of this book was published (1989), DNA sequencing was largely confined to the characterization of individual cloned genes; although some viruses had been completely sequenced, the concept of determining the complete sequence of a bacterial genome was still some way in the future. Even in the second edition (1994), the situation had only developed to the stage that 'Genome-sequencing projects are under way for several bacterial species . . . ' – the first complete genome sequences were to appear in 1995. In the third edition (1998), it was still possible to produce a list of the bacteria that had been completely sequenced, although with the caveat that it would be out of date very quickly. Now, it is no longer sensible even to attempt to produce such a list. The widespread use of automated, robotic methods means that new genome sequences are appearing weekly, if not daily. Coupled with this has been the advent of the polymerase chain reaction (PCR) – a 'recent development' in the first edition – and more recently microarrays and proteomics that exploit genome sequence data for global analysis of gene expression.

This technological explosion has largely relegated many of the classical techniques in bacterial genetics to the pages of history, and poses a considerable problem in writing, or updating, a book such as this. Excluding the older methods completely would mean we would lose all sense of how the subject developed to the stage we are now at. Furthermore, the limitations of the purely molecular approach should be appreciated. Sooner or later, in order to fully understand the roles that specific genes play in the biology of an organism, one has to return to studying the organism itself, rather than just analysing its DNA sequence on a computer. So the compromise approach that we have adopted has been to slim down the description of the classical methods, while hopefully still retaining enough to provide historical perspective. This allows space to provide a fuller introduction to the world of genomics and post-genomics (the study of genome sequences and the exploitation of that data for analysis of gene expression and other features), and also to expand the discussion of the ways in which our knowledge of the biological properties of the organisms themselves has developed through the study of bacterial genetics (whether by classical or molecular methods).

Molecular Genetics of Bacteria, 4th Edition by Jeremy Dale and Simon F. Park
 ISBN 0 470 85084 1 (cased) ISBN 0 470 85085 X (pbk)

Of course this is a compromise, and some will find it unsatisfactory. Why have we included some topics and excluded others? The choice is a very personal one. It should be remembered that this book is not even trying to be a comprehensive textbook of bacterial genetics, but rather to provide a manageable-sized distillation of the subject primarily for those students for whom bacterial genetics is one of a wide number of courses taken and for whom the sheer size of most genetics textbooks presents a daunting obstacle. At the same time, we hope that some of you will find the topics introduced in this book sufficiently interesting and exciting (as indeed they are) that you will want to enquire further.

Jeremy Dale
Simon F. Park

1 Nucleic Acid Structure and Function

In this book it is assumed that the reader will already have a working knowledge of the essentials of molecular biology, especially the structure and synthesis of nucleic acids and proteins. The purpose of this chapter therefore is to serve as a reminder of some of the most relevant points and to highlight those features that are particularly essential for an understanding of later chapters.

1.1 Structure of nucleic acids

1.1.1 DNA

In bacteria, the genetic material is double-stranded DNA, although bacteriophages (viruses that infect bacteria – see Chapter 4) may have double-stranded or single-stranded DNA, or RNA. The components of DNA (Figure 1.1) are 2′-deoxyribose (forming a backbone in which they are linked by phosphate residues), and four heterocyclic bases: two purines (adenine and guanine), and two pyrimidines (thymine and cytosine). The sugar residues are linked by phosphodiester bonds between the 5′ position of one deoxyribose and the 3′ position of the next (Figure 1.2), while one of the four bases is attached to the 1′ position of each deoxyribose. It is the sequence of these four bases that carries the genetic information.

The two strands are twisted around each other in the now familiar double helix, with the bases in the centre and the sugar-phosphate backbone on the outside. The two strands are linked by hydrogen bonds between the bases. The only arrangement of these bases that is consistent with maintaining the helix in its correct conformation is when adenine is paired with thymine and guanine with cytosine.

Molecular Genetics of Bacteria, 4th Edition by Jeremy Dale and Simon F. Park
 ISBN 0 470 85084 1 (cased) ISBN 0 470 85085 X (pbk)

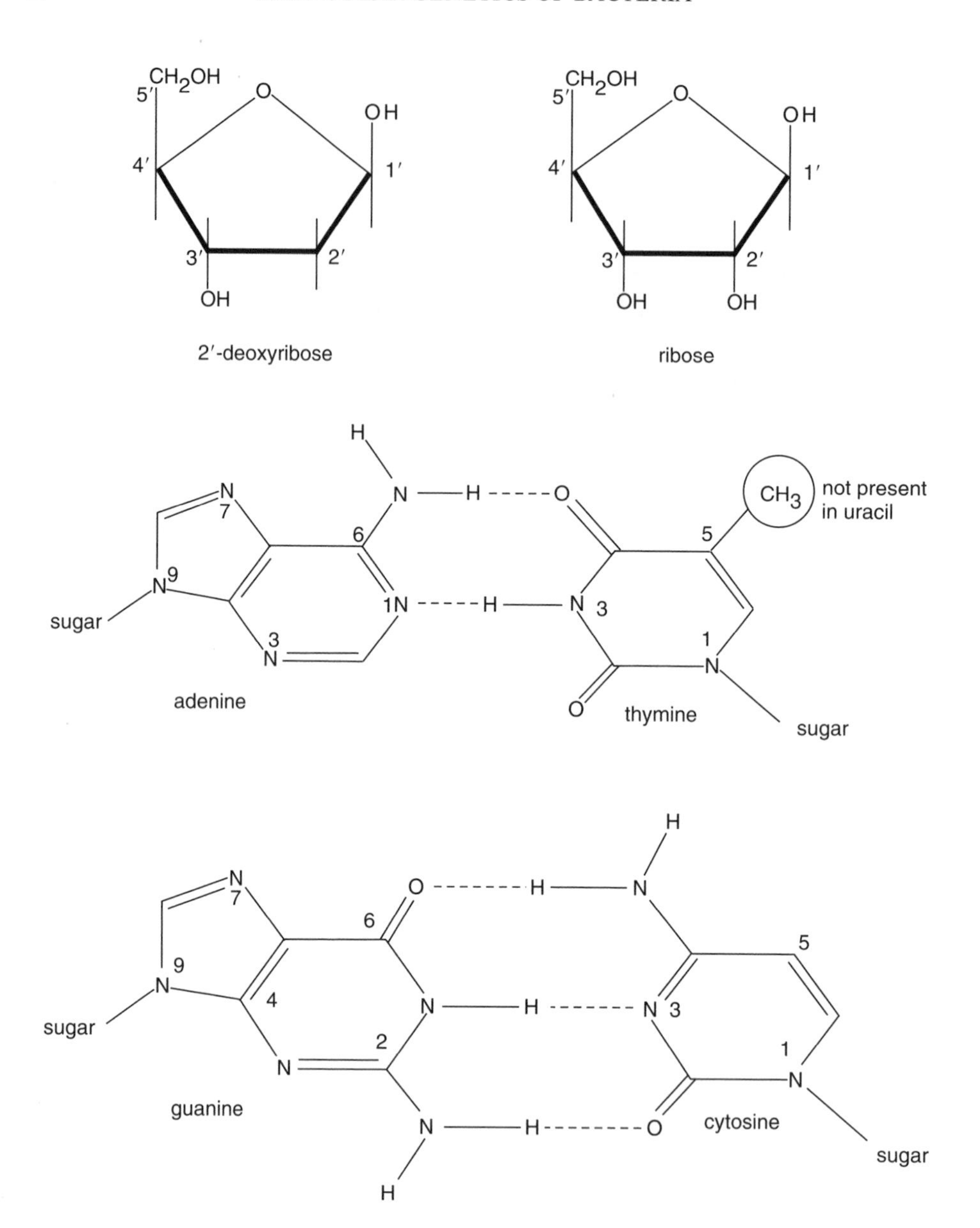

Figure 1.1 Structure of the basic elements of DNA and RNA. RNA contains ribose rather than deoxyribose, and uracil instead of thymine

One strand therefore consists of an image of the other; the two strands are said to be *complementary*. Note that the purines are larger than the pyrimidines, and that this arrangement involves one purine opposite a pyrimidine at each position, so the distance separating the strands remains constant.

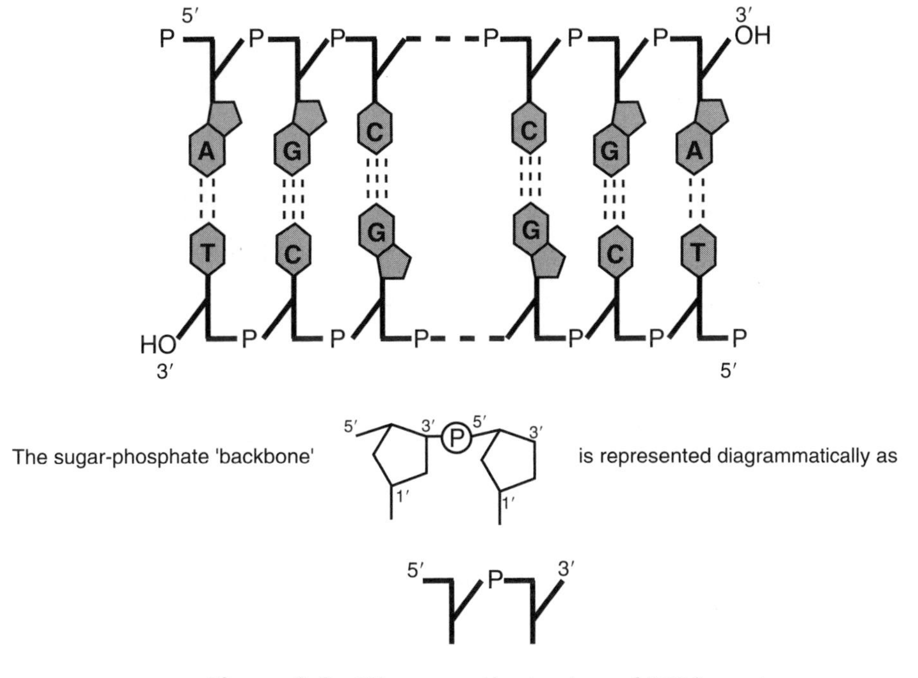

Figure 1.2 Diagrammatic structure of DNA

1.1.2 RNA

The structure of RNA differs from that of DNA in that it contains the sugar ribose instead of deoxyribose, and uracil instead of thymine (Figure 1.1). It is usually described as single stranded, but only because the complementary strand is not normally made. There is nothing inherent in the structure of RNA that prevents it forming a double-stranded structure: an RNA strand will pair with (hybridize to) a complementary RNA strand, or with a complementary strand of DNA. Even a single strand of RNA will fold back on itself to form double-stranded regions. In particular, transfer RNA (tRNA), and ribosomal RNA (rRNA) both form complex patterns of base-paired regions.

1.1.3 Hydrophobic interactions

Although geneticists emphasize the importance of the hydrogen bonding between the two DNA strands, these are not the only forces influencing the structure of the DNA. The bases themselves are hydrophobic, and will tend to form structures in which they are removed from the aqueous environment. This is partially achieved by stacking the bases on top of one another (Figure 1.3). The double-stranded

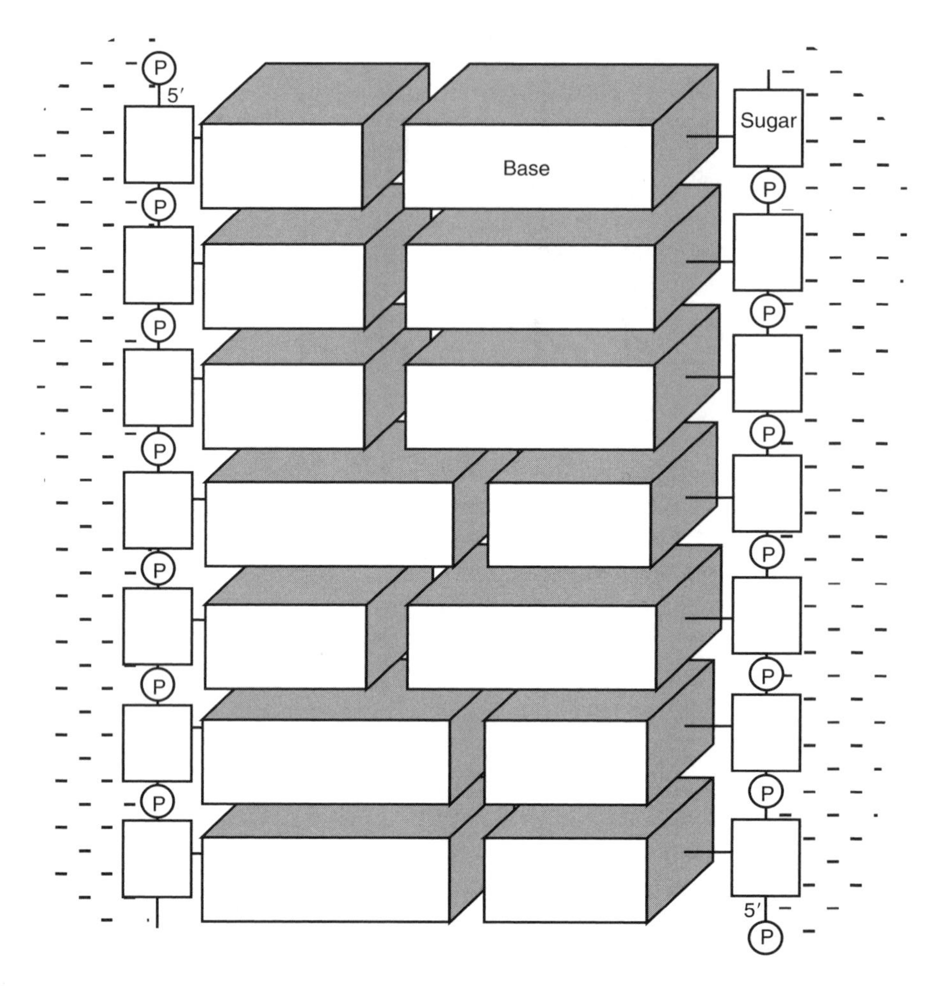

Figure 1.3 Hydrophobic interactions of bases in DNA. The hydrophobic bases stack in the centre of the helix, reducing their contact with water

structure is stabilized by additional hydrophobic interactions between the bases on the two strands. The hydrogen bonding not only holds the two strands together but also allows the corresponding bases to approach sufficiently closely for the hydrophobic forces to operate. The hydrogen bonding of the bases is however of special importance because it gives rise to the specificity of the base pairing between the two chains.

Although the bases are hydrophobic, and therefore very poorly soluble in water, nucleic acids are quite soluble, due largely to the hydrophilic nature of the backbone, and especially the high concentration of negatively-charged phosphate groups. This will also tend to favour a double helical structure, in which the hydrophobic bases are in the centre, shielded from the water and the hydrophilic phosphate groups are exposed.

1.1.4 Different forms of the double helix

A full consideration of DNA structure would be extremely complex, and would have to take into account interactions with the surrounding water itself, as well as the influence of other solutes or solvents. The structure of DNA can therefore vary to some extent according to the conditions. *In vitro*, two main forms are found. The Watson and Crick structure refers to the B form, which is a right-handed helix with 10 base pairs per turn (Figure 1.4). Under certain conditions, isolated DNA can adopt an alternative form known as the A form, which is also a right-handed helix, but more compact with about 11 base pairs per turn. Within

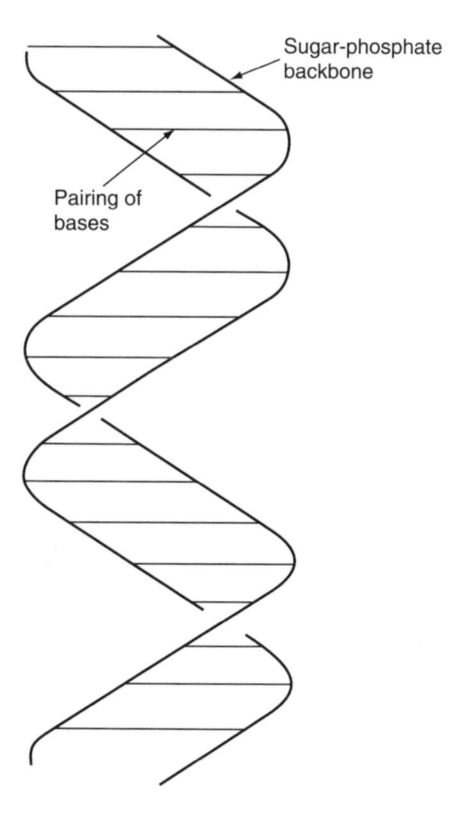

Figure 1.4 Diagrammatic structure of B-form DNA. The two anti-parallel sugar-phosphate chains form a right-handed helix with the bases in the centre, held together by hydrophobic interactions and hydrogen bonding

the cell, DNA resembles the B form more closely, but has about 10.4 base pairs per turn (it is *underwound*, see below)

Certain DNA sequences, notably those containing alternating G and C residues, tend to form a left-handed helix, known as the Z form (since the sugar-phosphate backbone has a zig-zag structure rather than the regular curve shown in the B form). Although Z DNA was originally demonstrated using synthetic oligonucleotides, naturally occurring DNA within the cell can adopt a left-handed structure, at least over a short distance or temporarily. The switch from left- to right-handed can have important influences on the expression of genes in that region.

1.1.5 Supercoiling

Within the cell, the DNA helix is wound up into coils; this is known as supercoiling. Figure 1.5 shows a simple demonstration of supercoiling, which the reader can easily try out for him/herself by taking a strip of paper and twisting one end to introduce one complete turn (i.e. the same side of the paper is facing the reader at each end). It will now look like the illustration shown in Figure 1.5a. The two ends should then be brought towards each other so that the conformation will change to that shown in Figure 1.5b which is a simple form of supercoiling. Not only has the strip of paper become supercoiled, but also the degree of twisting appears to have changed (in this example it now appears not to be twisted at all). If both ends have been held firmly, the twist of the strip cannot have

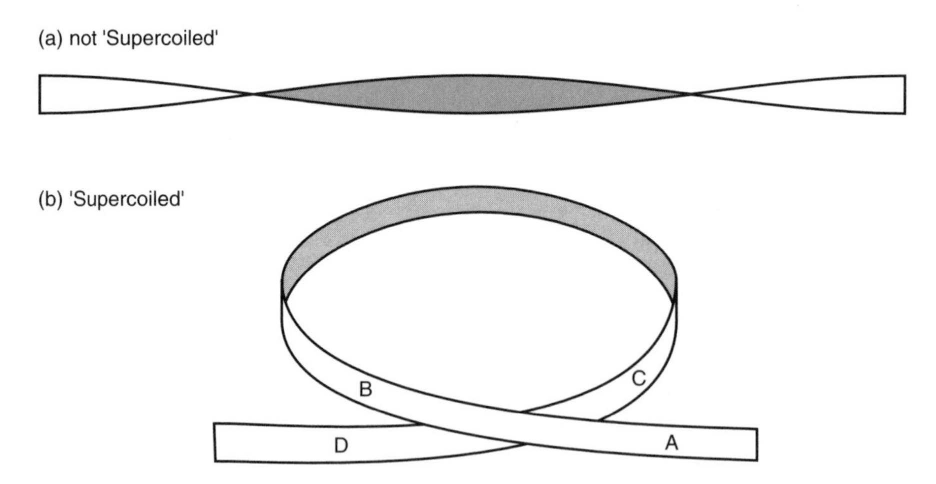

Figure 1.5 Interaction between twisting and supercoiling. (a) A ribbon with a single complete twist, without supercoiling. (b) The same ribbon, allowed to form a supercoil; the ribbon is now not twisted

disappeared completely; it has merely changed to a different form. If the ends are pulled apart again, it will change back to form a.

There are three parameters involved: twist (T), linking number (L) and writhe (W). The twist is the number of turns of the strip, while writhe (essentially a measure of the degree of supercoiling) can be considered as the number of times the strip crosses over itself in a defined direction. These two parameters vary according to the conformation: in (a), there is one twist ($T = 1$) but no supercoiling ($W = 0$), while in (b) there is no twist ($T = 0$) and the strip crosses itself once ($W = 1$). The sum of these two parameters is used to define the linking number, i.e. $L = T + W$. The linking number is therefore a measure of the overall twisting of the strip, or for DNA, the total number of times that the DNA strands wrap around one another.

If the ends of the strip are not free to rotate, then the linking number will remain constant. Most of the DNA molecules we will be considering are circular, and therefore do not contain ends that can rotate. Unless there is a break in the DNA, any change in the twist will be balanced by a change in supercoiling, and vice versa. This is illustrated by Figure 1.6.

In a, the strip (or DNA molecule) is not supercoiled ($W = 0$) but contains one complete twist ($T = +1$); the linking number (L) is +1. In b, the overall shape has been changed by rotating one end of the structure (i.e. introducing a degree of supercoiling). The strip crosses itself once, and by convention a crossover in this direction is assigned a negative value, so $W = -1$. At the same time the twist has changed; there are now two complete twists, so $T = +2$. Since $L = T + W$, we see that L remains the same (+1).

In (c), which is obtained by a further rotation of one end of the structure, there are two negative supercoils ($W = -2$), and three complete twists ($T = +3$). Once more, L stays the same. Changing the supercoiling alters the degree of twisting correspondingly, so that the linking number remains constant. The three structures shown in Figure 1.6 are interchangeable by rotating one end, without opening the circle. With an intact circle, the twist and the writhe can be changed jointly but not separately; any change in supercoiling will involve a compensating change in the twist (and vice versa) so that the linking number remains constant. *It is only possible to alter the linking number in circular DNA by breaking and rejoining DNA strands*, for example through the action of topoisomerases (see below).

Bacterial DNA is normally negatively supercoiled. Another way of putting it is to say that the DNA is underwound so that if the DNA was not supercoiled, the degree of twisting of the helix would be less than that seen in relaxed linear DNA. If the DNA is nicked (i.e. one strand is broken, leaving it free to rotate) it relaxes into an open circular, non-supercoiled form. Chromosomal DNA is usually broken into linear fragments during lysis of the cell, but bacterial plasmids (see Chapter 5) are usually small enough to be isolated intact in a supercoiled form.

The compact supercoiled structure of the DNA is also significant in that the chromosome, in its expanded state, would be a thousand times longer

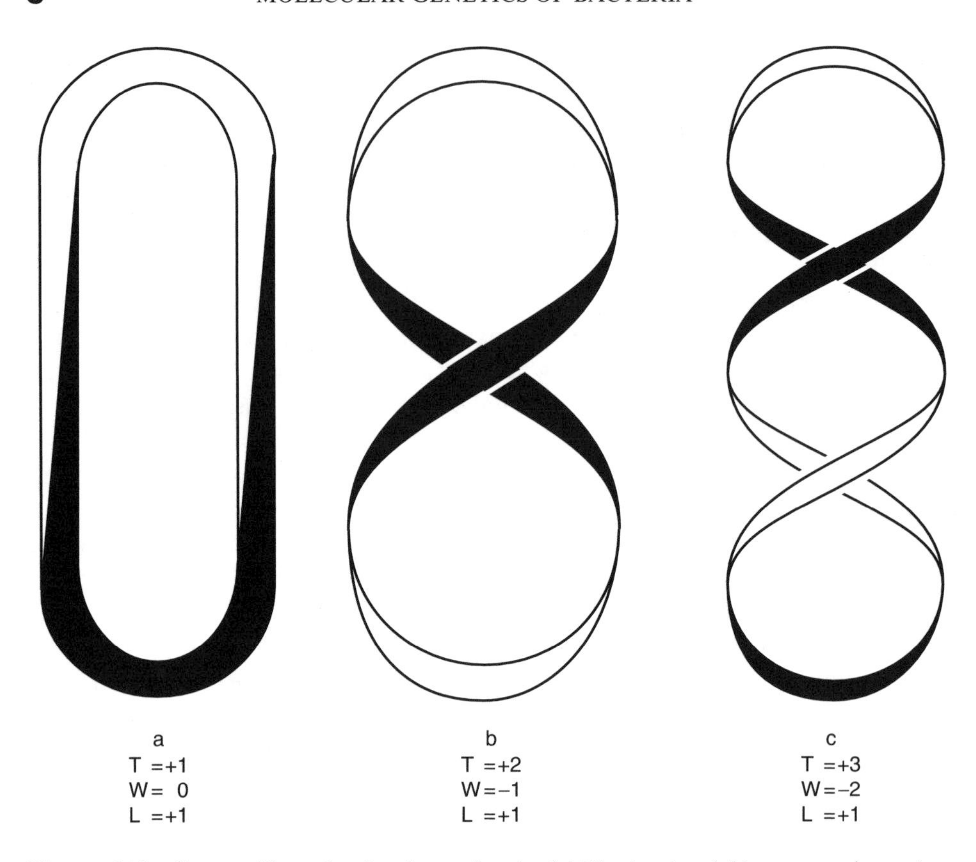

Figure 1.6 Supercoiling of a circular molecule. (a) The 'molecule' has one twist and no supercoils. Rotating one end of the molecule (b and c) introduces negative supercoils and increases the amount of twisting. Since the linking number (L) remains the same, the three forms are interchangeable without breaking the circle (see the text for further explanation)

(about 1 mm) than the bacterial cell itself. To put it another way, a bacterial operon of four genes, in its non-supercoiled B-form, would stretch from one end of the cell to the other. Supercoiling is only the start of story, as the bacterial chromosome consists of a large number of supercoiled loops arranged on a core to produce a highly compact and organized structure known as the nucleoid. Supercoiling (and other structural features) of the DNA are also important in the regulation of gene expression (see Chapter 3).

Action of topoisomerases

Supercoiling of bacterial DNA is not achieved by physically twisting the circular molecule in the way we illustrated in Figure 1.6. Instead the cell uses enzymes known as *DNA topoisomerases* to introduce (or remove) supercoils from DNA, by controlled breaking and rejoining of DNA strands.

DNA topoisomerases can be considered in two classes. Type I topoisomerases act on a segment of DNA by breaking one of the strands and passing the other strand through the gap, followed by resealing the nick. Since this increases the number of times the two strands cross one another, the linking number is increased by 1 which results in an increase in either T or W. The *E. coli* topoisomerase I acts only on negatively supercoiled DNA; the increase in the value of W means that the degree of negative supercoiling is reduced (the DNA becomes relaxed).

Type II topoisomerases break both strands and pass another duplex region through the gap. In Figure 1.7, the structure A looks at first glance to be

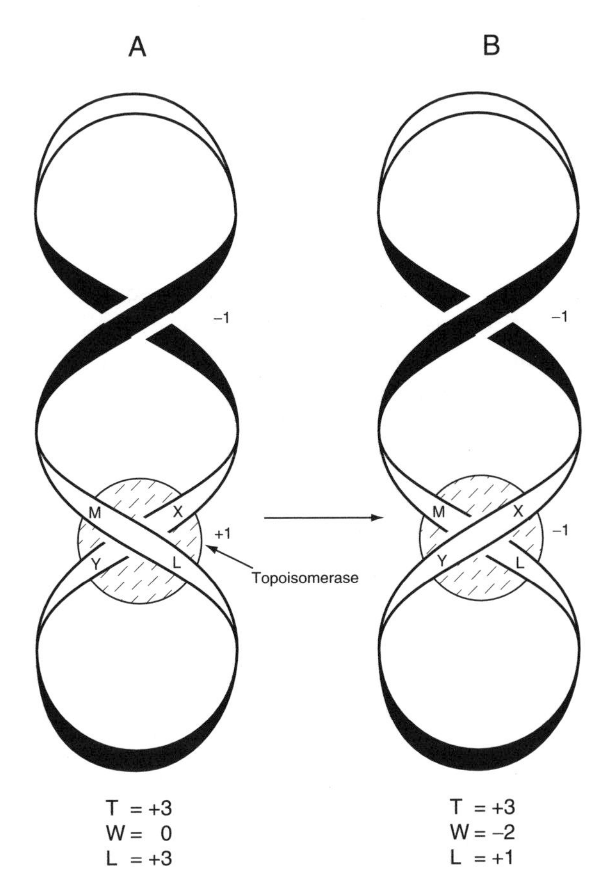

Figure 1.7 Action of Type II topoisomerase. Structure A is not supercoiled: the two crossing points are of opposite sign and cancel one another. The topoisomerase makes a double strand break between L and M, passes the X–Y region through the gap and reseals the break between L and M. This changes the sign of W at that point, so structure B is now negatively supercoiled

supercoiled, but closer inspection shows that the two crossover points are of opposite sign, and therefore cancel each other out. Structure A is a non-supercoiled circle that is drawn in this way to show the action of the topoisomerase. If both strands of the helix are broken between points L and M and the lower strands (X–Y) are moved through the gap followed by re-sealing the strands between L and M, structure B is formed. As a consequence of this reaction, the sign of W is changed at that point (changing W from +1 to −1), which has the overall effect of a change of −2 in the value of W and hence in the value of the linking number L. An important example of this type of enzyme is DNA gyrase, which is able to introduce negative supercoils into newly replicated DNA.

1.1.6 Denaturation and hybridization

Since the two strands of DNA are only linked by non-covalent forces, they can easily be separated in the laboratory, for example by increased temperature or high pH. Separation of the two DNA strands, *denaturation*, is readily reversible. Reducing the temperature, or the pH, will allow hydrogen bonds between complementary DNA sequences to re-form; this is referred to as *re-annealing* (Figure 1.8). If DNA molecules from different sources are denatured, mixed and allowed to re-anneal, it is possible to form hydrogen bonds between similar DNA sequences (*hybridization*). This forms the basis of the use of DNA probes to detect specific DNA sequences. The specificity of the reaction can be adjusted by altering the temperature and/or the ionic strength. Higher temperature, or lower ionic strength, gives greater *stringency* of hybridization. High stringency hybridization is used to detect closely-related sequences, or to distinguish between sequences with only small differences, while low stringency conditions are used to detect sequences that are only remotely related to the probe. This technique forms an important part of modern molecular biology.

Temporary separation of localized regions of the two DNA strands also occurs as an essential part of the processes of replication and transcription. Note that there are three hydrogen bonds linking guanine and cytosine while the adenine–thymine pairing has only two hydrogen bonds. The two DNA strands are therefore more strongly attached in those regions with a high G + C content. Because of this, such regions are more resistant to denaturation and conversely re-anneal more readily. The influence of base composition on the ease of separation of two nucleic acid strands may play an important role in the control of processes such as the initiation of RNA synthesis where an A-T rich region may facilitate the initial separation of the DNA strands (Chapter 3).

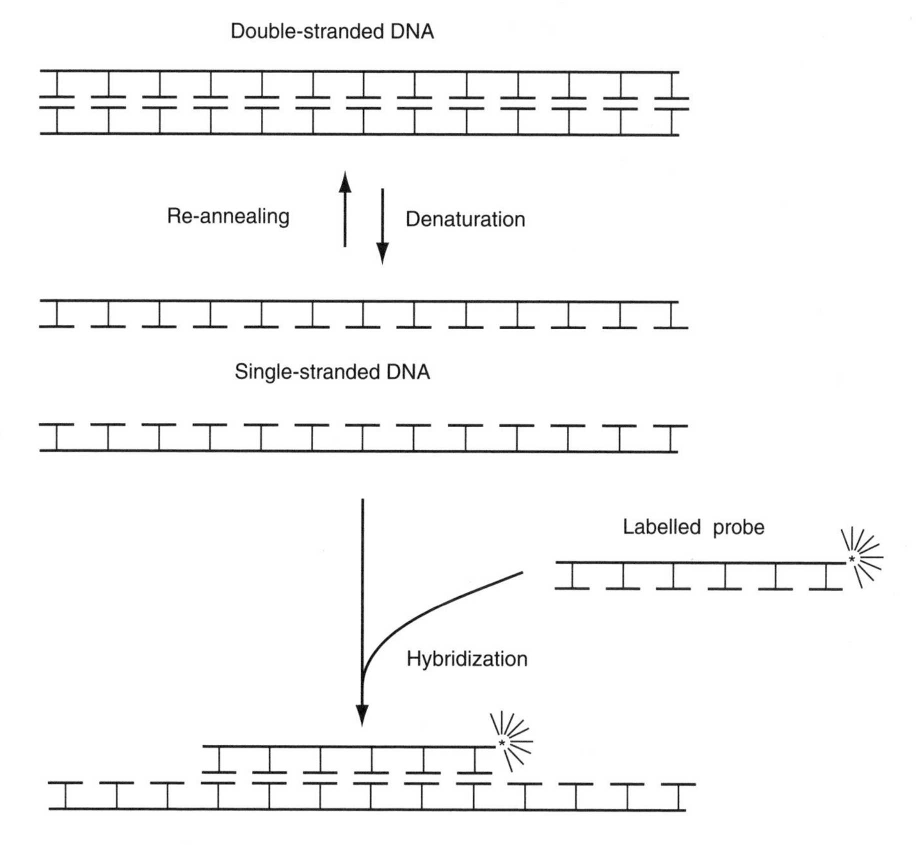

Figure 1.8 Denaturation and hybridization of DNA

1.1.7 Orientation of nucleic acid strands

A further noteworthy feature of the helix is that each strand can be said to have a direction, based on the orientation of the linkages in the sugar-phosphate backbone. Each phosphate group joins the 5′ position of one sugar residue to the 3′ position of the next deoxyribose. In Figure 1.2, the upper strand has a free 5′ group at the left-hand end and a 3′ OH group at the right-hand end. It is therefore said to run (from left to right) in the 5′ to 3′ direction. Conversely, for the lower strand, the 5′ to 3′ direction runs from right to left.

By convention, if a single DNA (or RNA) strand is shown, it reads in the 5′ to 3′ direction from left to right (unless otherwise stated). If both strands are shown, the upper strand reads (left to right) from the 5′ to 3′ end.

All nucleic acids are synthesized in the 5′ to 3′ direction. That is, the new strand is elongated by the successive addition of nucleotides to the free 3′ OH group of the preceding nucleotide. The phosphate to make the link is provided by the substrate which is the nucleoside 5′-triphosphate (ATP, GTP, CTP, UTP for RNA; dATP, dGTP, dCTP, dTTP for DNA).

1.2 Replication of DNA

A DNA strand can act as a template for synthesis of a new nucleic acid strand in which each base forms a hydrogen-bonded pair with one on the template strand (G with C, A with T, or A with U for RNA molecules). The new sequence is thus complementary to the template strand. The copying of DNA molecules to produce more DNA is known as *replication*; the synthesis of RNA using a DNA template is called *transcription*.

Replication is a much more complicated process than implied by the above statement. Some of the main features are summarized in Figure 1.9. The opposite polarity of the DNA strands is a complicating factor. One of the new strands (the 'leading' strand) can be synthesized continuously in the 5′ to 3′ direction. The enzyme responsible for this synthesis is DNA polymerase III. With the other new strand however, the overall effect is of growth in the 3′ to 5′ direction. Since nucleic acids can only be synthesized in the 5′ to 3′ direction, the new 3′ to 5′ strand (the 'lagging' strand) has to be made in short fragments (known as Okazaki fragments) which are subsequently joined together by the action of another enzyme, DNA ligase.

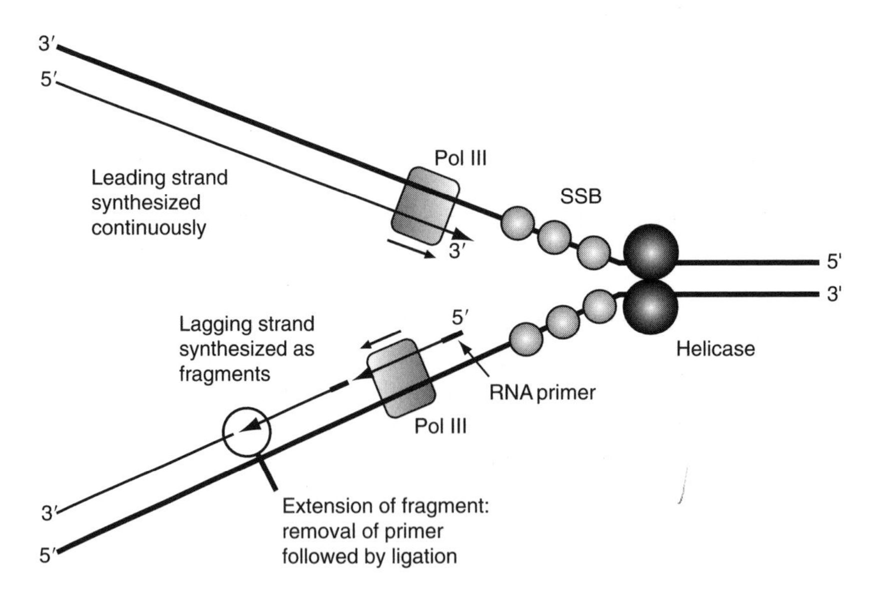

Figure 1.9 Simplified view of the main features of DNA replication. Note that the diagram does not show the helical structure of the DNA

Furthermore, DNA polymerases are incapable of starting a new DNA strand, but can only extend a previously existing molecule. This restriction does not apply to RNA polymerases, which are able to initiate synthesis of new nucleic acids. Each fragment is therefore started with a short piece of RNA, produced by the action of a special RNA polymerase (*primase*). This RNA primer can then be extended by DNA polymerase III. The primer is subsequently removed, and the gap filled in, by a different DNA polymerase (DNA polymerase I); this enzyme can carry out both of these actions since it has exonuclease as well as polymerase activity. After the gap has been filled, the fragments that have been produced are joined together by DNA ligase.

1.2.1 Unwinding and rewinding

Before any of these events can take place, it is necessary for the two strands to be separated, for a short region at least. This is achieved by enzymes known as *helicases* which bind to the template strand and move along it, separating the two strands. The separated strands are prevented from re-associating by the binding of another protein, the single-stranded DNA binding protein or SSB. A number of copies of the SSB will bind to the DNA strands, maintaining a region of DNA in an extended single-stranded form.

A further complication arises from the twisting of the two DNA strands around each other. DNA molecules within the cell cannot normally rotate freely. In bacterial cells for example the DNA is usually circular. Therefore it is not possible to produce a pair of daughter molecules by just separating the two strands and synthesizing the complementary strands, as is implied by the simplified representation in Figure 1.9. The strands have to be unwound to be separated. If they are not free to rotate, separating the strands at one point will cause overwinding further along. Unless this problem is overcome, the molecules would quickly become hopelessly tangled. (This can be demonstrated with the use of lengths of string!). The resolution of the problem requires the action of topoisomerases, as described earlier. By allowing the double helix to unwind ahead of the replication fork, they permit the strands to separate for replication. One topoisomerase, DNA gyrase, has the important role of introducing negative supercoils into the newly replicated DNA.

1.2.2 Fidelity of Replication: proof-reading

It is essential that the newly synthesized DNA is a precise (complementary) copy of the template strand. This does not arise simply by the nucleotides aligning themselves in the right position, but involves the specificity of the DNA polymerase in selecting nucleotides that are correctly aligned.

Most DNA polymerases are more complex enzymes than the name suggests, as they also possess exonuclease activity. We have already encountered one such activity: the removal of the RNA primer from the Okazaki fragments is achieved by means of the 5′ to 3′ exonuclease activity of the DNA polymerase (i.e. it can remove bases from the 5′ end of a chain) as it extends the following fragment. The fidelity of replication is enhanced by a second exonuclease function of DNA polymerases: the 3′ to 5′ exonuclease activity, which is able to remove the nucleotide at the growing end (3′ end) of the DNA chain. This is not as perverse as it sounds, since the 3′ to 5′ exonuclease only operates if there is an incorrectly paired base at the 3′ end. The DNA polymerase will only extend the DNA chain, by adding nucleotides to the 3′ end, if the last base at the 3′ end is correctly paired with the template strand. If it is not, polymerization will stop, and the 3′ to 5′ exonuclease function will remove the incorrect nucleotide, allowing a further attempt to be made (Figure 1.10). The reasons for the occurrence of errors in adding bases to the growing DNA chain are dealt with in Chapter 2.

This mechanism of correcting errors, known as proof-reading or error-checking, adds considerably to the fidelity of replication, thus reducing the rate of spontaneous mutation. There is a price to be paid however, as extensive error-checking will slow down the rate of replication. The balance between the rate of replication and the extent of error-checking will be determined by the nature of the DNA polymerase itself. Some DNA polymerases do not show efficient proof-reading and therefore result in a much higher degree of spontaneous errors. The rate of spontaneous mutation shown by an organism is therefore (at least in part) a genetic characteristic that is subject to evolutionary pressure.

The fidelity of replication is further enhanced by DNA repair mechanisms which are described later in this chapter.

1.3 Chromosome replication and cell division

Bacterial cells are generally regarded as having a single, circular chromosome. This is a simplification in several ways. Firstly, many bacteria often contain additional DNA molecules known as *plasmids* (see Chapter 5). In most cases these are additional, dispensable, elements, but in some bacterial species all strains carry two or more different DNA molecules, both (or all) of which appear to be essential for normal growth. These can equally well be regarded as essential plasmids or as additional chromosomes. Secondly, not all bacterial DNA is circular. Some bacteria (notably *Streptomyces*) have a linear chromosome and/or linear plasmids; these are also discussed further in Chapter 5.

More fundamentally, immediately before cell division there must be at least two complete copies of the chromosome, in order to ensure that both daughter cells acquire a copy. Therefore, the chromosome must replicate in tune with the cell division cycle, which means that at an intermediate time in the cycle

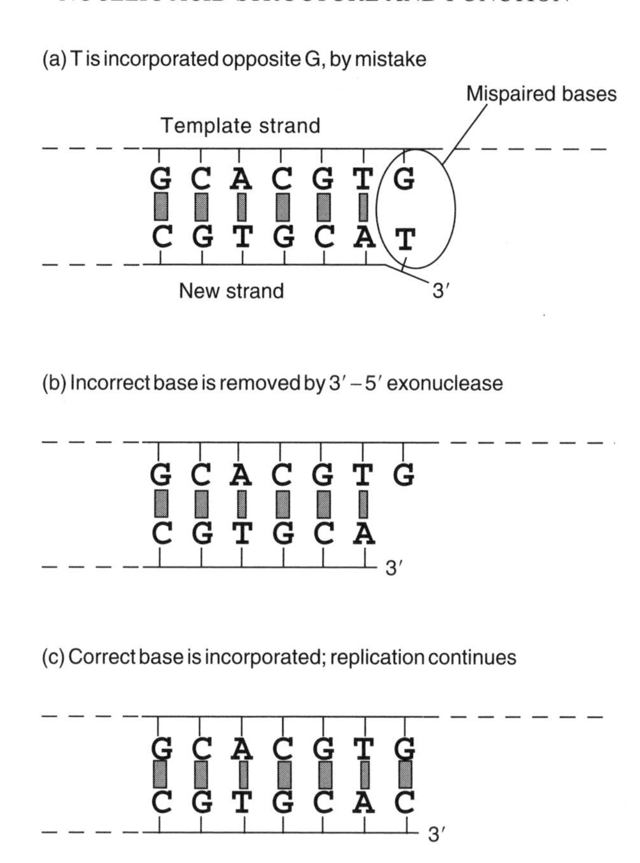

Figure 1.10 Elimination of mispaired bases by proof-reading. (a) An incorrect base has been added to the growing DNA strand; this will prevent further extension. (b) The mispaired base is removed by the 3′—5′ exonuclease action of DNA polymerase. (c) The correct base is added to the 3′ end; DNA synthesis continues

part of the chromosome will have been copied, with the consequence that there are at least two copies of this part of the chromosome.

Replication of a bacterial chromosome normally starts at a fixed point (the origin of replication, *oriV*) and proceeds in both directions to a termination point (*ter*) that is approximately opposite to the origin (Figure 1.11). In *E. coli*, this takes about 40 min. There is then a period of about 20 min before cell division, making a total of 60 min between the initiation of replication and cell division. If the growth rate of the cells is changed by using a richer or poorer medium, this time remains much the same. However under favourable conditions, *E. coli* will grow much faster than that – dividing perhaps every 20 min. How can the cells be dividing faster than the chromosome replicates and still allow every daughter cell to acquire a complete copy of the chromosome? The answer lies in the timing of the initiation of replication. Initiation is stimulated, not by cell division, but as a

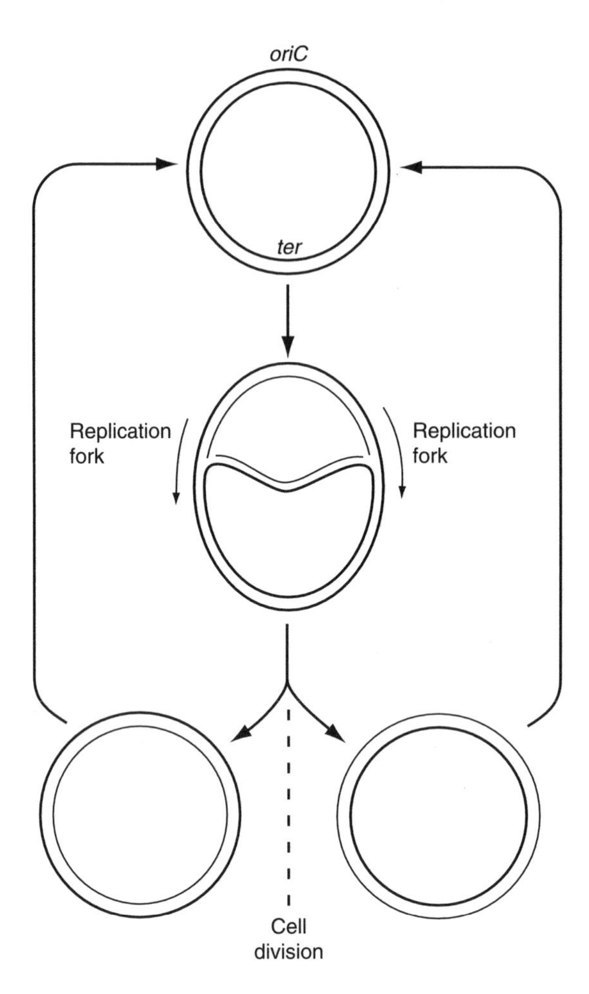

Figure 1.11 Chromosome replication. Bi-directional replication starts at *oriC* and continues to the termination site *ter*, producing two double-stranded molecules

function of the size of the cell. Consequently, when the cells are growing rapidly, there are several sets of replication forks copying the chromosome – so that when the cell is ready to divide there are not just a pair of completely replicated chromosomes, but each of these has in turn already been partly replicated by a second pair of replication forks (Figure 1.12).

There are two key regulatory points to be considered: the link between the completion of chromosome replication and subsequent cell division, and the control of the initiation of replication. Both are too complex (and still incompletely understood) to be considered fully here, but we can consider some aspects. On the first point, one (simplified) model is that the replicating chromosome

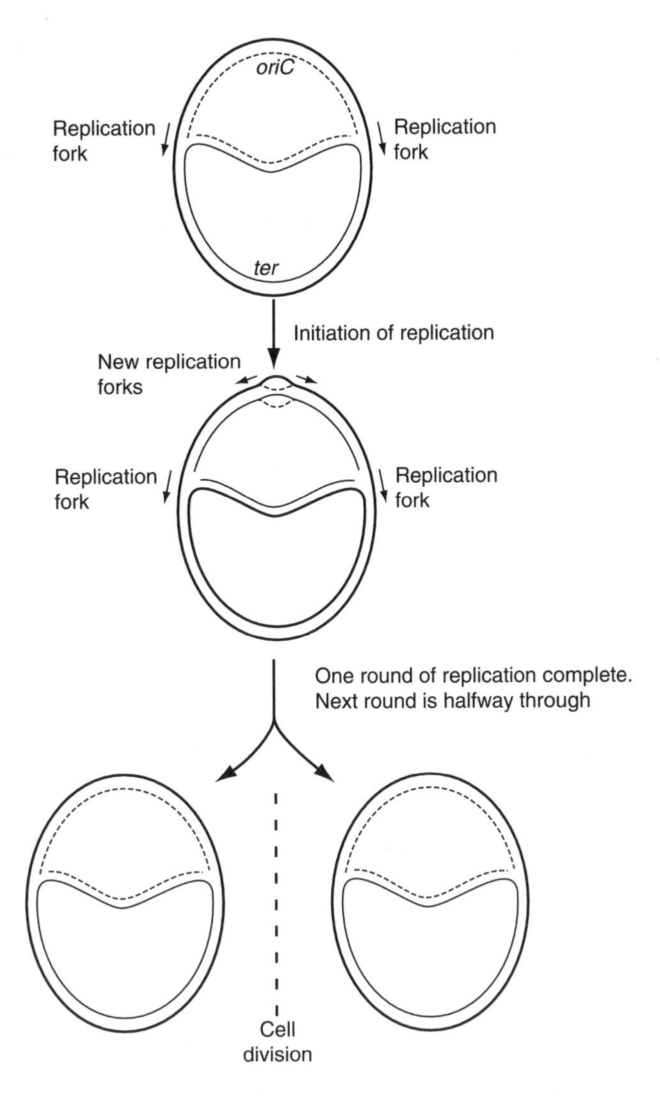

Figure 1.12 Chromosome replication at higher growth rates. When the interval between cell divisions is less than the time needed for replication of the chromosome, a new round of replication starts before the previous one has finished

occupies a region of the membrane at the midpoint of the cell which prevents the initiation of cell division at that point. When replication has finished, the two separate molecules can be pulled apart, towards the poles of the cell, thus freeing the site for cell division to start. We will look further at this in Chapter 9.

The second point, the control of initiation, is more difficult. Initiation of new rounds of replication is triggered when the cell reaches a critical mass. It is tempting to think that this means that an inhibitor of replication is diluted out as the cell grows, but it is far from that simple. However, we know that initiation requires a protein called DnaA which binds to specific DNA sequences known as *DnaA boxes*; the origin of replication contains a number of DnaA boxes. Wrapping the DNA around the aggregated DnaA proteins facilitates the separation of the strands that is necessary for the initiation of replication (Figure 1.13). However, the full story of the control of initiation is more complex and still incompletely understood.

One feature worth noting is that each replication origin fires only once at the time of initiation. Mature DNA is methylated by addition of methyl groups to adenine residues at certain positions. However, the newly synthesized DNA strand is not methylated and so the double-stranded DNA contains one methylated strand (the old strand) and one non-methylated one (the new strand). This *hemimethylated* DNA is refractory to the initiation of replication, and thus the newly replicated origin will not be available for another initiation event until the second strand has been methylated.

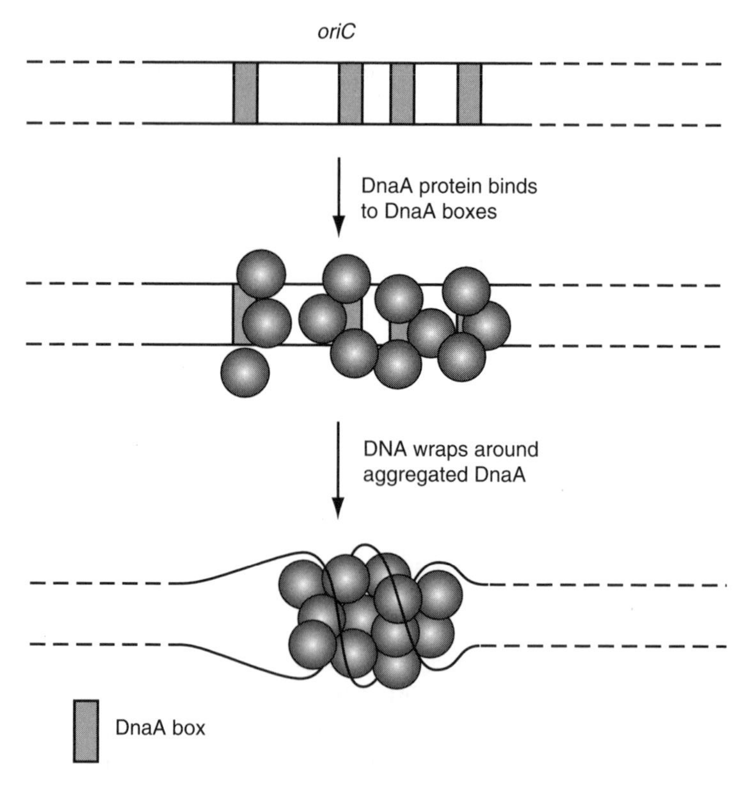

Figure 1.13 Binding of DnaA protein to DnaA boxes at *oriC*. The DnaA protein binds to several sites (DnaA boxes) at the origin of replication; wrapping the DNA around the bound DnaA helps to separate the strands, enabling the initiation of replication

1.4 DNA repair

In addition to the measures described earlier which enhance the fidelity of replication, the cell also possesses mechanisms to correct damaged DNA. This includes replication errors that have escaped the proof-reading process and also damage that may occur in non-replicating DNA.

1.4.1 Mismatch repair

The simplest of replication errors is one that leads to the wrong base being incorporated into the new strand. If this occurs, and is not dealt with by the proof-reading mechanism, it would lead to mutation. However, the cell has an effective mechanism for removing such mismatches and replacing them with the correct nucleotide. In order to do this, it has to know which of the two strands contains the correct information. The methylation of the DNA, as referred to above, identifies the new and old strands, at least until the new strand becomes methylated – hence the mechanism is referred to as *methyl-directed mismatch repair*. The system recognizes the mismatched bases, removes a short region of the non-methylated strand and fills in the gap.

1.4.2 Excision repair

Other types of DNA damage, in particular the formation of pyrimidine dimers by ultraviolet irradiation (see Chapter 2), give rise to distortion of the double helix, which can activate a repair mechanism known as excision repair.

The essence of this mechanism is summarized in Figure 1.14. The process is initiated by an endonuclease (a complex enzyme, coded for by genes known as *uvrA*, *uvrB* and *uvrC*, since mutations in these genes cause reduced UltraViolet Resistance). This enzyme cuts the DNA strand on either side of the damage, which exposes a 3′ OH group; this can be used as a primer by DNA polymerase I to replace the short region of DNA between the nicked sites (15–20 bases long). The final step is the joining of the newly repaired strand to the existing DNA by DNA ligase.

1.4.3 Recombination (post-replication) repair

There is another type of mutant that is abnormally sensitive to UV, although possessing a functional excision repair system. These bacteria are defective in a gene (*recA*) that is amongst other things responsible for general recombination (see Chapter 6). This indicates that excision repair is not the only mechanism for dealing with UV damage, but that there is a further repair mechanism involving

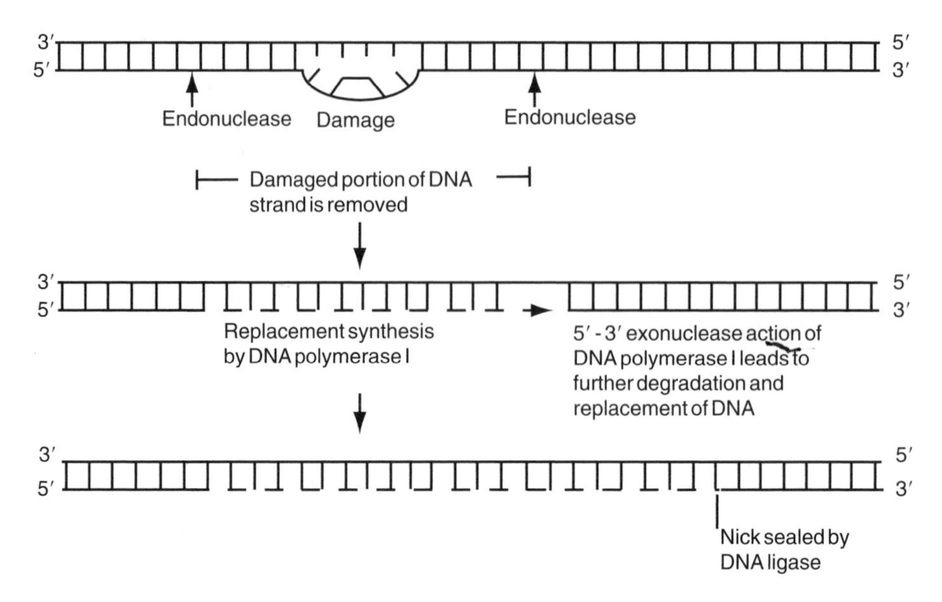

Figure 1.14 Mechanism of excision repair. Endonuclease cleavage removes a portion of the damaged strand. The gap is filled in by DNA polymerase I; the 5′—3′ exonuclease action of DNA polymerase I allows it to remove more DNA and replace it. Finally, the sugar-phosphate backbone is resealed by DNA ligase

general recombination. The double mutant (*uvrA recA*) is even more sensitive than either of the single mutants.

Forms of DNA damage that interfere with the base pairing between the strands will normally prevent replication, due in part to the requirement of the DNA polymerase for an accurately paired 3′ end. The replication fork will therefore pause. It is possible however for replication to restart beyond the lesion, thus leaving a gap in the newly synthesized strand (Figure 1.15). This portion of DNA cannot be repaired by excision repair, which requires one intact strand. The gap can however be filled using a portion of DNA from the other pair of strands by a recombination process (i.e. cutting and rejoining the DNA). Although this merely re-assorts the damage rather than directly repairing it, it does achieve a situation where the damage is repairable. The original damage can now be repaired by excision repair while the gap in the other DNA molecule can be filled in by DNA polymerase I and DNA ligase.

1.4.4 SOS repair

An alternative strategy, when faced with overwhelming levels of DNA damage preventing normal replication, is to temporarily modify or abolish the specificity of the DNA polymerase. This enables it to continue making a new DNA strand,

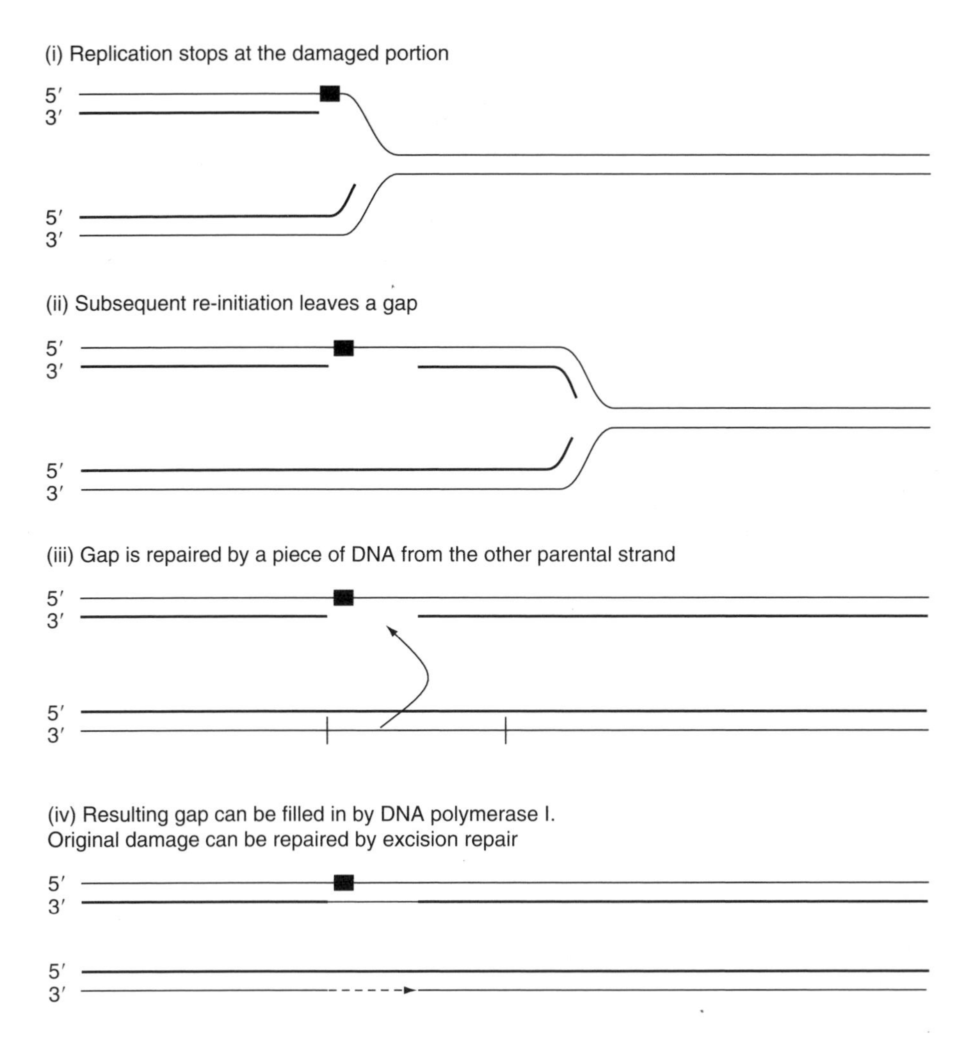

Figure 1.15 Post-replication repair. After replication stops at a damaged site (i), subsequent re-initiation leaves a gap in the new strand (ii). This can be repaired by exchange of DNA (iii), allowing the original lesion to be repaired by excision repair (iv)

despite the absence of an accurately paired 3′ end. Although the new strand is produced, it is obviously likely to contain many mistakes and the process is therefore described as 'error-prone'. SOS repair is the cause of mutations arising from ultraviolet irradiation and is considered further in Chapter 2.

1.5 Gene expression

The expression of the genetic material occurs for the most part through the production of proteins, involving two consecutive steps in which the information is converted from one form to another: transcription and translation. With those proteins that consist of several different subunits, each one is the product of a distinct region of DNA. The complete protein is thus the product of several different genes, mutation in any of which may lead to the absence of a functional product. The term *cistron*, meaning that region of the DNA that codes for a single polypeptide chain, is used where it is desirable to emphasize the distinction between a single polypeptide and a multimeric protein.

1.5.1 Transcription

The first step is the conversion of the information into messenger RNA (mRNA). This process (transcription) is carried out by RNA polymerase. As with DNA synthesis, the RNA strand is made in the 5′ to 3′ direction. However, there are major differences between transcription and replication. Firstly, only a comparatively short molecule is produced, and secondly, only one of the DNA strands is transcribed. (Some genes use one strand, and some use the other, but in general any specific region of DNA is only transcribed from one strand). Since only a single strand is made, it can be produced continuously using a single enzyme; there is no need for lagging strand synthesis. In addition the production of relatively short single-stranded RNA causes fewer topological problems: the enzyme and the RNA product can essentially rotate around the helix, so there is no need for the helicases and topoisomerases that are essential for replication. Furthermore, RNA polymerase can start synthesis from scratch – no primer is needed. Transcription is therefore considerably simpler than replication.

Since transcription results in the synthesis of comparatively short mRNA molecules (often just a few kilobases long, corresponding to a defined block of several genes) there must be a large number of signals around the chromosome that direct the RNA polymerase to start transcription at the required place and to stop when the block of genes has been transcribed. The start signals (*promoters*) also convey the information as to the direction in which transcription should proceed, or which strand to work from – which is another way of saying the same thing.

In *E. coli*, depending on growth conditions, 2000–5000 copies of RNA polymerase may be engaged on mRNA synthesis at any time. The basic structure of RNA polymerase consists of four polypeptides – two identical α chains plus two other chains (β and β') that are related to one another but are not identical. This structure ($\alpha_2\beta\beta'$) is referred to as the *core* enzyme. The specificity of promoter binding is due to a fifth subunit, the σ (sigma) factor; the complete structure including the σ factor is called the *holoenzyme*. As we will see in Chapter 3, there are different classes of promoters, recognized by different sigma factors, which allows selective expression of certain groups of genes.

After the RNA polymerase holoenzyme binds to the promoter region, the initial structure (the closed complex) is converted to an open complex

(Figure 1.16) in which there is localized separation of the two DNA strands. This exposes the bases of the coding strand, allowing base pairing of the ribonucleoside triphosphates for synthesis of the RNA. The first phosphodiester bond is formed and the σ factor dissociates from the complex. From now on, the core enzyme alone is required for extension of the RNA strand. A short region of the newly

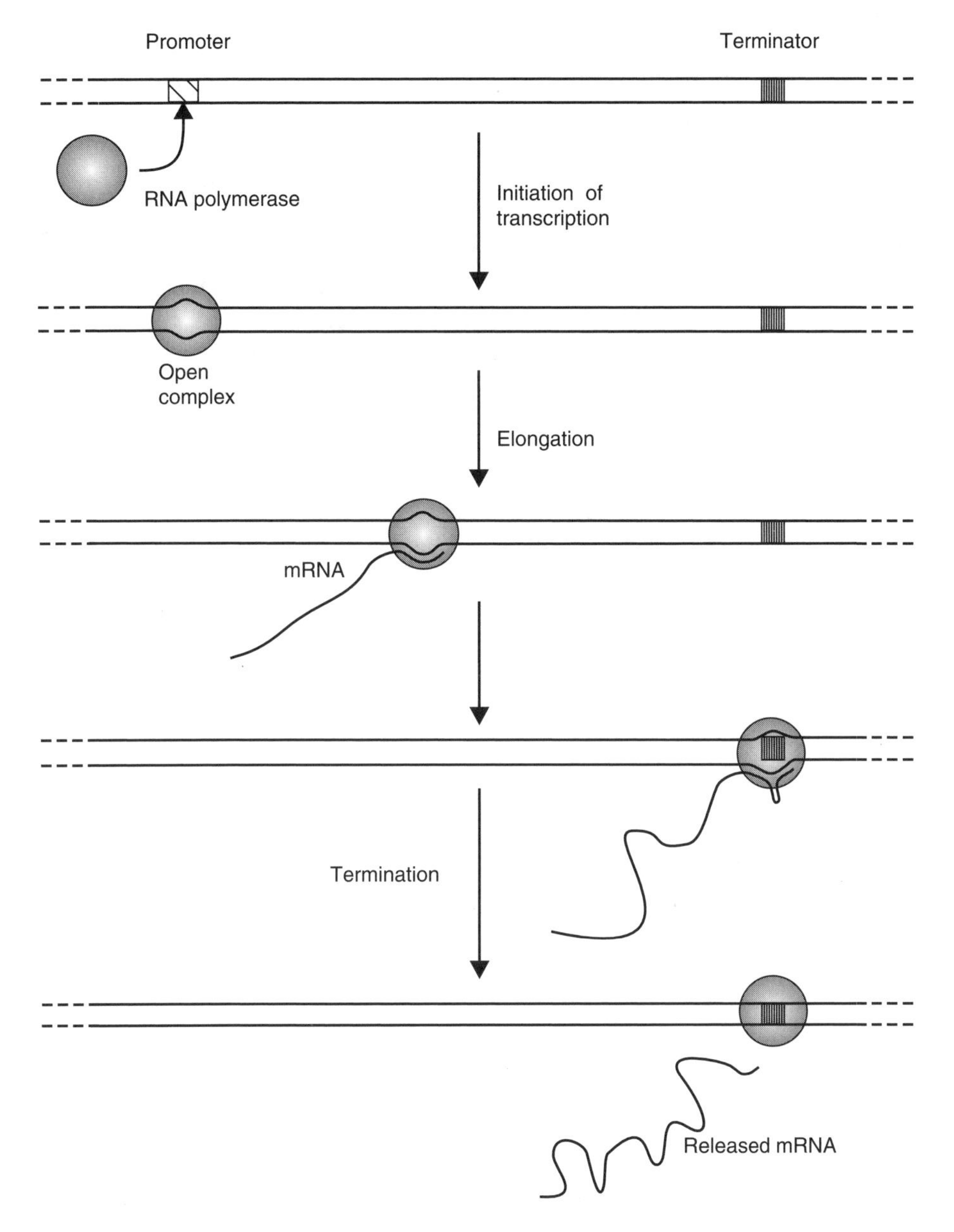

Figure 1.16 Main features of transcription

formed RNA remains base-paired to the DNA template, which keeps the DNA strands from re-associating, and therefore permits continued RNA synthesis, until a termination signal is reached when the mRNA and the RNA polymerase are released.

Transcriptional terminators

A characteristic feature of a transcriptional terminator is the presence of a short sequence that is complementary to the sequence just preceding it. When such a sequence is transcribed, the RNA formed can establish a stem and loop structure as shown in Figure 1.17. In most terminator sequences, the stem–loop structure is followed by a run of U residues.

A model that accounts for the termination of transcription is presented in Figure 1.18. RNA polymerase requires a short length of unwound DNA (about 17 bp) in which the two DNA chains are separated. However, the two DNA strands will tend to snap back together unless something prevents them.

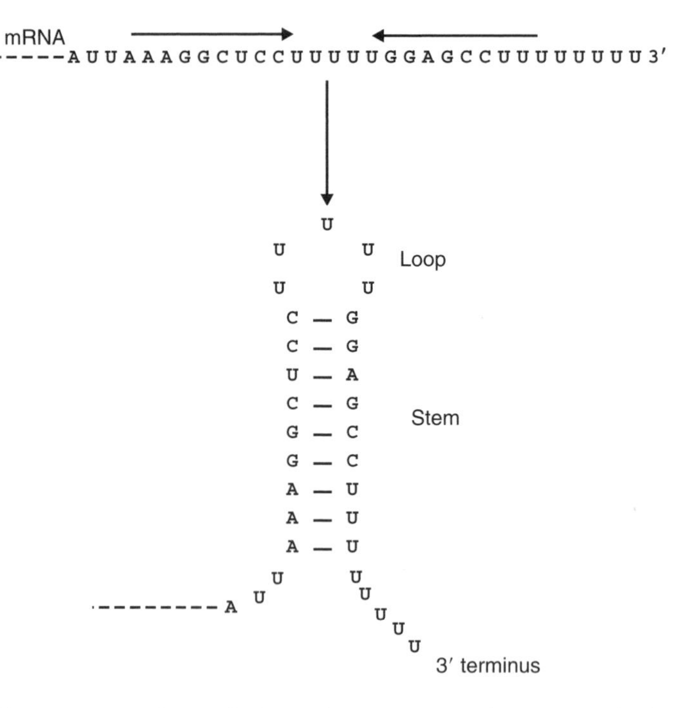

Figure 1.17 Structure of a typical terminator. The regions marked with arrows are complementary, and so can anneal together resulting in the formation of a stem–loop structure

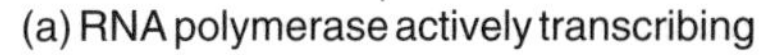

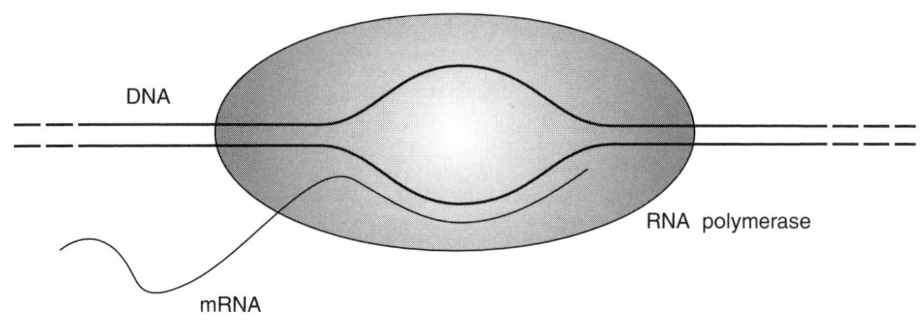

(b) Stem–loop structure forms in mRNA

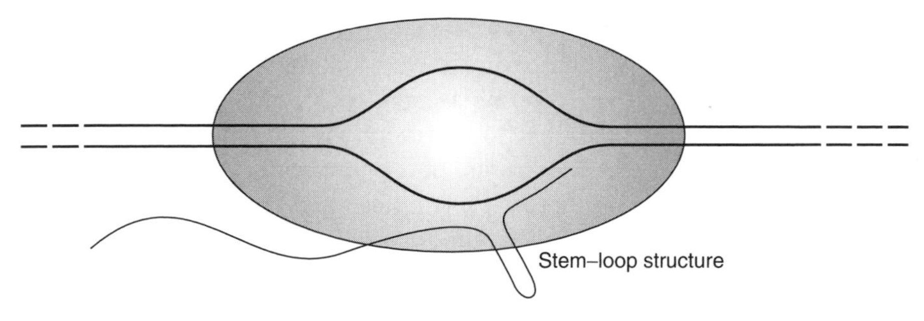

(c) DNA 'bubble' closes up; RNA polymerase pauses

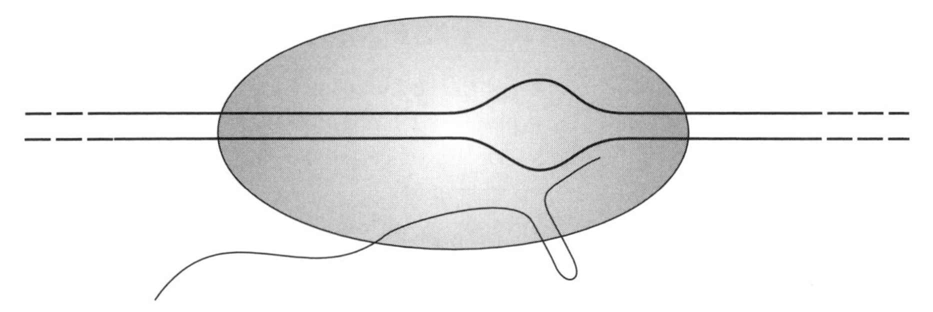

Figure 1.18 Model for transcription termination. (a) During transcription, the presence of the mRNA keeps the DNA strands separated over short region; (b) when the termination site is reached, the mRNA forms a stem–loop structure; (c) this allows the DNA strands to re-associate, leading to a pause in transcription

It is the mRNA itself that is responsible for keeping the DNA 'bubble' open, by remaining base-paired with the complementary DNA strand for a short while. Under physiological conditions, the RNA–DNA hybrid is more stable than the DNA–DNA pairing. The size of the bubble is limited however by topological constraints. The helix has to be unwound to some extent in order to allow the strands to separate, which causes stress in the molecule, since the two strands can only be separated by increasing the winding on either side. The larger the unwound region, the greater the stress. Beyond this point therefore the remainder of the mRNA molecule is dissociated from the template DNA (Figure 1.18a).

When the RNA polymerase encounters a transcriptional terminator sequence, it will transcribe it into RNA, with the consequent formation of a stem–loop structure which includes a portion of RNA that would otherwise be engaged on keeping the DNA bubble open (Figure 1.18b). The bubble will close up, which will hinder the activity of the RNA polymerase. The enzyme will therefore pause at this point, a few bases beyond the stem–loop structure (Figure 1.18c). Since the stem–loop structure in a typical termination sequence is followed by a string of U residues (Figure 1.17), all that keeps the mRNA attached at this stage is the relatively weak hydrogen bonding of the A-U base pairs. As a result, the RNA tends to dissociate from the DNA template, thereby terminating mRNA synthesis.

With some terminators, the stem–loop structure is not followed by a run of U residues. Although the RNA polymerase may pause at these sites, termination is dependent on the activity of another protein known as the rho factor. These are therefore known as rho-dependent terminators.

1.5.2 Translation

The genetic code

The mRNA carries the information for the sequence of amino acids in a protein in the form of the Genetic Code (see Appendix F) in which each occurrence of one of the 64 groups of three nucleotides (triplets or *codons*) codes for a specific amino acid (or for a stop signal).

The code is almost universal, in all species, although there are occasional minor differences, such as the use of UGA, which is normally a stop codon, to code for tryptophan or cysteine. This indicates that the code as we know it must have originated early in the evolutionary process and then became fixed because of the effect that any change would have on virtually every gene in the cell.

Ribosomes

Bacterial ribosomes typically consist of two subunits with sedimentation coefficients of 50S and 30S, the whole structure being referred to as a 70S ribosome (sedimentation coefficients are not additive). The larger (50S) subunit has two

RNA molecules (23S and 5S) plus 31 different polypeptides, while the smaller one (30S) contains a single RNA molecule (16S) and 21 polypeptides. Note that the structure of eukaryotic ribosomes is different in several respects.

The ribosomal RNA molecules form a very stable three-dimensional structure by extensive base-pairing, which allows them to perform a scaffolding role by attachment of the various ribosomal proteins. The role of the rRNA extends further than this, as it is involved in recognition of the mRNA (see below) and in the catalytic events leading to peptide bond formation.

In bacteria, the ribosomes attach to a specific sequence on the mRNA (the ribosome binding site, or RBS, also known as the Shine–Dalgarno sequence after the workers who first recognized its significance). This sequence is partly complementary to the 3′ end of the 16S rRNA (Figure 1.19), so that binding of the ribosomes can be mediated by hydrogen bonding between the complementary base sequences. This will normally occur as soon as the binding site is available, so the mRNA will start to be translated while it is still being formed (Figure 1.20).

However, it is not the complete ribosome that initiates these events. Ribosomes that are not involved in translation dissociate into their constituent 50S and 30S subunits. For translation to start, a 30S subunit binds to a ribosome binding site, and an initiator tRNA (see below) associates with an adjacent initiation codon (usually AUG, but sometimes GUG or even less commonly CUG). The 50S subunit can then attach to this initiation complex, and the process of translation can get underway. These steps also require the action of several non-ribosomal proteins known as initiation factors.

Since the mRNA is read in consecutive groups of three (with no punctuation), it could code for three completely different proteins, depending on where it starts, i.e. there are three potential reading frames. The position of the ribosome binding site and the initiation codon determines the reading frame. As we will see in Chapter 2, the addition or deletion of a single base will change the reading frame and the coding property of the subsequent message is totally different.

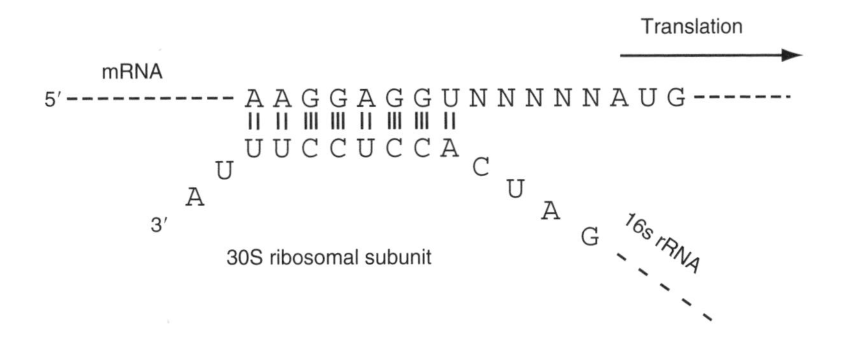

Figure 1.19 Ribosome binding site (Shine–Dalgarno sequence)

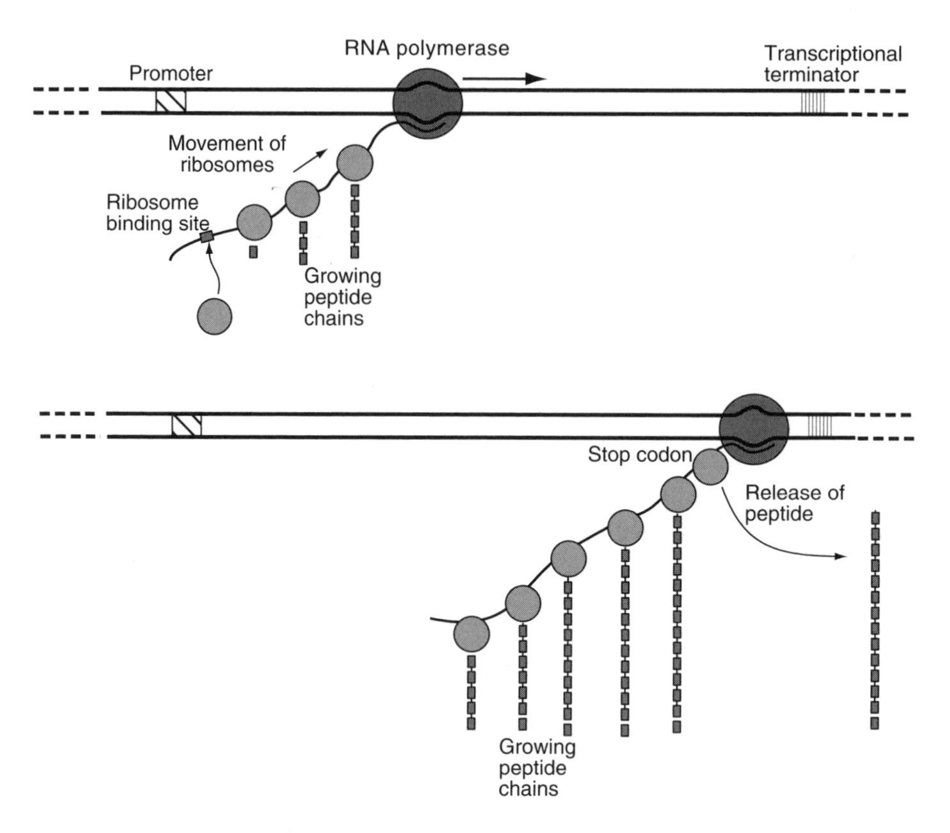

Figure 1.20 Translation of mRNA

Transfer RNA

Recognition of each triplet codon is mediated by small RNA molecules known as transfer RNA (tRNA). There is at least one tRNA species specific for each amino acid. However, they are all quite similar in their structure, consisting of a single RNA chain of 75–100 nucleotides folded back on itself in a form usually depicted as a cloverleaf structure (Figure 1.21); the actual three-dimensional structure is more complex and compact than this simplified two-dimensional diagram. Two parts of this molecule have clear functions in protein synthesis: the acceptor arm, formed by base-pairing of the 5′ and 3′ terminal regions, provides the site for attachment of an amino acid (by acylation of the 3′ end), and the anticodon arm which contains the bases (the anticodon) that recognize the triplet codon in the mRNA by base-pairing.

The appropriate amino acid is added to the tRNA by a specific enzyme (one of a number of aminoacyl tRNA synthetases) which has a crucial dual specificity: it is capable of recognizing a single tRNA species and also the correct amino acid with which that tRNA should be charged. Thus for example the codon

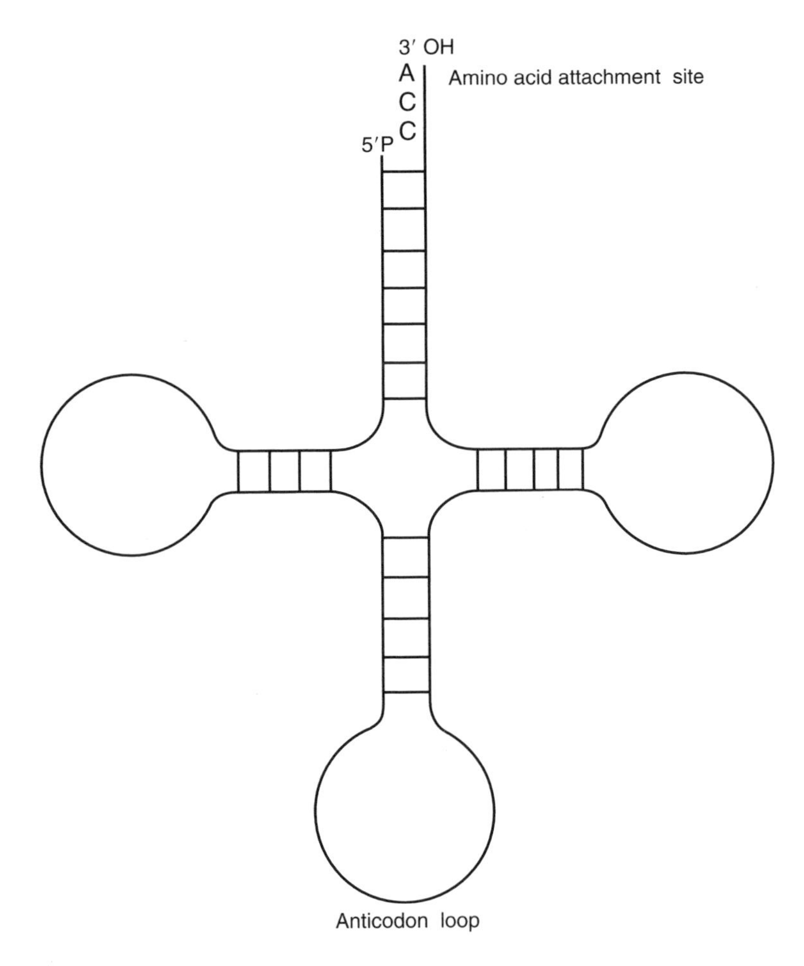

Figure 1.21 Diagrammatic structure of transfer RNA

UGG (which codes for tryptophan) will be recognized by a specific tRNA (designated $tRNA^{trp}$). This tRNA will be recognized by the tryptophanyl tRNA synthetase. This therefore ensures that the tRNA is charged with the appropriate amino acid.

So there are three separate elements to the specificity of this process: codon–anticodon interaction, recognition of the specific tRNA by the aminoacyl tRNA synthetase and recognition by the enzyme of the appropriate amino acid. Since tRNA molecules are all basically quite similar, and some amino acids (such as isoleucine and valine) are also similar to one another, it would not be surprising if mistakes were made occasionally. The low frequency of such errors (it has been estimated that one protein molecule in a thousand contains one incorrect amino acid) is due to the existence of an editing mechanism whereby the synthetase is able to cleave the amino acid from an incorrectly charged tRNA molecule.

If all three elements of specificity were absolute, then there would have to be at least 61 different tRNA species: one for each of the 64 codons less the three stop codons, for which there is no corresponding tRNA. For many of the amino acids there are indeed multiple tRNA species with different codon specificities. Some of these tRNA molecules are present at comparatively low levels in the cell, which would indicate that there could be a difficulty in translating that particular codon. This can be correlated to some extent with the frequency of occurrence of particular codons (codon usage): those codons that require a rare tRNA species tend also to occur less commonly, at least in highly expressed genes.

However, this is not the complete story. For many tRNA molecules, the codon–anticodon recognition is not absolutely precise; in particular, there is some latitude allowed in the matching of the third base of the codon. A rather complex set of rules (the wobble hypothesis) has been developed to account for the extent of allowable mismatching. So some tRNA molecules are able to recognize more than one codon. The number of tRNA species required for recognition of the complete set of codons is thus considerably less than 61 (commonly between 30 and 40).

Mechanism of protein synthesis

In bacteria, the initiation codon is recognized by a specific tRNA molecule, $tRNA^{fMet}$. After this tRNA molecule is charged with methionine, the amino acid is modified, to *N*-formylmethionine. Aminoacylated tRNA molecules normally bind to a site on the ribosome known as the A site (Acceptor), while their anticodon region pairs with the mRNA. Only after peptide bond formation is the tRNA able to move to a second site on the ribosome, the P (Peptide) site. The fMet-$tRNA^{fMet}$ (i.e. the $tRNA^{fMet}$ charged with formylmethionine) is unique in being able to enter the P site directly.

The $tRNA^{fMet}$ anticodon recognizes (forms base pairs with) the start codon on the mRNA, in association with binding of a 30S ribosome subunit to the nearby ribosome binding site. The 50S ribosome subunit then joins the complex (Figure 1.22). The charged tRNA corresponding to the second codon then enters the A site on the ribosome and peptide bond formation occurs by transfer of the fMet residue to the second amino acid. The $tRNA^{fMet}$, now uncharged, is released, and the ribosome moves one codon along the mRNA, which is accompanied by movement of the second tRNA molecule (now charged with a dipeptide) from the A site to the P site; this step is known as translocation. The A site is thus free to accept the charged tRNA corresponding to the third codon. The cycle of peptide bond formation, translocation, and binding of a further aminoacylated tRNA requires several additional non-ribosomal proteins (elongation factors) and is accompanied by the hydrolysis of GTP.

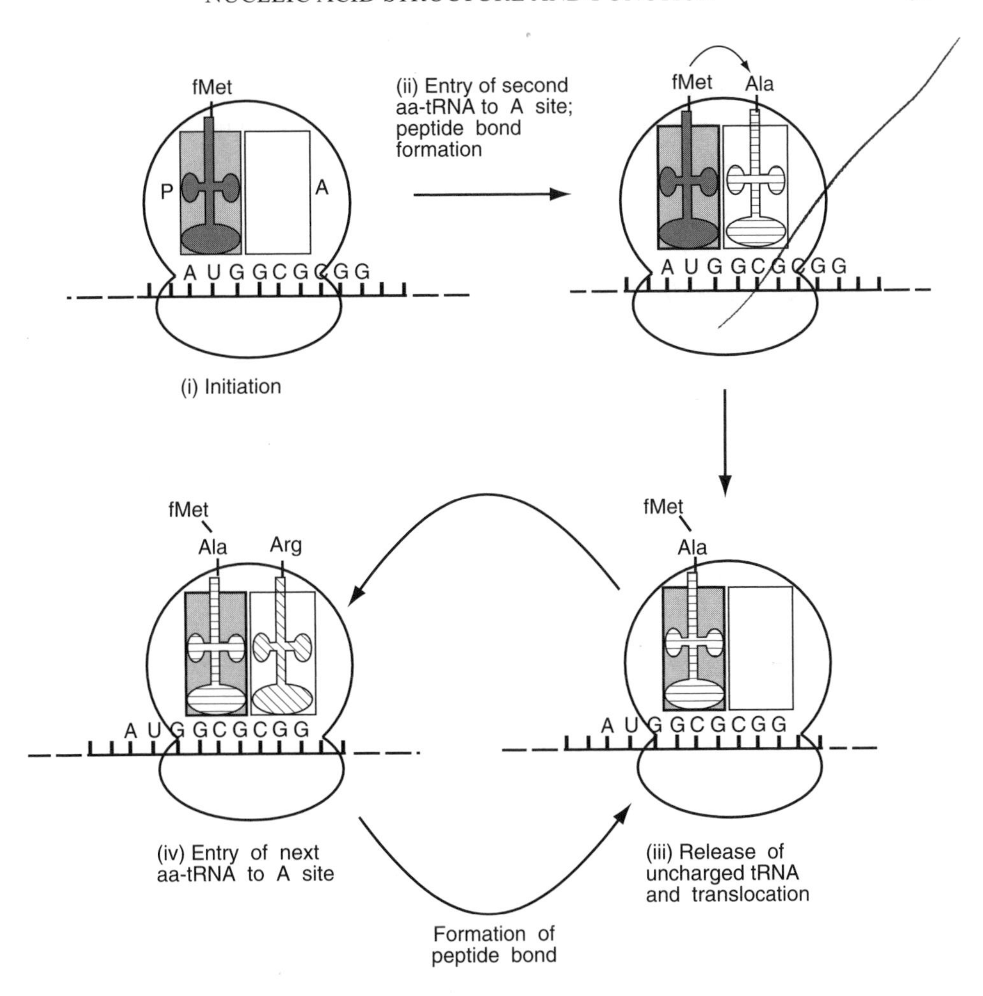

Figure 1.22 Outline of the mechanism of protein synthesis. (i) The initiation complex is formed by the binding of a 30S ribosomal subunit to the mRNA, followed by the initiating mRNA and a 50S subunit. (ii) The second tRNA enters the A site and the first peptide bond is formed by transfer of the N-terminal fMet to the free amino group of the second amino acid. (iii) The uncharged initiating tRNA is released and the ribosome moves along the mRNA. The second tRNA is now in the P site and the A site is free for the next tRNA. (iv) Entry of the next tRNA to the A site, ready for peptide bond formation as in step ii. Steps iii and iv are then repeated, until a stop codon is reached

When the ribosome has moved far enough, the ribosome binding site on the mRNA is exposed again and another ribosome can attach to it. A single mRNA molecule will therefore carry a number of ribosomes actively translating the sequence (Figure 1.20).

Each ribosome moves along the mRNA until a stop codon is reached. The absence of a corresponding tRNA species capable of recognizing this codon causes translation to stop at this point. The polypeptide chain is then released with the aid of proteins known as release factors and the ribosome dissociates from the mRNA.

1.5.3 Post-translational events

Translation of the information into a polypeptide does not complete the story. The formation of a biologically-active product involves several further steps. First of all, the protein has to fold correctly. There are three conformational levels in addition to the primary structure (which is the amino acid sequence itself). The secondary structure is the spatial arrangement of successive amino acids which may form regular structures such as α-helices or β-sheets, which are stabilized by non-covalent interactions (such as hydrogen bonds). Different regions of the protein will adopt different secondary structures, separated by turns or less-defined loops.

The various elements of secondary structure are in turn folded together to form the tertiary structure. This conformation is stabilized by non-covalent interactions and also by covalent disulphydryl bridges between cysteine residues. The tertiary structure may include two or more semi-autonomous regions known as *domains*. Many proteins consist of several (identical or different) polypeptide chains; the way in which these polypeptides are associated constitutes the quaternary structure.

It is not easy to predict the structure that will be adopted by a polypeptide chain, partly because of the large number of possibilities and partly because the polypeptide will tend to form secondary (and higher order) structures as it is being produced, thereby posing constraints on the structure adopted by the subsequent parts of the chain. The final structure is not necessarily the most thermodynamically stable form that can be adopted by refolding of the entire polypeptide.

Furthermore, the folding of the polypeptide is not entirely spontaneous. Cells contain proteins known as *molecular chaperones* which assist in obtaining the correct conformation of proteins, for example by interacting with the nascent polypeptide to prevent it from adopting an incorrect conformation until the complete protein is produced. Molecular chaperones can also play an important role in the refolding of denatured proteins, which can provide a degree of protection against heat and other stress conditions. Some of these molecular chaperones are specifically produced under conditions that lead to the accumulation of denatured protein and are known as heat-shock proteins or stress proteins. A further role of molecular chaperones is concerned with the assembly of polypeptide subunits into multimeric proteins or larger structures; for example, the assembly of bacteriophage heads (Chapter 4) may require the action of molecular chaperones.

Secretion

Many bacterial proteins have functions that require them to be present on the surface of the cell or in the extracellular environment. The first barrier to protein export is the cytoplasmic membrane and the most common mechanism for transport of proteins across this membrane is known as the general secretory pathway (GSP, sometimes called the Sec-dependent pathway). All proteins that utilize this system have a specific sequence at the N-terminus which targets the protein to this pathway and which is cleaved during transport. In Gram-positive bacteria, this mechanism is sufficient for export of proteins to the cell surface or to the surrounding medium. However, Gram-negative bacteria also have an outer membrane and the GSP by itself will deliver proteins, not to the outside of the cell, but to the region known as the periplasm, between the cytoplasmic and outer membranes.

Although Gram-negative bacteria are less prolific than Gram-positive bacteria in secreting proteins, they do have important secretion mechanisms. The most common of these, the Type II mechanism, is dependent on the GSP for transport of proteins to the periplasm and then uses a specific multiprotein complex to transport the protein across the outer membrane. Type V secretion systems are similar, in that they are dependent on the Sec machinery for transport to the periplasm, but the translocated protein has a specialized C-terminal sequence which is able to insert into the outer membrane to form a pore. The N-terminal sequence of the same protein is able to pass through this pore, after which it is cleaved from the C-terminal sequence releasing it into the extracellular environment. These proteins are often called 'autotransporters' because they mediate transport through the outer membrane themselves and do not require additional machinery.

Most Gram-negative bacteria also possess a number of separate secretion systems which are not dependent on the GSP and which are, therefore, termed Sec-independent. These include Type I, Type III and Type IV secretion pathways. The Type I pathway, is relatively simple, consisting of just three proteins, and targets proteins with a specific 60-amino acid secretion signal at the carboxy terminus. The Type IV secretory apparatus bears many similarities with the conjugal plasmid transfer systems described in Chapter 6 and can be used by a bacterium to introduce proteins into eukaryotic cells. Perhaps the most remarkable secretion system is the Type III pathway, which has been likened to a molecular syringe and which is used to inject proteins directly into the cytosol of eukaryotic cells (this type of system is described in more detail in Chapter 9).

Secretion is only triggered by direct contact between the bacterium and the host cell. Consequently, this pathway is widely used by pathogens to introduce effector molecules into host cells, thus subverting the normal function of the cell to the benefit of the bacterium.

Other post-translational modifications

In addition to the events described above, proteins may undergo a wide range of additional post-translational modifications, such as glycosylation, biotinylation, addition of lipids and proteolytic cleavage. The full range of these is too complex to be covered here, but the outcome is that the final structure can be influenced very strongly by the nature of the cell itself. Since the post-translational events may be essential in obtaining a product with full biological activity, the difficulty in obtaining accurate post-translational modification can severely affect the outcome of attempts to obtain functional gene expression in heterologous hosts (see Chapter 8).

1.6 Gene organization

In bacteria, genes with related functions are often (but not always) located together in a group known as an *operon* (Figure 1.23). An operon has a single promoter and is transcribed into a single polycistronic mRNA molecule, which carries the information for several proteins. This group of genes will therefore be coordinately controlled: growth of the cell under the appropriate conditions will

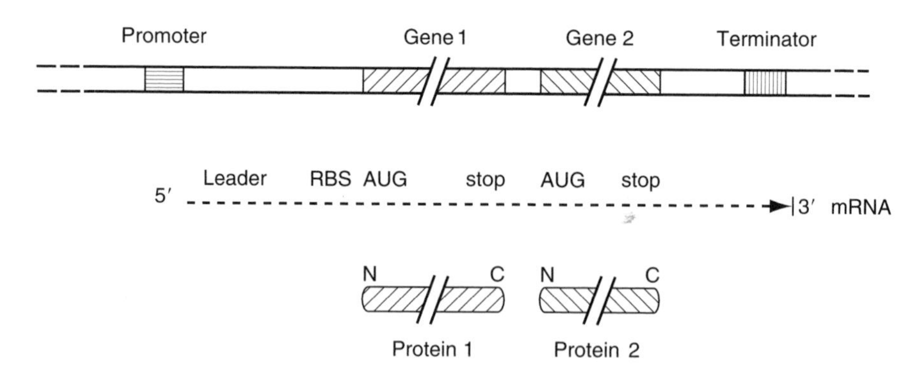

Figure 1.23 Structure and expression of a bacterial operon. A typical operon is transcribed from a single promoter into a polycistronic mRNA from which several independent polypeptides can be translated

induce all the genes in the operon simultaneously. This is discussed more fully in Chapter 3.

After the ribosome has translated the first cistron in an operon, it may dissociate, in which case translation of the next cistron will require attachment of ribosomes to another binding site adjacent to the initiation codon of the second cistron. In some cases the start codon for the second gene is very close to the stop codon of the first (in fact the sequences may actually overlap). If this occurs, then after the first polypeptide has been released, the ribosome may start translating again at the nearby start codon without dissociating.

Some major differences between bacteria and eukaryotes are worth noting. The mRNA in eukaryotes acts as a true 'messenger', being produced within the nucleus and migrating to the cytoplasm for translation to occur. In bacteria, transcription and translation occur in the same compartment and the ribosomes will attach to the mRNA as soon as a ribosome binding site is available. So, in bacteria, the mRNA is being actively translated while it is still being made. Most bacterial mRNA is extremely short-lived; it typically has a half-life of a few minutes only, which may be less than the time required for producing or translating it. This can only be achieved by the coupling of transcription and translation. A further difference related to the mechanism of ribosome binding is that eukaryotic mRNA (in general) codes for a single polypeptide only. The ribosome in eukaryotic cells attaches to the 5′ end of the mRNA and migrates until it reaches the start codon. (Internal ribosome entry sites do exist in eukaryotes, but they are the exception rather than the rule).

A further difference is that in eukaryotes, the initial product of transcription is a precursor of the mRNA. This precursor, which is found only in the nucleus, contains additional sequences (introns) that are removed by a process known as splicing or processing. In some cases the final size of the mRNA is less than 10 per cent of that of the original gene. Generally, bacterial genes do not contain introns, but there are a few examples of prokaryotic genes (mainly from bacteriophages) that do contain introns.

Finally, eukaryotic mRNA is often (but not always) polyadenylated, i.e. it has a run of adenine residues at the 3′ end. The presence of poly-A tails is often used as the basis of procedures for purifying mRNA from eukaryotic cells. Bacterial mRNA, on the other hand, is not consistently polyadenylated, although a small proportion of bacterial RNA molecules may carry a short oligo-A tail.

2

Mutation and Variation

The cornerstone of bacterial genetics has been, until recently, the isolation of specific mutants, i.e. strains in which the gene concerned is altered (usually in a deleterious fashion). This alteration shows up as a change in the corresponding characteristics of the organism. It is this change in the observable properties of the organism (the *phenotype*) that is used to follow the transmission of the gene. The genetic nature of the organism (the *genotype*) is inferred from the observable characteristics.

Later in the book we will look at the impact of gene cloning technology, which provides new dimensions to the study of genetics in that it is possible to isolate and study the structure of specific genes without previously obtaining mutants and also to make specific alterations to the structure of the gene *in vitro*. However, knowledge of the function of that gene in relation to the metabolism of the cell can only be completed by using *in vivo* analysis.

Any population of bacteria is far from homogeneous. A culture of *E. coli* (under optimal conditions) will grow from a single cell to its maximum cell density (commonly about 10^9cells ml^{-1}) in about 10–15 h, having passed through 30 generations. Within that culture, there will be some variation from one cell to another, which can be due to physiological effects or to genetic changes. Physiological variation means that, due to differences in the environment, growth history or stage of growth at any one moment, the cells may respond differently to some external influence. The key difference between physiological and genetic variation is whether the altered characteristic can be inherited.

2.1 Variation and evolution

Modern evolutionary theory is derived from the concepts advanced by Charles Darwin in *The Origin of Species* (1859), and independently by Alfred Russell

Molecular Genetics of Bacteria, 4th Edition by Jeremy Dale and Simon F. Park
 ISBN 0 470 85084 1 (cased) ISBN 0 470 85085 X (pbk)

Wallace, namely that species evolve by a process of natural selection, by which those individuals that are better adapted to their environment will survive and breed more effectively, thus passing on those characteristics to their progeny. There were two large gaps in this theory: firstly, the mechanism of inheritance was poorly understood, and secondly, it did not explain satisfactorily the nature of the variation in the population.

On the first point, the research of Gregor Mendel enabled him to propose that heredity was due to a combination of discrete units rather than a blend of the characteristics of the parent. This work, presented in 1865, was largely ignored until 1900 when the development of understanding of the role of chromosomes provided a basis for appreciation of Mendel's results.

As for the nature of variation, although it subsequently became a central dogma that variation is essentially a random process that occurs without being influenced by the environment, Darwin's ideas had much more in common with the inheritance of acquired characteristics (facetiously exemplified by the idea that giraffes have long necks because they stretch to reach the branches of trees), as formalized 50 years earlier by Jean Baptiste Lamarck. In the early part of the 20th century, evidence accumulated against Lamarckism and in favour of random variation. Nevertheless, a form of Lamarckism persisted in the USSR where Lysenko, who believed that environmental changes could cause permanent genetic alterations, held such a powerful position that all Soviet genetics was dominated by these views until 1964.

Our view of evolution therefore, although often described as Darwinian, is in reality an amalgam of ideas from many sources (especially Darwin and Mendel), and is more accurately referred to as the neo-Darwinian synthesis.

2.1.1 Fluctuation test

Lamarckism persisted amongst bacteriologists long after it had become discredited amongst most geneticists, largely because of the ease with which certain types of mutation could appear when the bacteria were exposed to selective conditions, which seemed to suggest that those conditions had actually induced the relevant mutations. Luria and Delbruck, in 1943, published results which were thought to be the final death blow to this view.

They examined the occurrence of mutations of *E. coli* that caused resistance to the bacteriophage T1. Such mutations are readily isolated, simply by exposing a bacterial culture to the phage. The question is: are the mutations already there before exposure to the phage (random variation), or do they develop after (and because of) exposure to the phage (directed mutation)?

In order to answer this question they set up a series of small cultures, inoculated with a small aliquot of a single starter culture (Figure 2.1). After incubation they determined the number of phage-resistant bacteria present in each small culture.

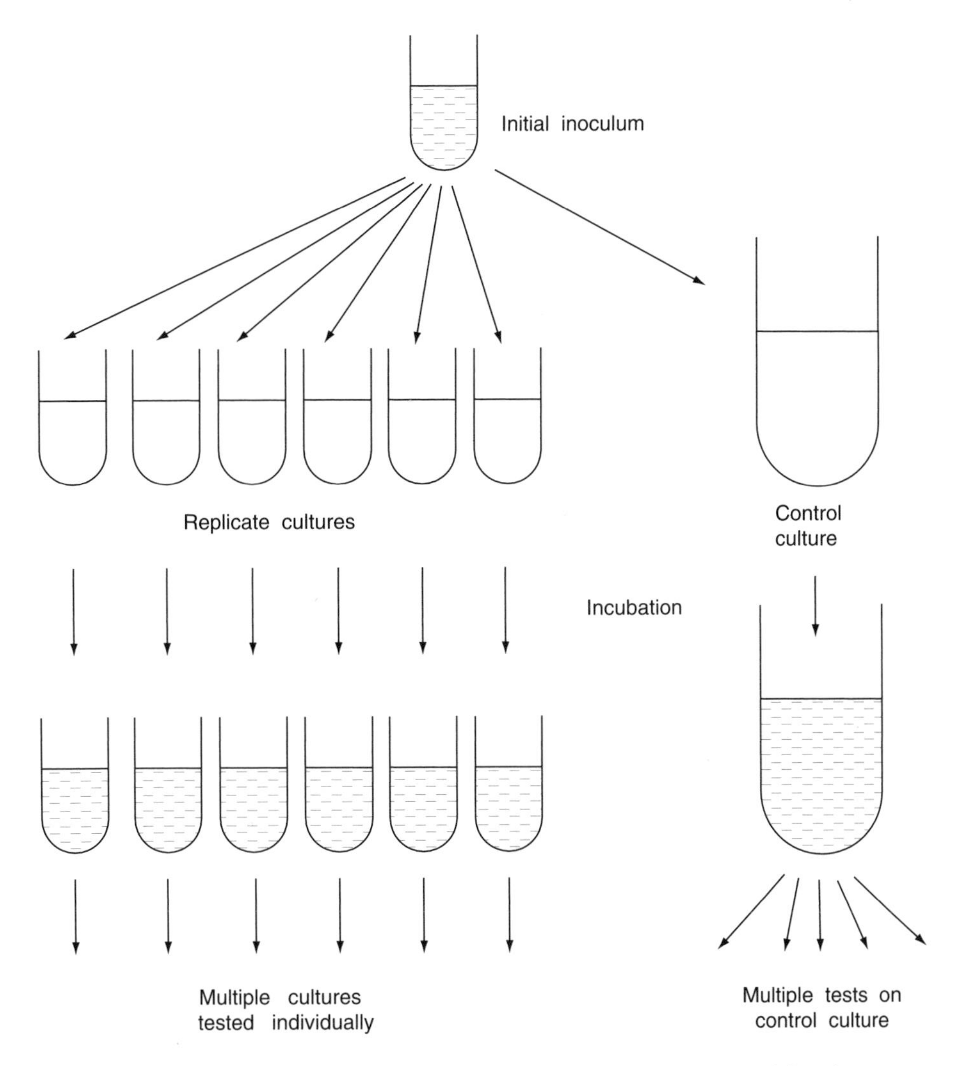

Figure 2.1 Fluctuation test. A number of replicate cultures are tested for the presence of a specific mutation. If mutation occurs randomly, the variation will be much greater than that seen in multiple tests of a single control culture (see Figure 2.2)

If the mutation occurred as a response to exposure to the phage, one would expect a similar number of mutants in each culture, subject to statistical and sampling variation. The number of mutants would be the same whether a number of samples from a single culture or one sample from each of a number of independent cultures had been selected.

On the other hand, if the mutation occurs at random, before exposure to the phage, then the number of resistant mutants should vary considerably from one tube to another, since the mutation can occur at any time during the growth of the cultures. This principle is illustrated in Figure 2.2 which shows that

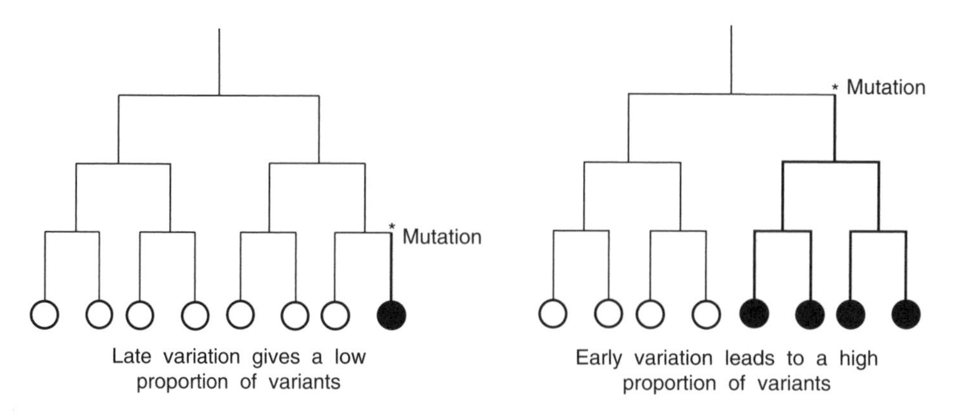

Figure 2.2 Principle of the fluctuation test. A random mutation may occur at any time during growth of a culture. If a mutation occurs at a late stage, the proportion of mutants in the culture will be low. If a mutation occurs earlier, the proportion of mutants will be much higher

mutations occurring at a late stage will give rise to a small proportion of mutants while those occurring earlier will yield more mutants. This was the result obtained: the variation between the independent cultures was considerably greater than between samples from the one tube, thus supporting the random mutation hypothesis.

2.1.2 Directed mutation in bacteria?

Despite the apparently conclusive results of the fluctuation test, there have been suggestions that something akin to directed mutation can be observed in bacteria under certain circumstances. This originated with work by Cairns and others which showed that a strain of *E. coli* that was unable to ferment lactose (Lac^-) reverted to Lac^+ at a higher frequency when supplied with lactose as the sole carbon source (i.e. conditions under which the ability to ferment lactose was advantageous) than when supplied with glucose (the ability to ferment lactose therefore being of no advantage).

It therefore seems that under these conditions, mutation was not totally random, but that advantageous mutations were occurring more readily than non-advantageous ones. This does not necessarily mean that Lamarck was right after all. This effect only seems to happen when the selective conditions prevent growth (i.e. not when the cells are actually killed, as was the case with the bacteriophage infection experiments), and it may be related specifically to what happens in cells that are unable to grow. The absence of net growth does not necessarily mean the absence of all metabolic activity, which may include some form of DNA replication (e.g. connected to DNA repair). There is some evidence that cells in such a state are more susceptible to mutation which could account for these observations. The interpretation of this effect is still highly controversial.

2.2 Types of mutations

2.2.1 Point mutations

There are many ways in which the structure of the genetic material may change. Much of the basis of genetics has been established using simple mutations (*point* mutations) in which the sequence of the DNA has been altered at a single position. Where this change consists of replacing one nucleotide by another, it is known as a *base substitution*. The consequence of such a change depends both on the nature of the change and its location. If the change is within the coding region of a gene (i.e. the region which ultimately is translated into protein), it may cause an alteration of the amino acid sequence which may affect the function of the protein. The alteration may of course have little or no effect, either because the changed triplet still codes for the same amino acid or because the new amino acid is sufficiently similar to the original one for the function of the protein to remain unaffected.

For example, the triplet UUA codes for leucine; a single base change in the DNA can give rise to one of nine other codons as shown in Figure 2.3. Two of the possible changes (UUG, CUA) are completely silent, as the resulting codons still code for leucine. These are known as *synonymous codons*. Two further changes (AUA and GUA) may well have little effect on the protein since the substituted amino acids (isoleucine and valine respectively) are similar to the

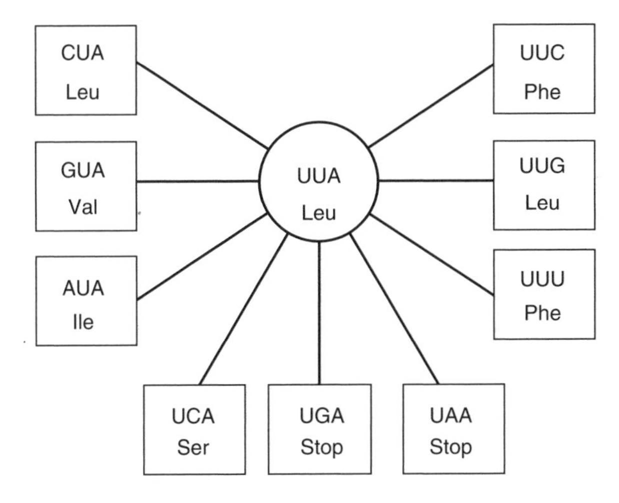

Figure 2.3 Codons arising by single base substitutions from UUA

original leucine (they are all hydrophobic amino acids). Phenylalanine (UUU or UUC codons) is also hydrophobic but is more likely to cause a significant change in the structure of the protein at that point. The significance of the change to UCA, resulting in the substitution of serine (which is considerably different) for leucine will depend on the role played by that amino acid (and its neighbours) in the overall function or conformation of the protein.

The final two changes (UAA, UGA) are referred to as stop or termination codons (as is a third codon, UAG), since they result in termination of translation; there is normally no tRNA molecule with the corresponding anticodon. The occurrence of such a mutation (also known as a *nonsense* mutation) will result in the production of a truncated protein; such a protein may or may not be functional, depending on the degree of shortening. The UAG codon was named 'amber' which is a literal translation of the German word 'Bernstein', the name of one of the investigators who discovered it; subsequently the joke was continued by calling the UAA and UGA codons 'ochre' and 'opal' respectively, although the latter two names are less commonly used.

A different kind of mutation still involving a change at a single position, consists of the deletion or addition of a single nucleotide (or of any number other than a multiple of three). This is known as a *frameshift* mutation, since it results in the reading frame being altered for the remainder of the gene. Since the message is read in triplets, with no punctuation marks (the reading frame being determined solely by the translation start codon), an alteration in the reading frame will result in the synthesis of a totally different protein from that point on. This point is illustrated in Figure 2.4.

In fact, protein synthesis is likely to be terminated quite soon after the position of the deletion. For most genes the two alternative reading frames are blocked by termination codons, which serves to prevent the production of aberrant proteins by mistakes in translation. Non-translated regions also often contain frequent stop codons, so that the existence of a long region of DNA that does not contain a stop codon in one reading frame (i.e. it has an Open Reading Frame or ORF) is used to identify hitherto unknown genes from the DNA sequence (see Chapter 10).

If a point mutation results in premature termination of translation, it may also affect the expression of other genes downstream in the same operon. This effect is known as *polarity*, and needs to be considered in genetic analysis (see Chapter 10).

2.2.2 Conditional mutants

There are many genes that do not affect resistance to antibiotics or bacteriophages, biosynthesis of essential metabolites or utilization of carbon sources. Some of these genes are indispensable and any mutants defective in those

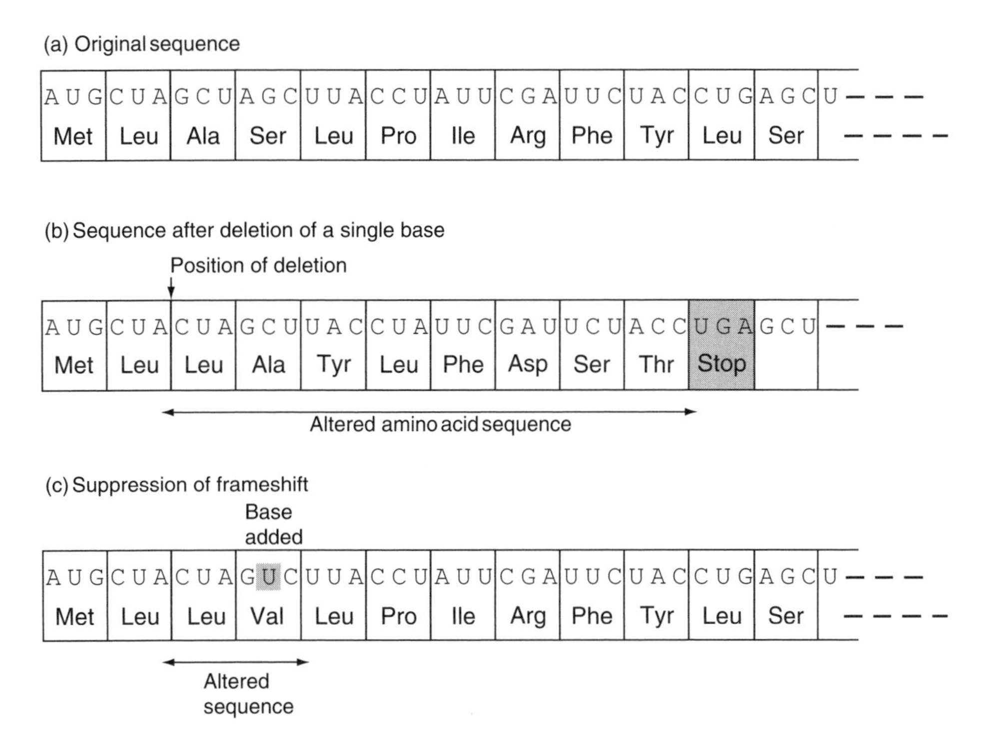

Figure 2.4 Frameshift mutation and suppression. (a) Initial (mRNA) sequence and translated product. (b) Deletion of a single base alters the subsequent reading frame producing a different amino acid sequence and encountering a stop codon. (c) Addition of a base at a different position restores the original reading frame and may suppress the mutation

activities would die (or fail to grow). Since this includes a wide range of genes that control the essential functions of the cell, such as DNA replication, it is important to be able to use genetic analysis to understand the role of these genes and their products. This can be done by using *conditional mutants*. This means that the gene functions normally under certain conditions while the defect is only apparent when the conditions are changed. One very useful type of conditional mutation confers temperature sensitivity on the relevant function. So, for example, a strain with a temperature-sensitive mutation in a gene needed for DNA replication would be able to grow normally at, say 30 °C (the *permissive* temperature) but would be unable to grow at a higher temperature, such as 42 °C. In a later section of this chapter we will consider a special class of mutants of bacteriophages (*amber* mutations) which are another example of conditional mutants, as they are only able to replicate in a host strain carrying an amber-suppressor mutation.

2.2.3 Variation due to larger scale DNA alterations

All the mutations described so far are point mutations. However, much of the variation that occurs in bacteria (and other organisms) is due to far more substantial alterations to the DNA structure.

The simplest of these are deletions where a substantial part of a gene (or even several genes) is totally absent. This obviously destroys the function of that gene and has the considerable advantage, for genetic analysis, that deletions do not revert. Genetic analysis often involves the study of very rare events (e.g. recombination between two closely linked genes) which happen perhaps once in 10^9 cells, and therefore relies on the stability of the genetic markers used. Simple point mutations may revert at a frequency in excess of this which makes the results difficult to interpret.

Other types of larger scale alterations are important in the generation of the natural diversity of micro-organisms. When an extraneous piece of DNA is inserted within a gene, it will usually inactivate that gene. Elements known as Insertion Sequences (IS) have a specific ability to insert into other DNA sequences, thus generating insertion mutations. A substantial proportion of spontaneous mutations may be due to inactivation of genes by insertion of a copy of an IS element rather than by replication errors.

Transposons are essentially similar to IS elements in that they have the ability to move (transpose) from one site to another; they differ in carrying one or more identifiable genetic markers. The most widely studied transposons carry antibiotic resistance genes and have played a key role in the evolution and spread of antibiotic resistance.

Instead of moving from one site to another, a region of DNA may flip round into the opposite orientation. If this invertible region contains the signals needed for expression of an adjacent gene, inversion will switch the associated gene on or off in a readily reversible (but still inherited) manner. The best studied example of this effect is the variation of the flagellar antigens of *Salmonella typhimurium*.

Insertion sequences, transposons, and invertible sequences are covered more fully in Chapter 7.

2.2.4 Extrachromosomal agents and horizontal gene transfer

In addition to alterations of the structure of the chromosomal DNA, variation in bacteria commonly occurs by the acquisition (or loss) of extrachromosomal DNA, either in the form of plasmids or bacteriophages. A wide range of characteristics, notably antibiotic resistance, can be encoded by these extrachromosomal elements. The transfer of genetic information from one strain to another is known as *horizontal gene transfer* and in some species may affect the structure of the chromosome itself, as well as the acquisition of extrachromosomal elements. Plasmids are considered in more detail in Chapter 5 and horizontal gene transfer is covered in Chapter 6.

2.3 Phenotypes

Mutations are initially characterized by means of their phenotypic effects – that is the effect they have on the observable characteristics of the cell. Many bacterial mutants fall into one of three phenotypic categories. Firstly, there are strains that require an additional growth supplement, such as a specific amino acid. These are referred to as auxotrophs, with the wild type, non-mutant strain being designated as a prototroph. Secondly, we have mutants that are defective in their ability to use a certain substrate. The best known of these are *E. coli* mutants that are unable to ferment lactose (Lac^-). The third type of commonly used mutation confers antibiotic resistance; for example, mutation in the gene coding for one of the ribosomal proteins can make the cell resistant to streptomycin. It is important to distinguish between antibiotic resistance caused by a mutation in a chromosomal gene and resistance that arises from the presence of a plasmid. Resistance to other chemical agents, or to bacteriophages, can also be valuable for genetic analysis. Mutation may of course result in a wide range of other changes to the phenotype and we will look at some examples in Chapter 9.

Unfortunately, observation of the phenotype does not always tell us much about the *genotype*, i.e. which gene is actually affected and the nature of the mutation in that gene. A particular phenotypic characteristic is often influenced by a number of genes. For example, the ability to synthesize tryptophan is determined by five genes that are linked together in the tryptophan operon. A defect in any one of these genes will lead to loss of the ability to make tryptophan and the strain would thus be designated Trp^-. This does not tell us which gene is at fault, but further analysis may show which gene has been altered. It is important to be able to describe the characteristics of a particular strain in a way that can be interpreted unambiguously by an informed observer. For this reason, a standard nomenclature has been devised for describing the phenotype and genotype of bacterial strains (see Box 2.1). The individual genes are denoted with, if necessary, an additional letter to indicate which gene is referred to. Thus the *trpA* and *trpB* genes code for the constituent subunits of the enzyme tryptophan synthase which is responsible for the final stages of tryptophan production i.e. the conversion of indole-3-glycerol phosphate to tryptophan.

However, designation of a strain as, for example, *trpA* does not necessarily mean that the strain is actually auxotrophic, merely that the *trpA* gene has been somehow altered. There are a number of ways in which a gene can be altered without abolishing the activity of its product. For example, the mutation may confer temperature sensitivity on the product of the gene concerned.

Box 2.1 Genetic nomenclature

There is a standard system for naming bacterial genes and their products. In all cases, three letter abbreviations are used; a list of gene designations used in this book will be found in Appendix E.

The phenotype is described using 'ordinary' letters (i.e. not italics), with the initial letter in upper case; a suffix is often added to clarify the nature of the defect. For example:

> Amp^r means the strain is resistant to ampicillin
> His^- denotes a histidine auxotroph
> Lac^- indicates inability to ferment lactose

Note the possible confusion that can arise between mutants of the second and third types. Thus a Lac^- strain is unable to use lactose, *not* unable to synthesize it. An element of common sense is needed to distinguish these possibilities.

For the genotype, a three-letter code is also used, but in italics and without the initial capital letter; in general, an additional upper case letter is added to indicate a specific gene. For example, inability to ferment lactose may be due to a defect in the β-galactosidase gene; the genotype would be described as *lacZ*. In the description of a genotype, it is not necessary to include a minus sign, a mutation being implied, but sometimes minus or plus signs are included to clarify the description.

This convention is also used to denote specific genes and their products. So 'the *recA* gene' is referred to, whereas the protein coded for by that gene would be called 'the RecA protein'.

A gene that has been inactivated by insertion (especially of a mobile element such as a transposon or insertion sequence) is shown as e.g. *hisA*::Tn*10*., indicating that the mutation in the *hisA* gene is due to insertion of the transposon Tn*10*.

The description can also show that the strain carries a plasmid which contributes to the phenotype. So a strain described as $hisA/F'hisA^+$ has a mutation in the chromosomal *hisA* gene, but contains an F′ plasmid which carries a functional *hisA* gene (indicated by the plus superscript).

This convention is not universally applicable; in this book it has been shown that many bacteriophage genes do not have three letter codes and in eukaryotic organisms the naming of genes is even more variable. However, the use of italics for genes and non-italic letters for gene products is a standard feature.

There are other genes with a *trp* designation that are not part of the tryptophan operon and where mutation does not lead to auxotrophy. For example, the *trpT* gene codes for tryptophan-specific tRNA and the *trpS* gene for the corresponding tryptophanyl-tRNA synthetase, while the TrpR protein is a repressor protein involved in the regulation of the *trp* operon. A *trpR* mutation may therefore actually increase the expression of the tryptophan operon or reduce the ability of the cell to turn off expression of the *trp* operon in the presence of tryptophan.

2.4 Restoration of phenotype

2.4.1 Reversion and suppression

Since a point mutation arises by a change at a single point in the DNA, an event which can occur at random, it follows that there is a good chance of a second mutation occurring which will restore the original DNA sequence (or a sequence that is for practical purposes indistinguishable). In the example shown in Figure 2.3, a mutant strain with the UUU codon (phenylalanine) may undergo a further mutation which restores the UUA codon (a true *back mutation*) or one that substitutes UUG or CUU (both of which are also leucine codons). This strain will now have the same properties as the original one and is said to have *reverted.*

The effect of a mutation can also be negated by a second, unrelated mutation; this effect is known as *suppression*. This can take a wide variety of forms, most of which are specific to the particular gene involved and occur when alteration of a second gene can counteract the deleterious effects arising from the loss of the first gene function. There are two types of suppression that are of more general importance.

The first occurs with frameshift mutations. These may revert by the restoration of a deleted base for example, but can also be suppressed by the addition or deletion of further bases (not necessarily at the same place as the original mutation) so that the total number of bases added or lost is a multiple of three (see Figure 2.4). In this way the original reading frame is restored leaving a limited number of altered codons. Whether this altered product has sufficient biological function to result in observable suppression of the original mutation will of course depend on the size and nature of this altered sequence and its effect on the function of the protein.

Another important type of suppression occurs with 'nonsense' mutations, where a stop codon has been created within the coding sequence. These result in termination of translation largely because there is no corresponding tRNA to recognize them. However, tRNA molecules are themselves coded for by genes, which are of course susceptible to mutation. It is therefore possible for an existing tRNA gene to be changed in such a way that the tRNA it codes for will now recognize one of the stop codons rather than (or as well as) the codon it normally recognizes. For example, in Figure 2.5 the original mutation changes a glutamine codon (CAG) to a stop codon (UAG). Suppression of this mutation can occur by alteration of the glutamine tRNA gene so that its anticodon now pairs with the amber codon. Glutamine will therefore be inserted into the growing peptide chain and the final product will be identical with the wild-type protein. Since there is more than one glutamine tRNA gene, the cell does not lose the ability to recognize genuine glutamine codons.

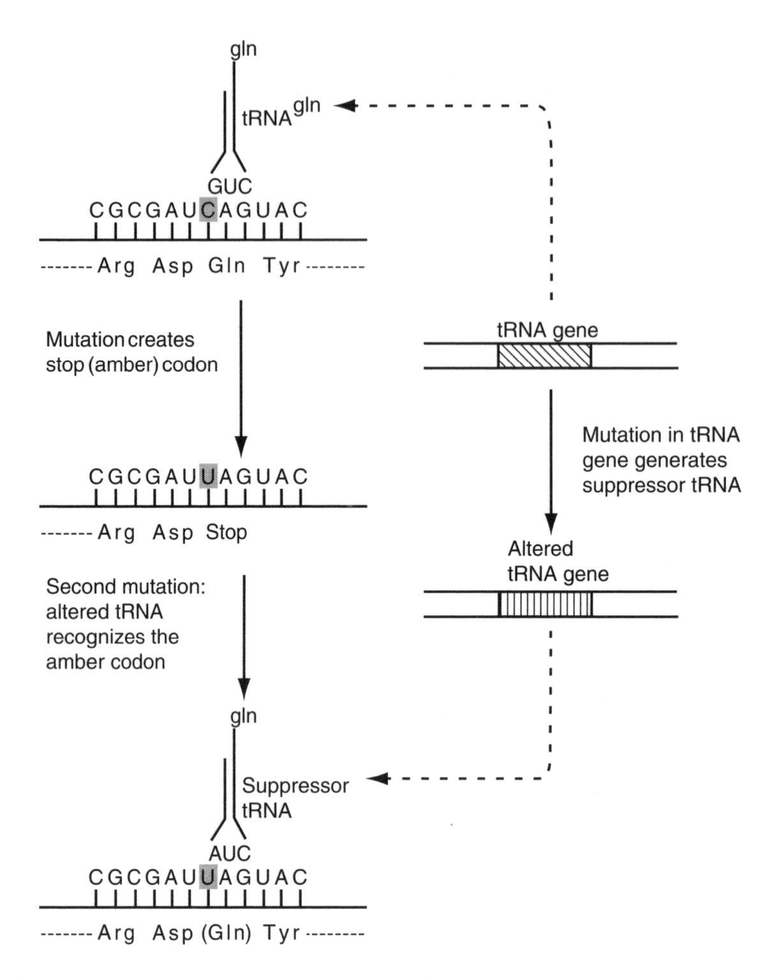

Figure 2.5 Suppression of a nonsense mutation. A base substitution changes CAG to the stop codon UAG, causing premature termination of translation. This can be suppressed by a separate mutation in a tRNA gene, giving rise to a tRNA that can recognize the UAG codon

It might be expected that this suppressor tRNA would now prevent normal chain termination at the end of the translated region which would clearly be very damaging to the cell. However, translational termination is reinforced by the action of release factors which means that suppression is far from absolute (it may range from 10–50 per cent for an amber suppressor). In addition, at the genuine termination site, there are often multiple termination codons which will lead to efficient termination even in the presence of a suppressor tRNA.

The importance of this type of suppression is that the tRNA mutation is able to suppress any corresponding mutation, not just the original one it was selected for. Of course, each of the stop codons can arise by mutation of a number of different codons (and there are also a number of possible suppressor tRNAs with different amino acid specificities) which means that in many cases the activity of the suppressor tRNA will result in the insertion of an amino acid other than the correct one. So, although a full-length product will be obtained, it may not be fully functional. Provided this can be overcome, the effect is extremely useful in bacteriophage genetics where amber mutations are commonly employed. Phages carrying an amber mutation will show a mutant phenotype (or fail to grow) on a normal bacterial host, but will show a wild-type phenotype when an appropriate amber suppressor host is used. The latter host is therefore referred to as the permissive host. This is another example of the use of *conditional mutants*, as discussed earlier.

2.4.2 Complementation

Another way in which a mutant phenotype can be converted back to the wild type is by acquisition of a plasmid that carries a functional version of the affected gene. For example, a Lac^- strain of *E. coli* will become able to use lactose again following introduction of a plasmid carrying the relevant genes. In this case, we say that the plasmid has *complemented* the chromosomal defect. This only works if the functional version is effective in the presence of the mutated gene, i.e. the mutation is recessive.

Traditionally, genetics has followed the route described above: isolating variants on the basis of the altered phenotype and then attempting to identify the nature of the genetic change responsible. The advent of gene cloning technology, and especially of genome sequencing, has opened up a different route. Using these techniques, we commonly know the sequence of many, or all, of the genes of an organism, but without knowing their functions. If we have some idea of the possible function of a specific gene (for example by comparing its sequence with that of known genes from other sources), we can test this hypothesis by examining the ability of the cloned DNA fragment to complement a characterized mutant.

2.5 Recombination

Recombination, in the sense of re-assorting the observable characteristics in the progeny of a cross, has been a fundamental feature of genetics since long before its inception as a formal discipline. The term 'recombination' can be used in an analogous fashion in bacterial genetics, but is also used to refer to the physical breaking and joining of DNA molecules.

At the simplest level, we can consider two linear DNA molecules: breaking both molecules at a single point, crossing them over and rejoining them will produce two recombinant DNA molecules, both of which have a part of each of the parental

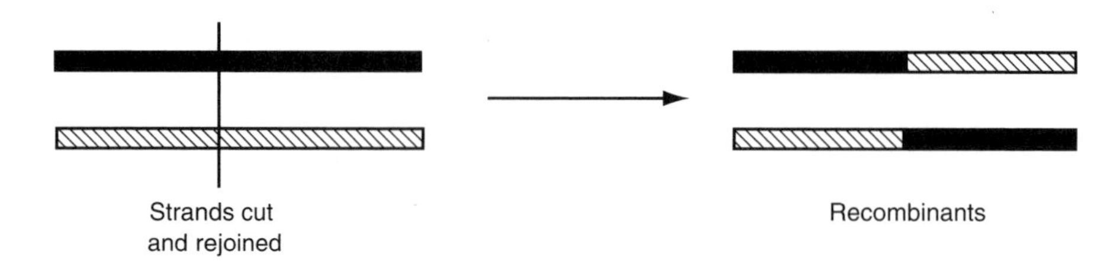

Figure 2.6 Recombination between two linear DNA molecules

molecules (Figure 2.6). This general concept applies to a variety of recombinational mechanisms, of which the principal one is known as *general* or *homologous* recombination; this requires a substantial degree of homology between the sequences to be recombined but will work with any two pieces of homologous DNA. In contrast, *site-specific* recombinational mechanisms require little or no homology, but (as the name implies) operate only within specific sequences. The RecA protein is required for homologous recombination, but not for site-specific processes. Recombination mechanisms are considered further in Chapter 6.

2.6 Mechanisms of mutation

2.6.1 Spontaneous mutation

Spontaneous mutation occurs through errors in the replication of DNA. However, the model presented for the structure and replication of DNA (Chapter 1) does not leave room for the existence of errors so it is necessary to consider how this may happen. Within the normal structure of the double helix of the DNA molecule, the only base-pairing combinations that are allowed are A-T and G-C; any other combinations would result in a distortion of the helix, and such distortions will be removed enzymically and repaired (see Chapter 1).

This is the basic dogma. However, it makes assumptions about the structure of the bases. If they exist only in the form in which they are usually drawn, then the statement is true. However, each of the bases can exist in alternative tautomeric forms with different hydrogen bonding capabilities. Two examples of tautomerism, the change from an amino to an imino form and a keto-enol tautomerism, are shown in Figure 2.7. (Other forms of tautomerism also exist, but will not be discussed here). The consequences for base pairing are illustrated in Figure 2.8. The normal pairing of adenine and thymine (Figure 2.8a) occurs with adenine in the amino form and the keto form of thymine. The imino form of adenine (Figure 2.8b) however, will base pair with cytosine rather than thymine, while the enol form of thymine will hydrogen bond to guanine (Figure 2.8c). These alternative forms of the bases are thermodynamically unfavoured; if considered as an equilibrium between the two states, then about 1 in $10^4 - 10^5$ molecules will be in the

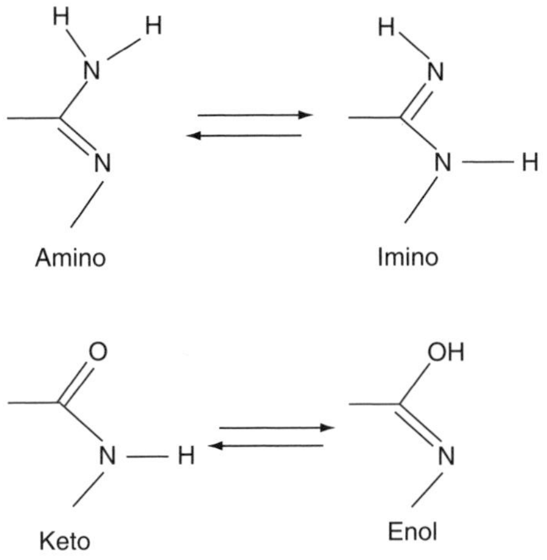

Figure 2.7 Examples of tautomerism

alternative state at any one point in time. Incorporation of one of these into the DNA will result in a mutation at that point, which would thus be expected to occur with a frequency of the order of $10^{-4} - 10^{-5}$ per base per generation.

The problem with this explanation is that the observed frequency of mutation is in fact many orders of magnitude lower than this (with *E. coli* at least). The discrepancy is accounted for by the proof-reading activity of the DNA replication machinery, arising from the 3′–5′ exonuclease function of the DNA polymerase, as described in Chapter 1, which enables it to remove incorrectly paired bases at the 3′ end of the growing strand. The abnormal tautomer can be considered as reverting rapidly to the more energetically favoured form, in which state it will be unable to pair correctly with the opposing base in the original strand. It will therefore be removed by the proof-reading mechanism. Spontaneous mutations will then only occur when this mechanism fails.

In fact, failure of proof-reading is not sufficient for the mutation to become established, since the proof-reading mechanism is only the first of several mechanisms for the repair of damaged or altered DNA (see Chapter 1 and later in this chapter). These DNA repair mechanisms are an important component of the cell's defences against a variety of agents that damage the DNA. The fundamental point here is that it is the existence of proof-reading and other repair mechanisms that keeps the mutation rate low. It should be noted that mutation rates can differ from strain to strain. In some strains, known as *mutator* strains, the rate at which mutations accumulate may be many thousand times higher than normal. These strains often contain primary mutations in the proof-reading activities of DNA polymerase (see Chapter 1). In addition, some genes contain highly variable regions known as homopolymeric tracts which serve to turn gene expression on or off (see Chapter 7).

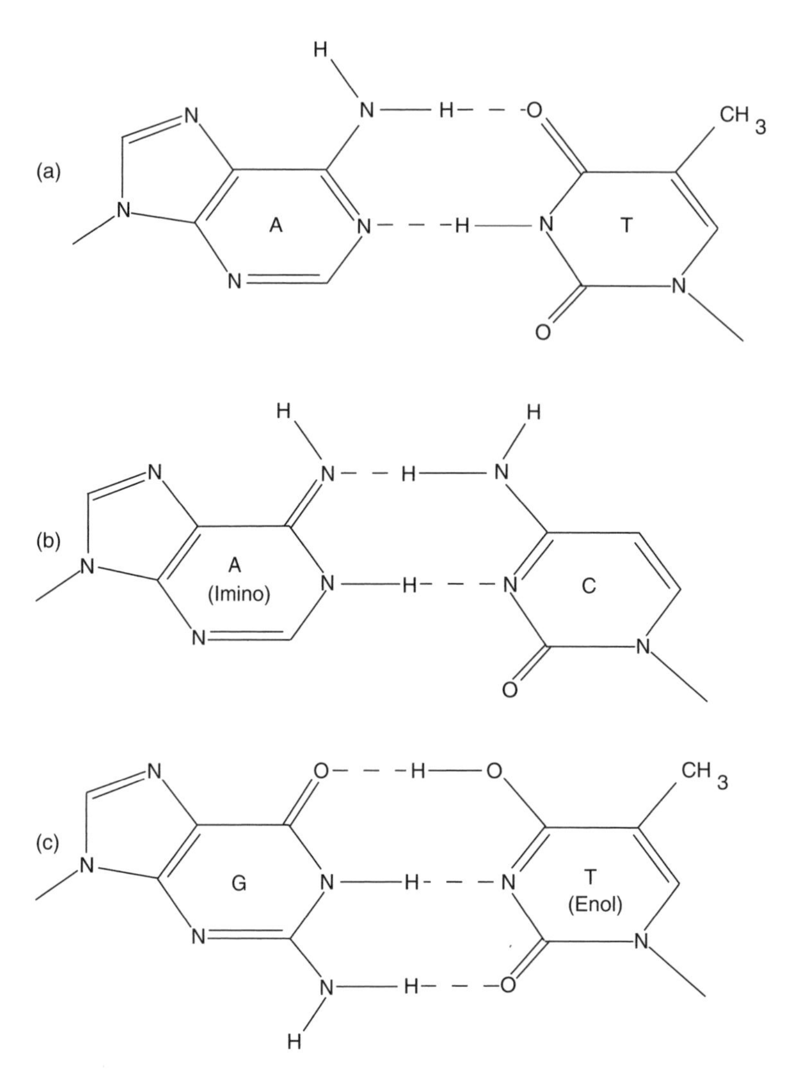

Figure 2.8 Tautomerism of bases leads to mispairing. (a) Normal pairing of adenine (amino form) and thymine (keto form). (b) Adenine in the imino form pairs with cytosine. (c) Thymine in the enol form pairs with guanine

2.6.2 Chemical mutagens

The natural rate of spontaneous mutation is much too low for convenient isolation of most types of mutants (apart from a handful of easily selected mutations such as antibiotic resistance). Ways must be found of enhancing that frequency. It is often possible to use *in vitro* mutagenesis or transposon mutagenesis (see

Chapter 10), but there are still many situations where chemical or physical procedures are preferred or essential.

Many different chemical agents interact with DNA or the replication machinery so as to produce alterations in the DNA sequence. Of these, the simplest to understand are those agents that act by chemically modifying a base on the DNA so that it resembles a different base. For example, nitrous acid causes an oxidative deamination in which amino groups are converted to keto groups and thus cytosine residues for example will be converted to uracil (Figure 2.9). Uracil is not a normal base in DNA and the cell contains enzymes that will remove it. However, if it persists through to replication it will be capable of pairing with adenine, thus causing a change from a C-G pair to U-A and ultimately T-A. Similarly deamination of adenine creates the base hypoxanthine which will base-pair with cytosine.

Nitrous acid can react directly with isolated DNA, although to produce mutations the DNA must of course be reintroduced into a bacterial cell. Treatments of this sort are especially useful for producing alterations of bacteriophages or plasmids where isolation and reintroduction of the DNA is readily achieved and where it is desirable to mutate the phage or plasmid without exposing the bacterial cell itself to the mutagenic agent. This ensures that any mutations selected are a consequence of changes to the DNA of the bacteriophage or plasmid, rather than being due to alterations in a chromosomal gene.

Some types of chemical agents act against the DNA within cells, rather than against isolated DNA. Alkylating agents such as ethyl methane sulphonate (EMS) and 1-methyl-3-nitro-1-nitroso-guanidine (MNNG) are extremely

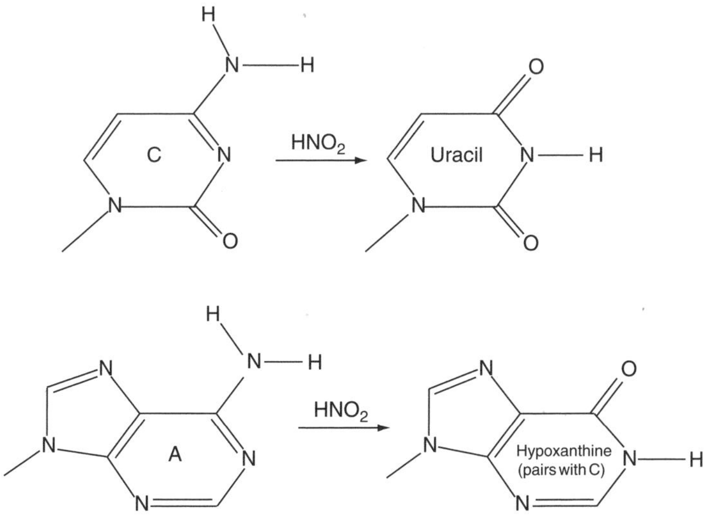

Figure 2.9 Nitrous acid causes oxidative deamination of bases

powerful mutagens – so much so that the latter in particular is extremely hazardous to use. They act by introducing alkyl groups onto the nucleotides at various positions, especially the O_6 position of guanine, and tend to cause multiple closely linked mutations in the vicinity of the replication fork.

The intercalating agents, such as acridine orange and ethidium bromide, have a different mechanism of action. These molecules contain a flat ring structure (Figure 2.10) which is capable of inserting (intercalating) into the core of the double helix between adjacent bases. The consequences of this are the addition (or sometimes deletion) of a single base when the DNA is replicated, giving rise to a frameshift mutation. These dyes (ethidium bromide in particular) are also much used in molecular biology in the detection of DNA, since the complex formed with DNA is fluorescent.

Another type of agent that acts only against growing cells (but with a very different mechanism) consists of the base analogues such as 5-bromouracil (Figure 2.11). Despite its name, this is an analogue of thymine in which the methyl group is replaced by a bromine atom which is a similar size. 5BU can be incorporated into DNA in place of thymine, since it will form base pairs with adenine residues on the template strand. However, the tautomerism referred to above is much more pronounced with 5BU. Therefore, in subsequent rounds of replication, it may pair with guanine rather than with adenine, thus giving rise to an A-T to G-C mutation.

2.6.3 Ultraviolet irradiation

Any agent that damages DNA can in principle lead either to the death of that organism or, amongst the survivors, to mutation. This is true of irradiation as

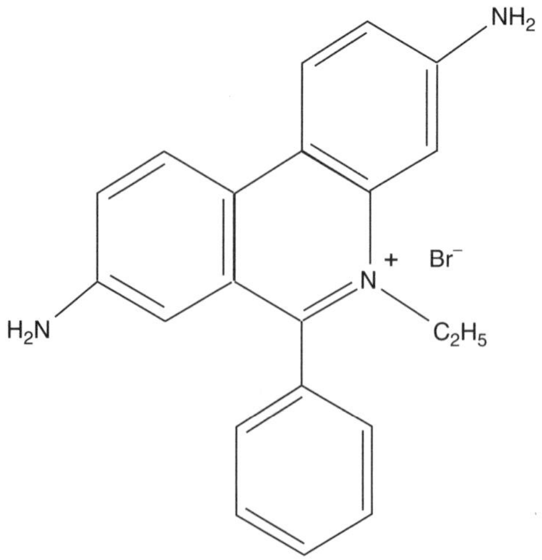

Figure 2.10 Ethidium bromide

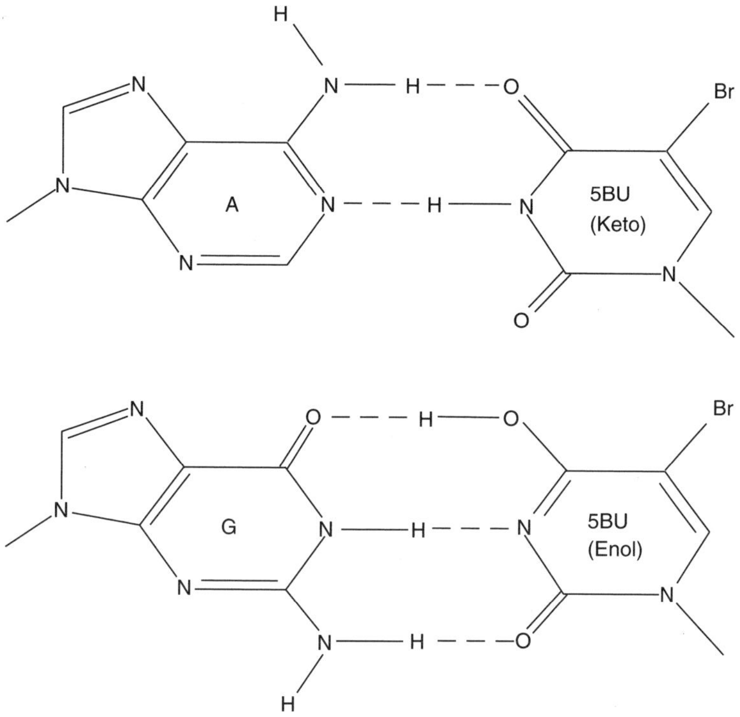

Figure 2.11 Enhanced tautomerism by the base analogue 5-bromouracil

well as of chemical agents. Many types of irradiation have been used to generate mutations. The higher energy rays such as X-rays and gamma rays however require expensive apparatus and safety equipment and are not really suitable for routine use in a microbiology laboratory. In addition, they produce an excessive amount of gross chromosomal damage that is not easily repaired by the micro-organism. Ultraviolet irradiation on the other hand is easily controlled (although eye and skin protection is necessary) and requires only comparatively inexpensive equipment.

The principal effect of UV irradiation with which we are concerned is the production of *pyrimidine dimers* (commonly referred to as thymine dimers, although the effect can also occur with cytosine). Where two pyrimidine residues are adjacent on the same DNA strand (Figure 2.12) the result of UV irradiation is the creation of covalent links between them. These pyrimidine dimers cannot be replicated and are therefore lethal to the cell unless it is able to repair the damage.

It is the attempts to repair the damage caused by ultraviolet irradiation that can lead to mutagenic effects. Although most repair mechanisms are reasonably accurate (error-free repair), in the event of these mechanisms being unable to cope with the damage an additional defence comes into play. This error-prone

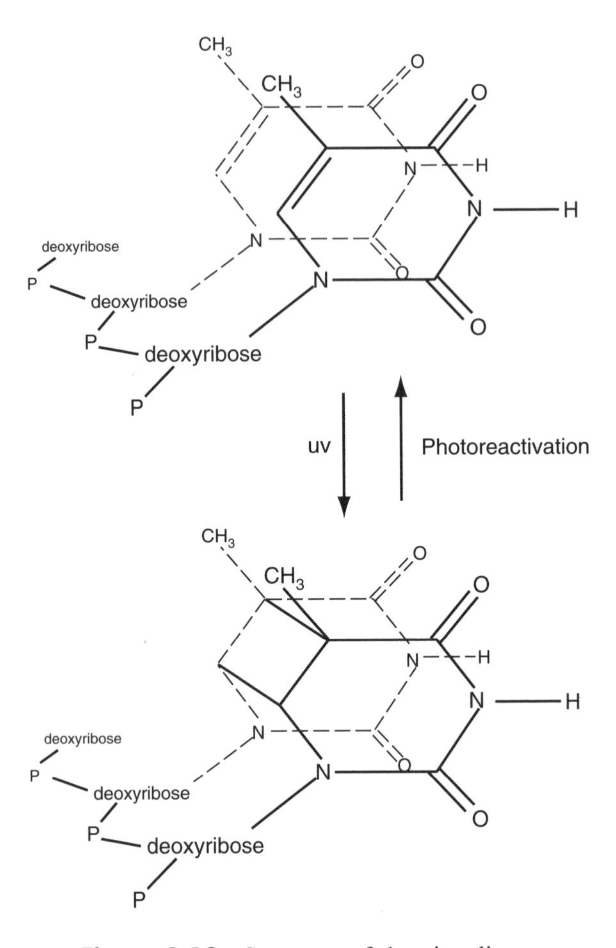

Figure 2.12 Structure of thymine dimers

repair mechanism (part of the SOS response) results in the incorporation of incorrect bases in the DNA.

Photoreactivation

The best defence that is mounted against damage by UV irradiation is known as *photoreactivation*. This is catalysed by an enzyme (photolyase) within the cells that in the presence of visible light can break the covalent bonds linking the two pyrimidine residues, thus re-establishing the original nature of the base sequence at that point. This method is very efficient and clearly does not lead to the establishment of mutations. For this reason, when UV mutagenesis is being performed in the laboratory, it is necessary to exclude light from the cultures (e.g. by wrapping the bottles in foil) while the cells are recovering from the UV treatment.

SOS repair

Photoreactivation is an error-free process, and therefore does not lead to mutation. If light is excluded, and photoreactivation prevented, the cell still has recourse to other error-free repair processes such as excision repair and recombination repair (post-replication repair), as described in Chapter 1. It is only when these processes are overwhelmed (or if we use mutant strains that lack these error-free repair mechanisms) that significant numbers of mutations result. This is due to yet another repair mechanism that is error-prone which is part of the so-called SOS response. (Although we are considering this effect in relation to UV irradiation, it is also involved in repair of other types of DNA damage and is related to the more general stress response systems).

In the presence of damaged DNA, the expression of a number of genes involved in DNA repair is induced; these genes include the excision repair genes *uvrA* and *uvrB*, the *recA* gene and the genes involved in error-prone repair. This inducibility of the repair pathways can be demonstrated by irradiating lambda bacteriophage and testing its ability to form plaques on irradiated and non-irradiated host cells. More plaques are obtained if the host cells are also subjected to a low dose of UV irradiation before infection by the phage due to the induction of the repair enzymes in the host by the pre-existing DNA damage.

The mechanism of induction involves the products of two genes: *recA* and *lexA*. The LexA protein acts as a repressor of the genes of the SOS response, including both *recA* and *lexA* itself. The LexA protein also has the ability to cleave itself but only after the RecA protein binds to it. The RecA protein has a co-protease function in stimulating the self-proteolysis of LexA. This activity of RecA arises after it binds to single-stranded DNA which arises as a consequence of DNA damage. This causes a conformational change in the protein that enables it to bind to LexA, resulting in cleavage of LexA and expression of the SOS genes.

Two of the SOS genes in particular, *umuC* and *umuD*, are involved in mutagenesis, since strains that are defective in these genes not only have increased UV sensitivity but also are not susceptible to UV-induced mutagenesis. Certain plasmids carry analogous genes (*mucA* and *mucB*); the presence of these genes increases resistance to UV and also results in an increased level of mutagenesis by UV and many other mutagenic agents. The UmuCD complex is able to act as a DNA polymerase, taking over from the normal polymerase (DNA polymerase III) when that enzyme stalls due to the presence of DNA damage. The low specificity of the UmuCD polymerase enables it to continue synthesis of DNA past the lesion, but at the expense of producing errors in the new DNA strand.

2.7 Isolation and identification of mutants

2.7.1 Mutation and selection

With the techniques described above, it is usually easy enough to alter the DNA randomly; the real art comes in devising means of isolating mutants in which specific genes have been altered. If possible, this is most effectively done using growth conditions in which either the original strain or the mutant is not able to grow. The choice of a method for selection of a mutant depends on the nature of the mutation and its consequences for the cell. We can consider mutants of the three types referred to earlier:

(1) mutants that are resistant to antibiotics or to specific bacteriophages, toxic chemicals or any other agents that are usually lethal or inhibitory to the parent cell;

(2) auxotrophs, i.e. mutants that require some additional growth factor, such as an amino acid;

(3) mutants that are unable to use a particular growth substrate (usually a sugar).

Resistant mutants can be obtained for any type of bacterium (provided it can be grown in the laboratory). Mutants of the second and third types cannot be isolated easily unless the bacterium can grow on a simple defined medium. This is a major reason why bacterial geneticists concentrated initially on *E. coli* and a limited range of other bacteria such as *Salmonella typhimurium* and *Bacillus subtilis*, although the range of organisms studied has now expanded substantially.

Antibiotic resistant mutants are, in principle, easy to isolate. The procedure simply involves plating the culture on agar containing the antibiotic at a concentration that will inhibit the parent strain. These selective conditions are extremely powerful so it is often not necessary to use any mutagenic agent. Very large numbers of bacteria can be used (10^8–10^9 cells or 0.1–1 ml of an overnight culture of *E. coli*), so mutations that occur at a very low frequency can be readily detected. If the mutation frequency is 10^{-8} and 1 ml of culture containing 10^9 cells is plated out, then (on average) 10 resistant mutants on each plate would be expected.

Auxotrophic mutants cannot usually be selected directly in this way, since it is not possible to devise conditions where the mutant will grow but the parent will not. For example, wild type *E. coli* will grow on a minimal medium consisting of a buffered solution of inorganic salts plus glucose as a carbon/energy source and ammonium ions as a source of nitrogen. A histidine auxotroph will not grow on this medium but will require the addition of histidine, since it is defective in one of the enzymes of the histidine biosynthetic pathway. The parental strain will also grow quite happily in the presence of histidine. So, although it is not possible to devise a medium on which only the auxotrophic mutant will grow, the converse (a medium on which the auxotroph will not grow) is easy.

If the mutation rate was high enough, it would be possible to pick colonies one by one and test each one to see if it has lost the ability to grow on the unsupplemented minimal medium. Such a test procedure is indeed common practice for many purposes in a microbial genetics laboratory. Each colony is sampled by stabbing it with a sterile toothpick and its ability to grow on a different medium (or a range of media) is tested by stabbing or touching the surface of each test plate in turn. A number of colonies (usually up to 100) can be tested on each plate, provided they are arranged systematically on a numbered grid so that one plate can be compared with another. This procedure becomes extremely tedious with more than a few hundred colonies (although automated procedures involving robotic workstations are now available). The frequency with which a desired mutation occurs in a culture, even after treatment with a mutagenic agent, is very low, so that recovering the required mutant would usually involve testing thousands (or even millions) of colonies.

2.7.2 Replica plating

Replica plating (Figure 2.13) provides a method for testing large numbers of colonies more rapidly. In this procedure, the mutagenized culture is plated out to obtain single colonies on a nutrient medium on which mutants and parents will grow. After incubation, a sterile velvet pad is pressed lightly onto the surface of the plate so that a tiny impression of each colony is transferred to the velvet. This is used to inoculate first a minimal agar plate and then a similar plate to which the appropriate supplement (in this case, histidine, since we are looking specifically for histidine auxotrophs) has been added. Histidine-requiring auxotrophs will be unable to grow on the first plate, but will grow on the second one, from which they can be picked off for further work.

The parent strain will grow on both plates, while other types of auxotrophic mutant will not grow on either. The second plate thus serves to eliminate other types of auxotrophs that are not wanted and also serves as a control to ensure that the absence of growth on the first plate was not due to failure to pick that colony up on the velvet pad initially.

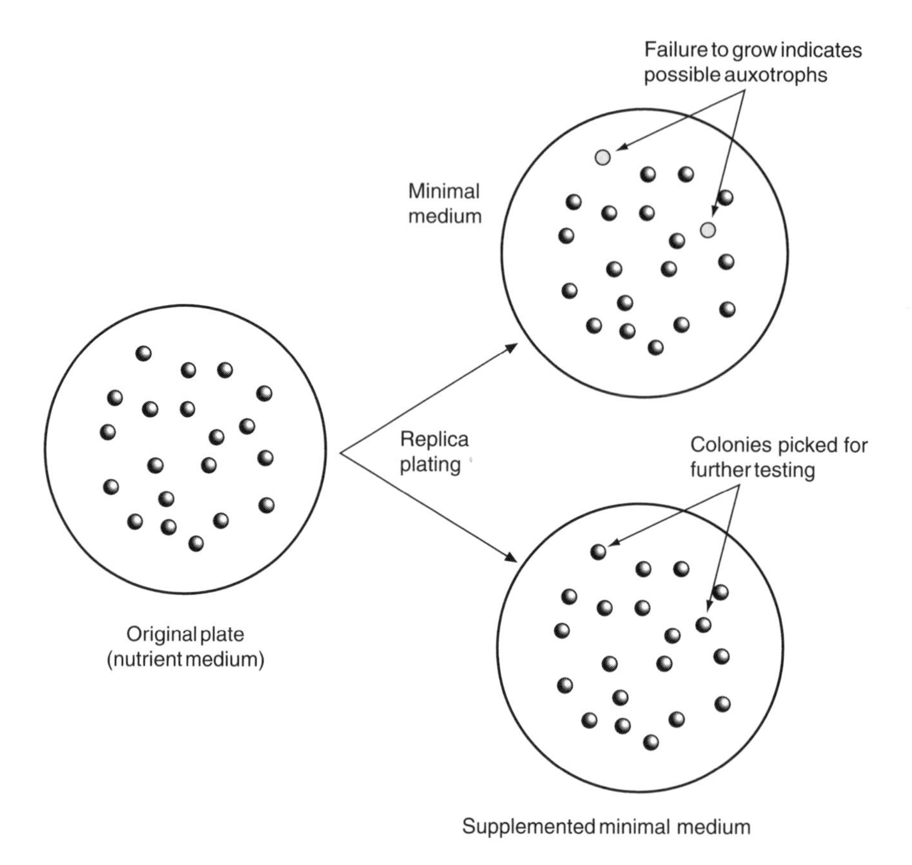

Figure 2.13 Replica plating to isolate auxotrophic mutants

Even with replica plating, the number of colonies that can be handled is still limited. If the original plate contains more than a few hundred colonies it is very difficult to identify colonies that have failed to grow. If the required mutation occurs at a frequency of 10^{-6} (a high frequency for a spontaneous mutation in *E. coli*) it would be necessary to replicate over 1000 plates to find it. Pre-treatment with a mutagenic agent will increase the proportion of mutants in the population so that they can be isolated from a manageable number of plates.

Replica plating can also be used to test whether mutation occurs at random, without exposure to the selective agent, rather than being induced by selection (see the earlier discussion in this chapter). For example a master plate containing a very large number of colonies can be used to replicate colonies onto an agar plate containing streptomycin. This will indicate which regions of the master plate contain streptomycin-resistant mutants. A pool of colonies can be recovered from those parts of the plate and subjected to further rounds of replica plating, until eventually a pure culture of streptomycin-resistant bacteria which have never been exposed to the antibiotic, is obtained. In this case, the mutations must have been random rather than directed.

2.7.3 Penicillin enrichment

One procedure to increase the proportion of auxotrophs relies on the fact that some antibiotics (notably penicillin) are only active against bacteria that are growing. Cells that have stopped growing, for whatever reason, are relatively unsusceptible. Penicillin enrichment is therefore carried out by resuspending mutagenized bacteria in a minimal medium and incubating to allow growth to start (Figure 2.14). Once the culture is growing well the penicillin can be added. The auxotrophic mutants will of course be unable to grow because of the absence of, in our example, histidine, so they will not be affected by the penicillin. The parental prototrophs, which are able to grow in this medium, will be killed by the action of the antibiotic.

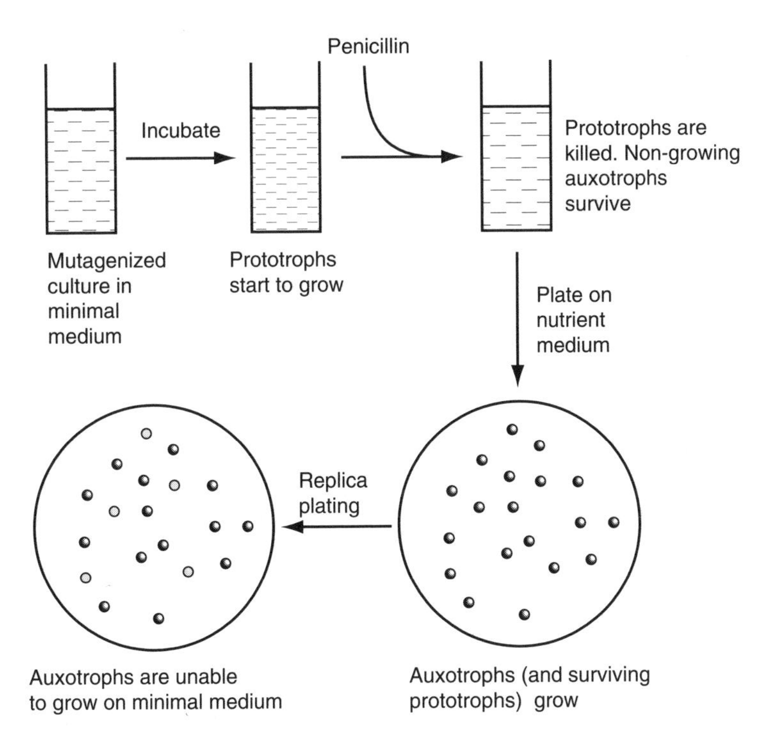

Figure 2.14 Penicillin enrichment for the isolation of auxotrophic mutants

2.7.4 Isolation of other mutants

Mutants that are not able to utilize a particular carbon source (lactose, for example) can be isolated in a similar way, i.e. by comparing the ability of colonies to grow on plates containing lactose or glucose as the carbon/energy source. Alternatively, with *E. coli* for example, advantage can be taken of the fact that sugar utilization often leads to acid production which can be detected by incorporating a pH indicator in the agar. *E. coli* mutants that are not able to ferment lactose (Lac^- mutants) can be detected using a medium with a nutrient base (so that both mutants and parents can grow), which also contains lactose and a pH indicator. Bacteriologists often use MacConkey agar for this purpose. This medium contains lactose and a pH indicator so that Lac^+ colonies will be red and Lac^- colonies will remain a pale colour that can be easily distinguished.

A more convenient and sensitive test for distinguishing Lac^+ from Lac^- colonies is to use a chromogenic substrate, i.e. a substrate that shows an easily detectable colour change when acted on by the enzyme concerned. In this case the enzyme is β-galactosidase, which catalyses the hydrolysis of lactose into its constituent sugars glucose and galactose. A commonly used chromogenic substrate for β-galactosidase is 5-bromo-4-chloro-3-indolyl-β-D-galactoside, more popularly known as X-gal. This is a synthetic analogue of the natural substrate, containing a dye linked to galactose. X-gal itself is colourless; the colour of the dye is only manifest when it is released by hydrolysis of the linkage by β-galactosidase. Lac^+ colonies will be blue on a medium containing X-gal while colonies that do not produce β-galactosidase will be white.

2.7.5 Molecular methods

The concepts and techniques described so far in this chapter relate to the behaviour of the bacteria themselves. However, some of the molecular methods that will be increasingly important in later chapters, especially the polymerase chain reaction (PCR), gene probes and Southern blotting, and the sequencing of DNA, also have a role in the characterization of mutants, so it is appropriate to introduce them briefly here. In later chapters, we will consider the impact of gene cloning and related techniques which provide highly specific and versatile ways of manipulating the genetic composition of a bacterial cell.

Gel electrophoresis

Central to many of these techniques is the ability to separate DNA fragments on the basis of their size by electrophoresis through an agarose gel (or for very small fragments, an acrylamide gel) – see Box 2.2. At the pH used (commonly 8.3),

Box 2.2 Agarose gel electrophoresis

Fragments of DNA can be separated by electrophoresis through an agarose gel (or for very small fragments, an acrylamide gel). The samples are applied to slots or wells in the gel and a voltage is applied across the gel to separate the DNA fragments. The gel is then stained with ethidium bromide. The ethidium bromide–DNA complex fluoresces under ultraviolet illumination.

Within certain limits, the mobility of a DNA fragment is determined by its size; larger fragments move more slowly through the gel. At the pH used – usually about 8.3 – the DNA is negatively charged and there is essentially no variation in charge between different fragments. Fragment sizes can be estimated by calibrating the gel with a mixture of known DNA fragments and plotting the logarithm of the size against the distance moved.

The relationship between size and mobility is only true for molecules with the same conformation. A gel that is calibrated with linear DNA fragments can only be used for sizing linear fragments.

Gel electrophoresis can also be used preparatively, by cutting the required band out of the gel and eluting the DNA from it.

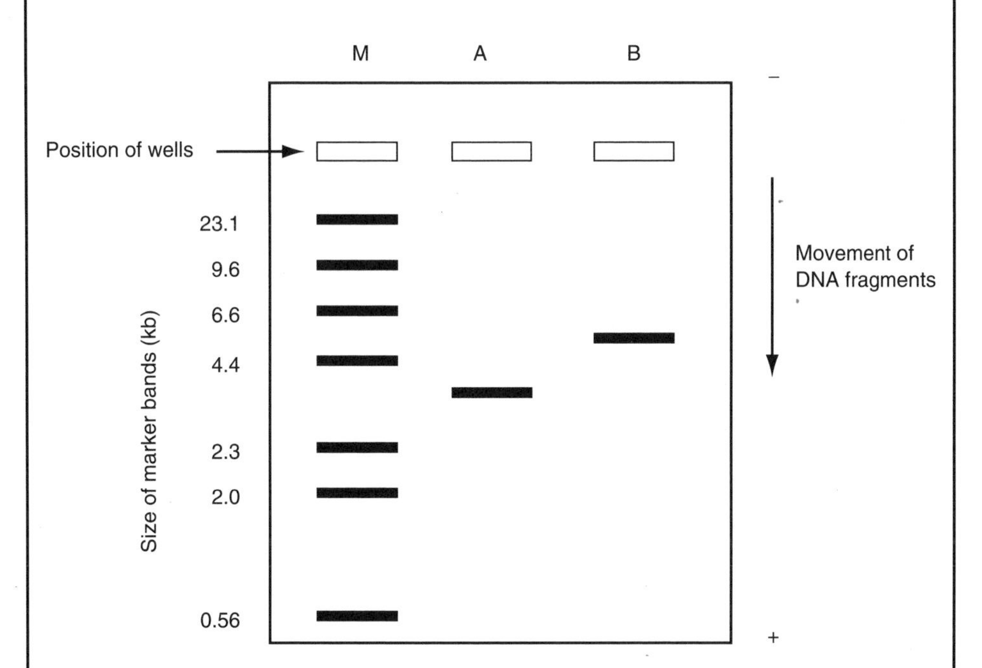

Diagrammatic representation of a gel after staining with ethidium bromide. M is a set of standard DNA fragments (a *Hin*dIII digest of λ DNA), with sizes shown in kilobases; A and B are the fragments whose size is to be determined

DNA is negatively charged, and will therefore move towards the positive electrode. Since smaller fragments move faster than larger ones, gel electrophoresis can be used to estimate the sizes of specific DNA fragments. In Chapter 9 we will consider further how this can be used to examine variation between bacterial strains.

Polymerase chain reaction

The principle of the polymerase chain reaction is outlined in Box 2.3. It depends on the fact that DNA synthesis requires primers base-paired to the template DNA. So using synthetic oligonucleotides that are complementary to the DNA on either side of the gene being studied, specific synthesis of the target gene can be achieved. Repeated cycles of heating the mixture (to dissociate the DNA strands), and then allowing the primers to associate to the new strands followed by further DNA synthesis, result in an exponential increase in the amount of the target DNA. Twenty cycles give a 1 million-fold increase in the amount of the specific target. If the mutation being studied has caused a significant change in the size of a specific gene (such as an insertion or a deletion) this can be detected by a change in the size of the PCR product which can be determined by using gel electrophoresis as described above.

Although conventional gel electrophoresis can only be used to detect changes in the *size* of a DNA fragment, rather than sequence changes, there are some specialized electrophoretic techniques that can detect small differences (even single base changes) between two PCR products. These are considered in Chapter 10.

Gene probes and Southern blotting

The use of a labelled DNA fragment (a *gene probe*) to detect similar sequences by hybridization was introduced in Chapter 1. If we use a short oligonucleotide as a probe, under conditions of high stringency (high temperature and/or low ionic strength), small changes in the sequence of a gene, even single nucleotide substitutions, can be detected.

A common method of using gene probes for looking at similarities, or differences, in the structure of a gene, or in a PCR product, is the technique known as *Southern blotting*. As shown in Box 2.4, this involves separating fragments of DNA by electrophoresis in an agarose gel (see above) and transferring them to a filter which can then be hybridized with the labelled probe. Not only can we confirm that we have amplified the correct PCR product, but by using highly specific probes we can detect differences in the sequence. Furthermore, Southern

Box 2.3 Polymerase chain reaction

The polymerase chain reaction (PCR) produces large amounts of a specific DNA fragment by enzymic amplification. Two synthetic oligonucleotide primers that will hybridize to the DNA sequence of interest are mixed with the DNA sample (template), together with a heat-stable DNA polymerase and dNTP substrates. When this mixture is heated to denature the template DNA and then cooled, the primers anneal to the template. The enzyme uses the primers to start synthesis of new DNA strands, complementary to each strand of the template, thus doubling the amount of the DNA fragment. The cycle of heating (denaturation), cooling (annealing) and synthesis is then repeated; each cycle again doubles the amount of the target DNA, so that after 20 cycles a single molecule of DNA will be amplified to about 1 million copies.

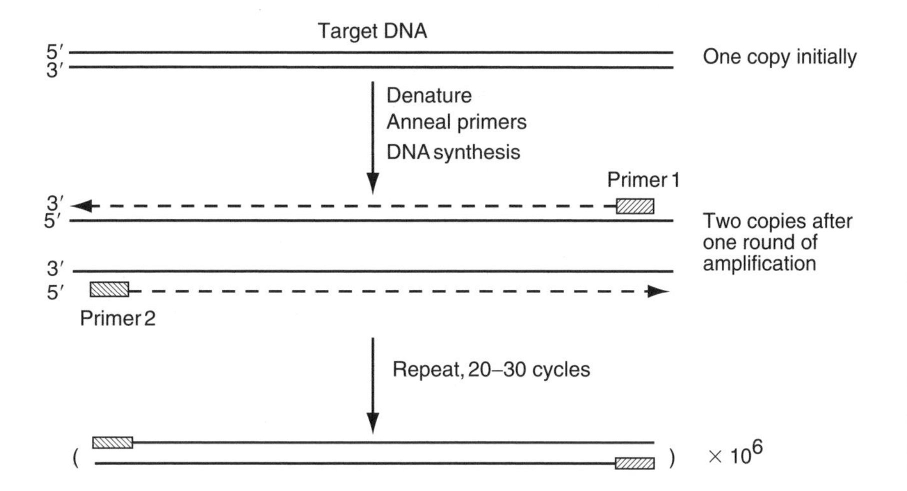

There are many applications of the polymerase chain reaction. These include

- amplification of gene probes
- generation of DNA fragments for cloning
- amplification of a specific chromosomal fragment, e.g. for sequence analysis of mutations
- detection of specific pathogens (molecular diagnosis)
- *in vitro* mutagenesis (using primers with an altered sequence)
- analysis of gene expression by detection of mRNA (reverse transcript PCR)

blotting and hybridization allows us to examine specific genes within the complex mixture of DNA fragments that are generated by digestion of complete chromosomes since only the fragments we are interested in will be detected by the probe.

Box 2.4 Southern blotting

Southern blotting enables the discriminatory power of gel electrophoresis to be combined with the specificity of DNA hybridization. The name has no geographical connotations, but is named after Ed Southern who first developed it.

In the original version of this technique (as shown in the diagram below), a filter is laid on top of the gel in which the DNA fragments have been separated by electrophoresis and a stack of dry blotting paper is used to draw buffer through the gel and the filter by capillary action. The DNA fragments travel with the buffer and are trapped on the filter where they can be detected by hybridization with a labelled DNA probe followed by autoradiography.

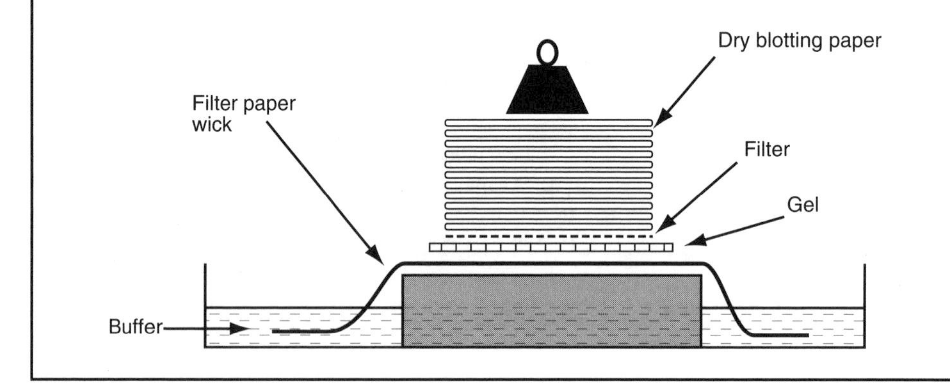

This forms the basis of widely-used procedures for analysing variation between bacteria, as considered in more detail in Chapter 9.

DNA sequencing

In order to establish exactly what changes have occurred in the gene under study, it will be necessary determine the DNA sequence of that gene. This is now a routine procedure in molecular biology laboratories, extending not just to individual genes or fragments but to the complete sequence of the DNA (the *genome*) of the organism. The methods for doing this and the impact this has had on the study of bacteria are described in Chapter 10.

3

Regulation of Gene Expression

Adaptability is a crucial characteristic of bacteria that are able to prosper in a wide variety of environmental conditions. The most versatile organisms contain a large reservoir of genetic information, encoding mechanisms that enable the bacterium to cope with a variety of challenges. Genome sequence data enables us to estimate the total number of genes in a variety of bacteria – in the sequences known so far this ranges from less than 700 to nearly 6000, although only 600–800 are needed at any one time. From the genome sequence information, we can also identify those genes that are responsible for regulating gene expression. In terms of ATP consumption, gene expression is expensive (~3000 ATP molecules per protein), so this process has to be controlled precisely to prevent wasteful synthesis of unnecessary materials. *Pseudomonas aeruginosa* for example, which can survive in a wide range of environments, has 468 proteins that regulate its response to different stimuli, whilst *Helicobacter pylori*, which is specifically adapted to the human stomach has just 18 such proteins.

In addition to controlling gene expression in response to environmental or other stimuli, a bacterial cell needs to be able to produce some proteins (e.g. structural proteins, ribosomal proteins) at very high levels, while other proteins (such as some regulatory proteins) are only produced at a very low level. Although these levels may go up and down in response to environmental changes, or at different stages of growth, the maximum potential expression of genes is fixed at different levels. Fortunately, the mechanisms used for fixed and variable controls are similar, so we can consider them together.

Looking at the flow of information from the structure of the gene to the activity of the enzyme as the final product (Figure 3.1), we can see that control is achieved at three main stages: production of mRNA, translation of that message into protein and control of the enzymic activity of that protein. Within this framework, there are a number of potential regulatory factors.

Molecular Genetics of Bacteria, 4th Edition by Jeremy Dale and Simon F. Park
 ISBN 0 470 85084 1 (cased) ISBN 0 470 85085 X (pbk)

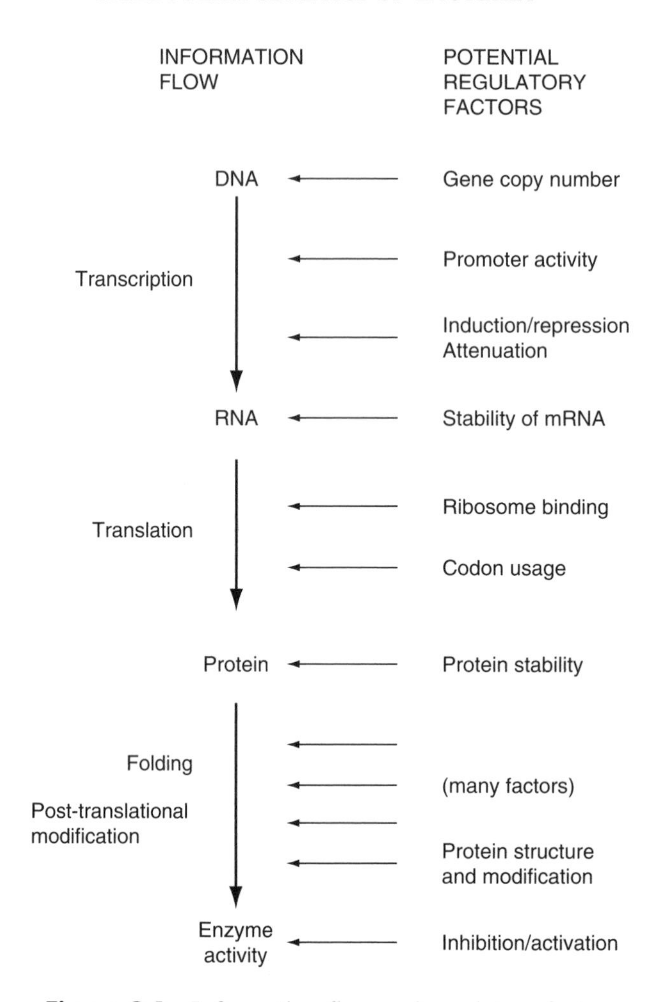

Figure 3.1 Information flow and regulatory factors

(1) The number of copies of the gene. In general, if there are several copies of a gene the level of product is likely to be higher (although the relationship is not necessarily linear).

(2) The efficiency with which the gene is transcribed, which is mainly determined by the level of initiation of transcription by RNA polymerase (promoter activity). In bacteria this is the major factor influencing the expression of individual genes, whether we are considering fixed or variable controls.

(3) The stability of the mRNA. It is important to recognize that the amount of specific mRNA will be determined by the combined effect of the rate at which it is

produced and the length of time each molecule persists in an active state in the cell. Most bacterial mRNA is very short-lived, typically being degraded with a half-life of about 2 min. The instability of bacterial mRNA is a key feature in the rapidity with which bacteria can respond to changes in their environment. However, some bacterial mRNA species are more stable than others, in some cases with a half-life as long as 25 min. Other forms of RNA (rRNA, tRNA) are also considerably more stable, which can be ascribed to the high degree of secondary structure possessed by these molecules.

(4) The efficiency with which the mRNA is translated into protein. This will be influenced by the efficiency of initiation (ribosome binding) and also by factors that affect the rate of translation (mainly codon usage).

(5) The stability of the protein product. As with the mRNA, the amount of protein reflects both its rate of production and its stability. Different proteins vary in their stability to a very marked degree, as might be expected from their different functions: a protein that forms part of a cellular structure is likely to be more stable than one that transmits a signal for switching on a transient cellular event.

(6) Post-translational effects. This includes a wide variety of events such as protein folding which is necessary for conversion of polypeptide chains into a biologically active conformation, as well as covalent modifications that can influence the activity of the protein. Phosphorylation is especially important as a mechanism for regulating the function of specific proteins. In addition, to complete the picture, control of the metabolism and physiology of the cell is influenced by temporary effects such as inhibition and activation of individual enzymes.

We can now consider some of these factors in more detail.

3.1 Gene copy number

Most genes on the bacterial chromosome are present as single copies (with a few notable exceptions such as the genes for ribosomal RNA); gene copy number is not therefore an important method of control for most of the normal metabolic activities of a bacterial cell. It does become important however when we consider plasmid-mediated characteristics, particularly with reference to the cloning and expression of heterologous DNA. Some plasmids are present within the cell in very high numbers, running into thousands of copies (see Chapter 5) and this is reflected in the enhanced level of expression of the genes they carry.

3.2 Transcriptional control

3.2.1 Promoters

The principal method of control of gene expression in bacteria is by regulating the amount of mRNA produced from that gene, which is primarily determined by the affinity of RNA polymerase for the promoter (see Chapter 1). Strong promoters are highly efficient and lead to high levels of transcription, while others (weak promoters) give rise to low levels of transcription. The nature of the promoter is therefore of major importance as a fixed level of control that determines the potential level of expression of different genes.

From the comparison of hundreds of these regions a consensus can be established (Figure 3.2). Most *E. coli* promoters, for example, have two key parts (*motifs*) that are involved in the recognition by the RNA polymerase and resemble TTGACA and TATAAT at positions that are centred at 35 bases and 10 bases before (upstream from) the transcriptional start site and are hence referred to as the -35 and -10 positions respectively. The latter is also known as the Pribnow box. Strong promoters, which can direct transcription of genes every 2 s, tend to have a sequence close to this ideal consensus, whilst weak promoters may have

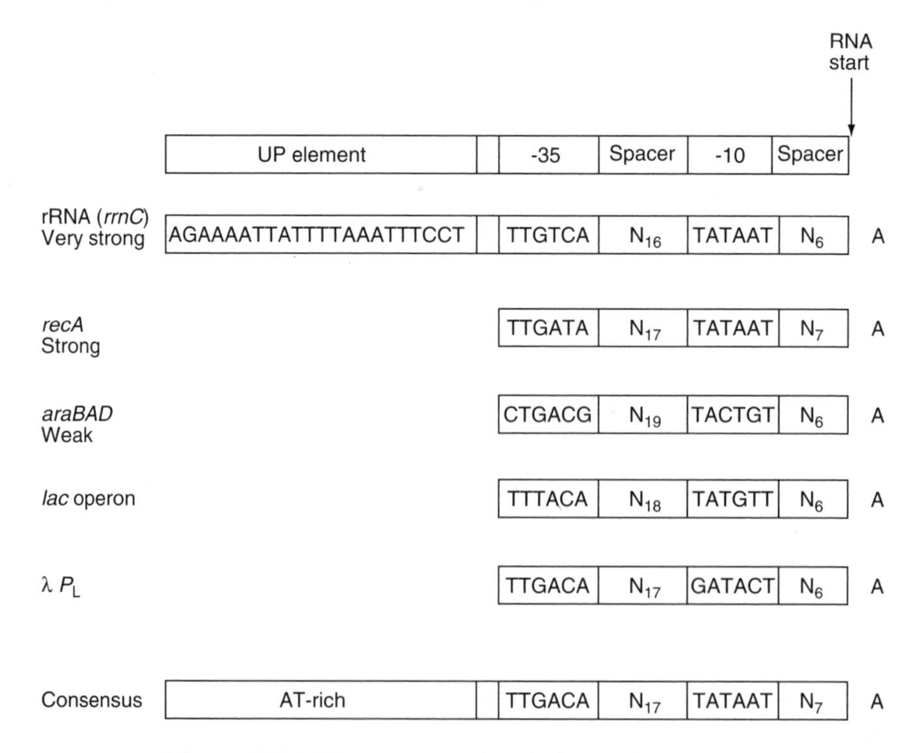

Figure 3.2 The structure of typical *E. coli* promoters

base changes in these regions or may differ in the spacing between the two motifs. Consequently, these may only be transcribed once every 10 min or so.

While the requirement of the enzyme for specific sequences as recognition points is not surprising, the restriction on the distance separating the two motifs may be less obvious. RNA polymerase is large compared to the promoter region and a few bases one way or the other might be expected to make little difference to the ability of the enzyme to reach both sequences. However, because the two strands form a double helix, an increase (or decrease) in the distance between the two motifs will mean that they are no longer on the same side of the DNA (Figure 3.3) and the RNA polymerase will be less able to make contact with both motifs simultaneously.

Regions of the promoter sequence other than the -35 and -10 motifs may also affect its efficiency. Promoters which direct transcription of ribosomal RNA operons are the most active promoters, in *E. coli* accounting for more than 60 per cent of all transcription in the cell, and direct the production of more mRNA than all of the other 2000 or so promoters in the cell added together. In addition to the -10 and -35 motifs, these promoters have a third element, an AT-rich sequence termed an UP element (Figure 3.2) which increases transcription 30–90-fold and

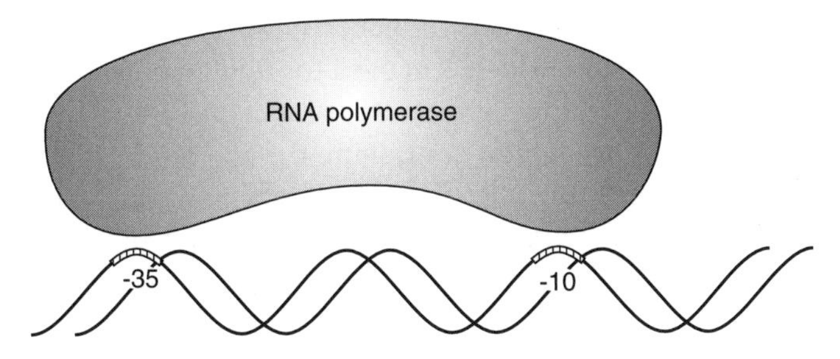

(b) Increased distance between -35 and -10 means they are no longer on the same side of the DNA.
RNA polymerase is unable to contact both regions

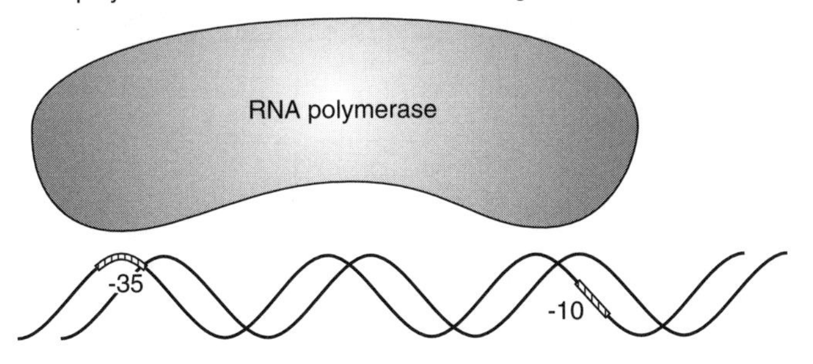

Figure 3.3 Importance of the distance between the -35 and -10 regions of a promoter

contributes to the remarkable strength of these promoters. Furthermore, as we will see later on, the activity of a promoter may be influenced by the local structure of the DNA; supercoiling and bending of the DNA in particular affect the ability of the RNA polymerase to open up the helix to allow transcription to proceed.

It should be noted that this discussion refers to those promoters that are used under 'normal' conditions. Under certain conditions, such as heat shock, alternative promoters may be used that bear little or no resemblance to the sequences for the -10 and -35 described above (see below).

Operons and regulons

As mentioned in Chapter 1, bacterial genes may be part of a single transcriptional unit known as an *operon*. The whole operon is transcribed, from a single promoter, into one long mRNA molecule from which each of the proteins is translated separately. Transcriptional regulation can therefore apply to the operon as a whole. In consequence, the genes in the operon are coordinately regulated. This way of organizing genes appears to be unique to bacteria. In *E. coli* for example, whilst there are 4289 genes in the genome, many of these are organized into 578 known operons.

One of the best known examples of an operon is the *lac* operon of *E. coli* (Figure 3.4). This consists of the structural genes *lacZ* (coding for β-galactosidase), *lacY* (coding for a permease which is needed for uptake of lactose) and *lacA* (coding for an enzyme called thiogalactoside transacetylase). To the 5′ side of these genes is found a regulatory region containing the promoter site and also an overlapping sequence known as the *operator*. The operator and the separate *lacI* gene, which codes for a repressor protein, are connected with the inducibility of the operon which is described later in this chapter.

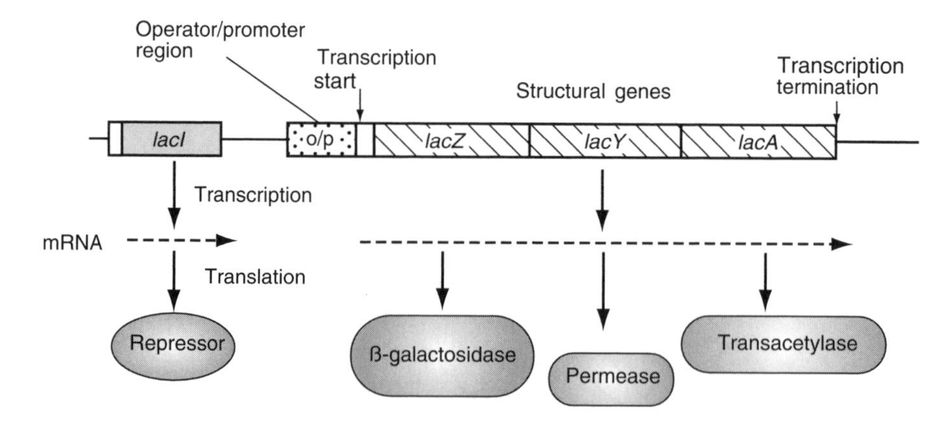

Figure 3.4 Structure of the *lac* operon

In some cases, co-ordinated control of several genes is achieved by a single operator site that regulates two promoters facing in opposite directions (Figure 3.5). In one example, the genes *ilvC* (coding for an enzyme needed for isoleucine and valine biosynthesis) and *ilvY* (which codes for a regulatory protein) are transcribed in opposite directions (they are *divergent* genes), but transcription of both genes is controlled by a single operator site. Since there is a single operator, this arrangement is also referred to as an operon, even though there are two distinct mRNA molecules. This provides an exception to the general rule that genes on an operon are transcribed into a single mRNA.

(a) Operon 1: single operator/promoter controlling a set of adjacent genes

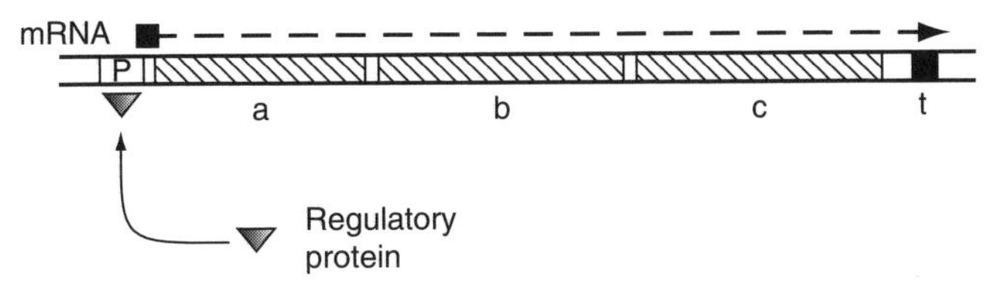

(b) Operon 2: single operator controlling divergent promoters

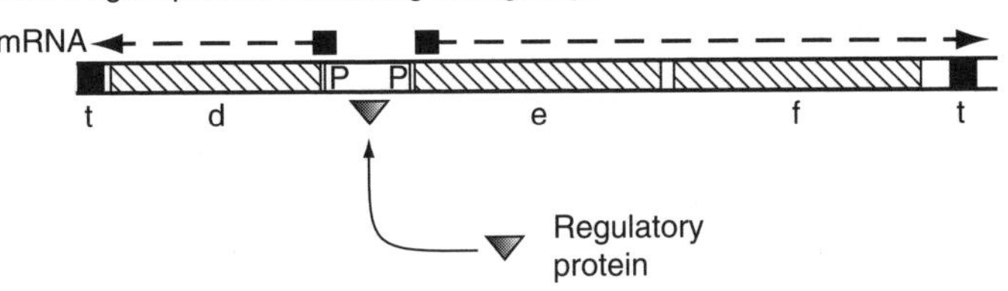

(c) Regulon. Regulatory protein interacts with several operators, controlling genes on different parts of the genome

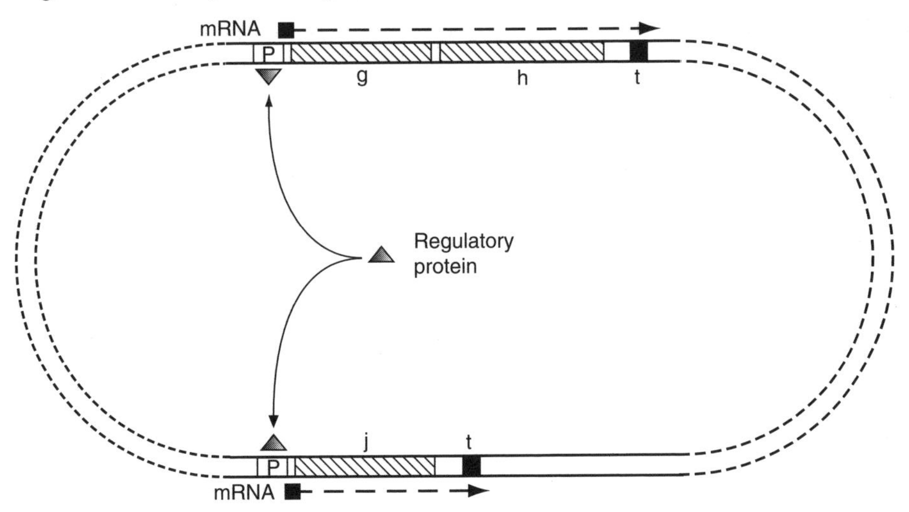

Figure 3.5 Structure of operons and regulons

Not all coordinately controlled genes are arranged in operons. In some cases, groups of genes at different sites on the chromosome are regulated in a concerted fashion. Such a set of genes or operons, expressed from separate promoter sites but controlled by the same regulatory molecule is called a *regulon* (Figure 3.5). For example, arginine biosynthesis requires eight genes (*argA-H*), but (in *E. coli*) only three of these (*argC, argB, argH*) form an operon with a single promoter. A fourth gene (*argE*) is divergently transcribed from an adjacent promoter (thus providing another example of a divergent operon as described above) while the remaining three genes (*argA*, *argF*, and *argG*) are found at different sites on the chromosome, each with its own promoter. Despite this scattering of the arginine biosynthesis genes, they are coordinately controlled.

Alternative promoters and σ-factors

We saw in Chapter 1 that RNA polymerase consists of a core protein, composed of four subunits ($\alpha_2\beta\beta'$) with a fifth dissociable subunit called a σ-factor (sigma-factor). It is the σ-factor that allows the polymerase to recognize specifically the two conserved nucleotide motifs in the promoter region. Thus the sigma factor determines the specificity of the enzyme.

The promoter consensus described above is recognized by the primary σ-factor, commonly referred to as σ^{70} (because it is about 70 kDa in size in *E. coli*). This subunit is responsible for recognition of the promoters used for transcription of most of the genes required in exponentially growing cells. These are sometimes called 'housekeeping' genes since they encode essential functions needed for the cell cycle and for normal metabolism such as glycolysis, the TCA cycle and DNA replication. Most bacterial species, however, have several different σ-factors. Replacement of the primary σ-factor with a different subunit radically changes the recognition of promoter sequences by the RNA polymerase (Figure 3.6). These alternative σ-factors allow the bacterium to bring about global changes in gene expression in response to particular environmental stresses. For example, in many bacteria, an abrupt temperature increase triggers the expression of heat-shock proteins which can counter the detrimental effects of high temperature. The promoters of about 30 heat shock genes are only recognized by RNA polymerase containing σ^{32} (also called the heat shock σ-factor). This normally has a short half-life but following exposure to high temperatures, it is not degraded and thus stimulates transcription of target genes. Consequently, this σ-factor allows the bacterium to express a subset of genes whose products are required to counter heat stress. Because many of these genes are scattered around the genome, this system constitutes a heat shock regulon. Other examples of σ-factors, their regulatory activity and recognition sequences are given in Figure 3.6. We will encounter further examples in Chapter 4.

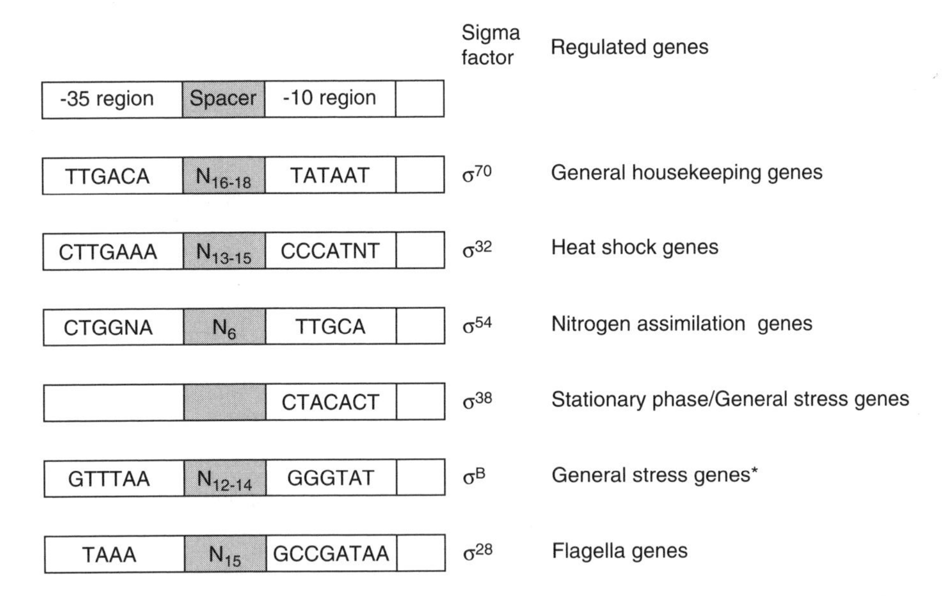

Figure 3.6 Alternative sigma factors and promoter recognition sequences. *In *B. subtilis*. All others are in *E. coli*

When the Gram-positive bacterium *B. subtilis* encounters stress, one of the survival mechanisms it can evoke is sporulation, resulting in highly resistant endospores. This is governed by a complex transcriptional regulatory programme that controls the expression of more than 100 genes and involves the sequential activation of six different σ-factors. A simplified view is shown in Figure 3.7. At the onset of starvation, the primary σ-factor (σ^A) and a low abundance factor called σ^H direct the transcription of a set of genes whose products cause an asymmetric invagination of the membrane, thus separating the forespore from the mother cell. During this process a copy of the chromosome is partitioned into each of the two compartments. Another σ-factor, σ^F, is present before the septum forms, but is inactive (due to the presence of an anti-σ-factor, see below). When the septum forms, σ^F becomes active but only in the forespore, where it switches on a new set of genes. Following this, a third sporulation specific σ-factor, σ^E, also becomes active, but only in the mother cell. A different set of genes is thus activated in the mother cell. The forespore then becomes surrounded by a second membrane (engulfment). The subsequent condensation of the spore chromosome is dependent on gene expression mediated by yet another sigma factor, σ^G, which becomes active in the forespore following engulfment. The changes in the forespore that are brought about by the genes expressed using σ^G lead to the activation of a further sigma factor, σ^K, in the mother cell. The products of the

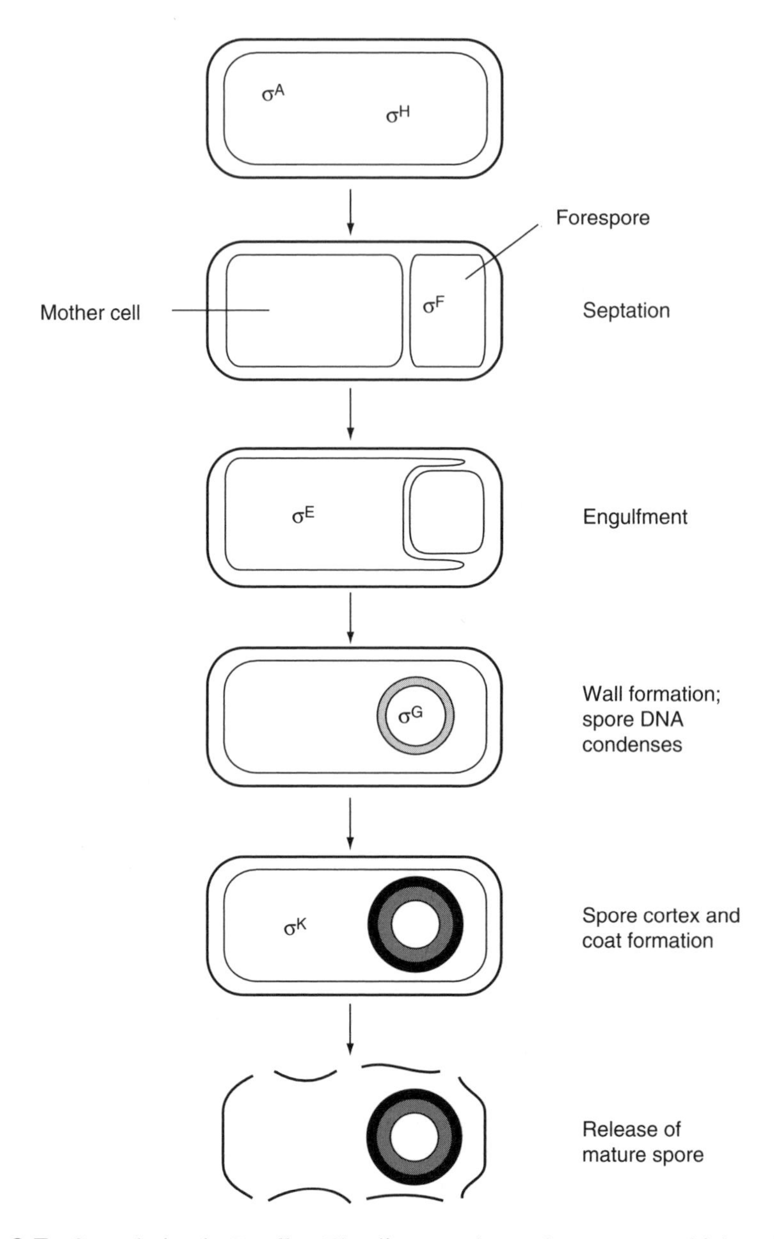

Figure 3.7 Sporulation in *Bacillus*. The diagram shows the stages at which each sigma factor first becomes active

genes expressed at this stage are involved in the final stages of sporulation including the production of the cortex and coat layers that encase the spore, leading to a structure that is highly resistant to desiccation and also to chemical agents such as disinfectants.

An important and intriguing part of this process is that the two compartments of the cell have to develop in parallel. Each stage in the mother cell has to be matched by corresponding stages in the forespore and vice versa. Without this co-operation, the proper development of the spore would not occur. As each stage develops, there is a form of signalling between the two compartments, known as *cross-talking*, that stimulates the development of the next stage. This provides one of the best understood examples of developmental control in bacterial systems.

Anti-σ-factors

The ability of σ-factors to activate sets of genes in response to various stimuli has to be controlled so that they are inactive in the absence of that stimulus. One way of doing this (as in the example of σ^F above) involves anti-σ-factors – proteins that bind to a specific σ-factor and prevent it interacting with the core RNA polymerase. Anti-σ-factors also provide an additional layer of transcriptional regulation as demonstrated by the regulation of flagella biosynthesis in *Salmonella typhimurium* (Figure 3.8). The flagellum is amongst the most complicated structures built by bacterial cells and requires the coordinated expression of more than 52 genes (1.5 per cent of the genome). One of the major regulators is the flagella-specific σ-factor σ^{28}. Bacterial flagellar assembly occurs sequentially, with the inner substructures being laid down first, followed by the construction of the external structures. Consequently, one of the problems faced by a bacterium is how to sense when the inner substructures have been completed so that it can start production of the external filament which is attached to this. The FlgM anti-σ-factor provides a mechanism for doing just this. Before completion of the inner substructures, FlgM binds to σ^{28} to prevent its action and thus prevent filament production. However, once the intracellular portion (hook-basal body) of the flagella is completed, it serves as a hollow channel and export pathway, not only for the filament protein but also for FlgM. Thus, at the same time as the hook-basal body is completed, FlgM is exported out of the cell and in its absence, σ^{28} can activate expression of the filament genes.

3.2.2 Terminators, attenuators and anti-terminators

The structure and action of transcriptional terminators, which stop transcription at the end of an operon, was described in Chapter 1. Transcriptional termination can also play a role in adjusting the level of expression of different genes within an operon.

The genes forming an operon are transcribed from a single promoter and are therefore switched on or off simultaneously. However, although the cell requires all these products at the same time, it does not necessarily need all of

(a) Before construction of the hook basal body, FlgM binds to σ^{28} to prevent it binding to RNA polymerase (RNAP)

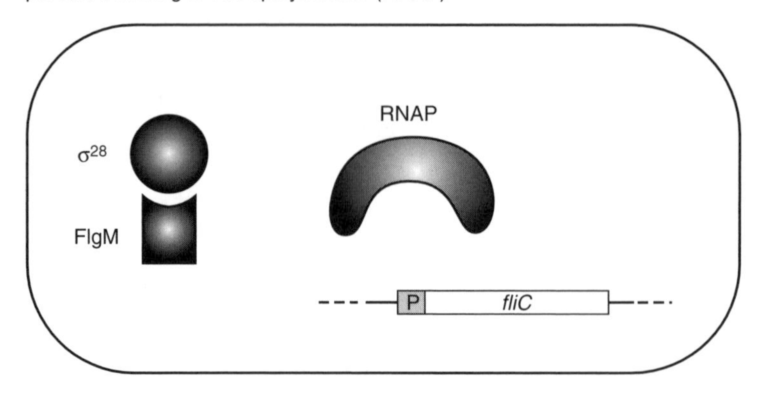

(b) After construction of the hook basal body FlgM is exported through a pore formed by the basal body. In its absence, σ^{28} is able to bind to RNA polymerase and initiates expression of the flagella filament protein FliC

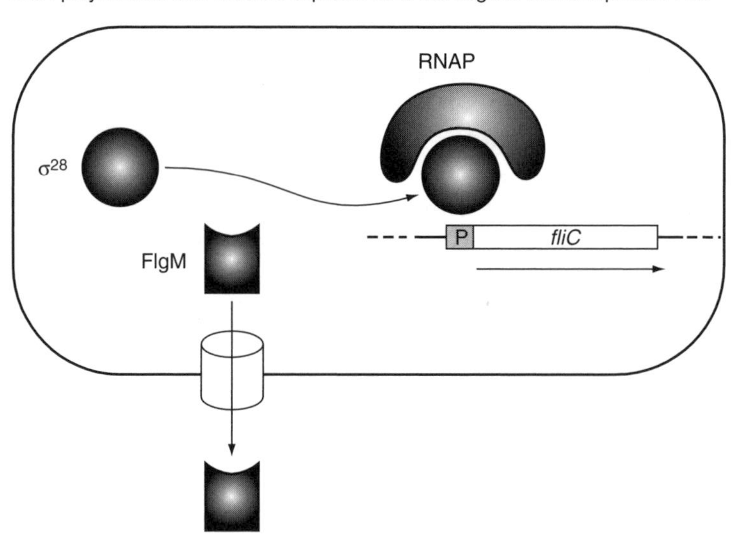

Figure 3.8 The regulation of flagella filament production (FliC) by the anti sigma-factor FlgM

them at the same level. Transcriptional termination provides one way of doing this.

In the example shown in Figure 3.9, the products of genes a and b are required at a higher level than those of genes c and d. The presence of a termination site (labelled t1) located between genes b and c will cause the RNA polymerase to pause. In this case, however, termination will only occur in a proportion of transcripts, while in the remainder transcription will be resumed. This will

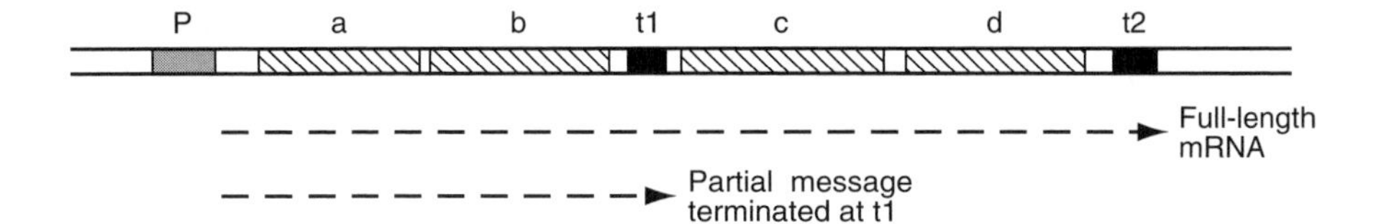

Figure 3.9 Attenuation within an operon. The presence of a weak transcriptional termination site (t1) within an operon leads to reduced expression of the distal genes (c and d). The strong terminator t2 causes termination at the end of the full-length mRNA

lead to a mixture of partial and full-length mRNA molecules. This process is referred to as attenuation; the related but distinct process of attenuation in the regulation of the *trp* operon is described later in this chapter.

In some cases, termination within an operon can be overridden by the action of proteins known as anti-terminators, which modify the RNA polymerase to allow it to ignore the relevant terminators. The best example of anti-termination comes from the control of bacteriophage lambda (see Chapter 4).

3.2.3 Induction and repression: regulatory proteins

The lac *operon*

One of the main ways in which the cell can switch transcription on or off uses protein repressors to bind to a site (the operator) which is adjacent to, or overlaps with, the promoter. The binding of the repressor protein to the operator prevents the initiation of transcription. In the case of the *lac* operon, this repressor protein is produced by a separate gene (*lacI*) that is not part of the *lac* operon (see Figure 3.4). The operator region, to which the *lac* repressor binds, overlaps with the promoter and with the first 20 or so bases that are transcribed (Figure 3.10). The relative position of operator and promoter sites is not the same for all operons; in some cases the operator overlaps the upstream end of the promoter. The end result

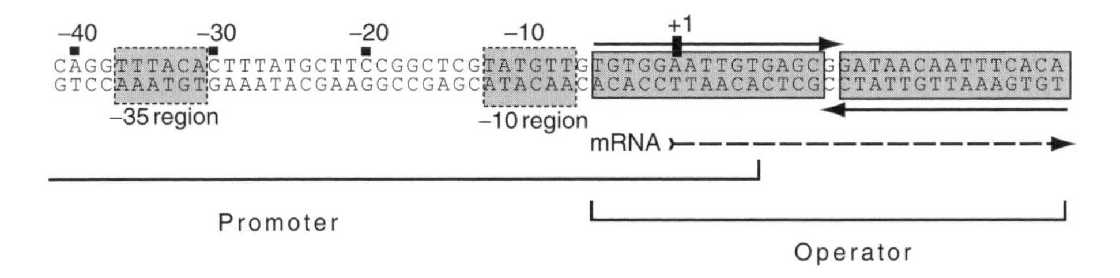

Figure 3.10 Structure of the operator/promoter region of the *lac* operon

is the same: binding of the repressor to the operator prevents the RNA polymerase from obtaining access to the promoter or from initiating transcription.

The *lac* repressor is a multimeric protein, consisting of four identical subunits, showing a secondary structure feature consisting of two α-helices separated by a few amino acids that place the two α-helices at a defined angle to each other. This conformation, known as a helix-turn-helix motif, is characteristic of DNA-binding proteins and enables this part of the protein to fit into the major groove of the DNA and make specific contacts with the operator DNA. The *lac* operator site exhibits dyad symmetry. In other words, part of the sequence (shown boxed in Figure 3.10) is repeated (imperfectly) in the inverse orientation. The tetrameric structure of the regulatory protein enables each half of it to recognize half of the binding site, thus increasing both specificity and affinity for the operator (Figure 3.11).

The *lac* repressor protein also has affinity for allolactose (a derivative of lactose). Binding of allolactose to the repressor causes an allosteric change in the protein so that it is no longer able to bind to the operator site. So, in the absence of lactose, the repressor is active, binds to the operator and prevents transcription of the *lac* operon (Figure 3.12). In the presence of lactose, the repressor is inactivated, the operon is expressed, and the cell produces the enzymes that are necessary for the metabolism of lactose.

The natural situation in *E. coli* is therefore that lactose fermentation is an inducible characteristic, i.e. it is only expressed in the presence of lactose. In the laboratory, lactose is an inconvenient inducer, since it will start to be metabolized as soon as induction is effective, thus tending to diminish the effect. However, the ability to act as an inducer and the ability to act as a substrate are separate characteristics: the former depends on recognition by the repressor protein and the latter depends on recognition (and breakdown) by β-galactosidase. It is therefore possible to design analogues of lactose that can still act as inducers, since

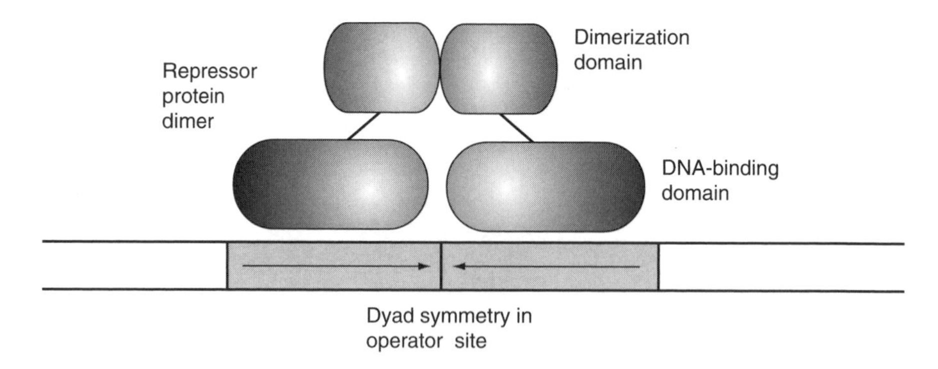

Figure 3.11 Binding of a dimeric regulatory protein to an operator site with dyad symmetry

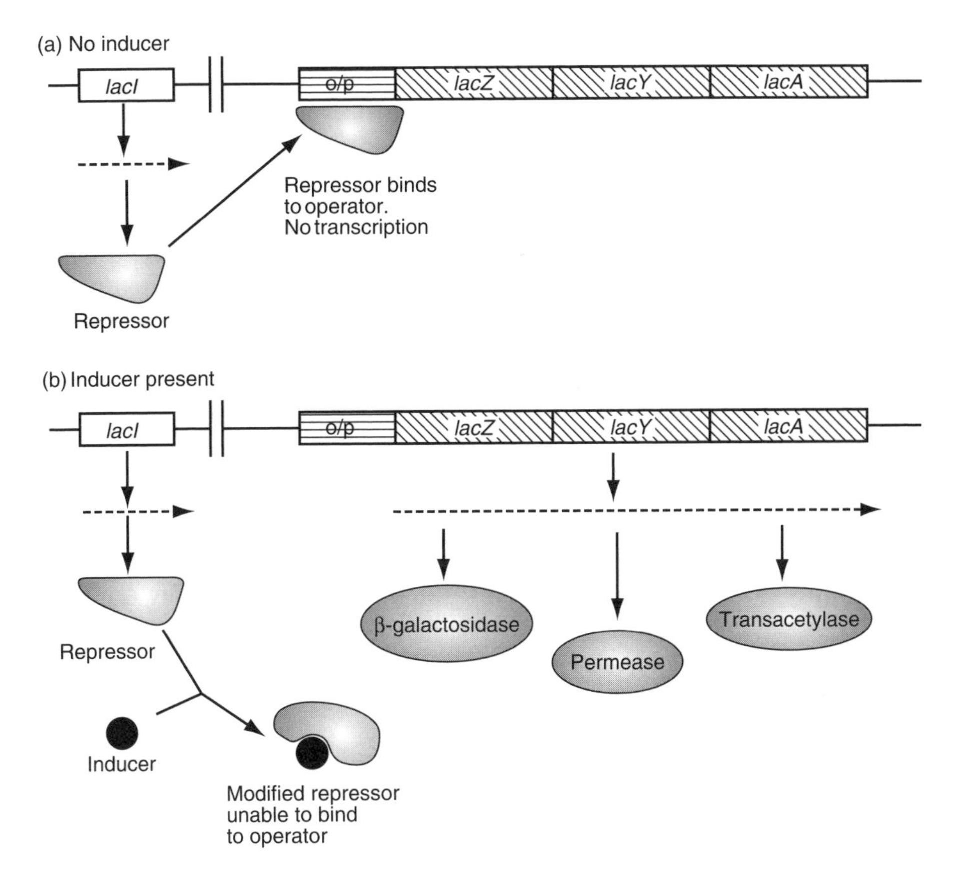

Figure 3.12 Regulation of the *lac* operon

they will bind to the repressor, but cannot be destroyed by the β-galactosidase. One such *gratuitous inducer* is the synthetic analogue iso-propyl-thiogalactoside (IPTG).

The converse is also true: some compounds are substrates for breakdown by β-galactosidase but are not able to act as inducers since they are not recognized by the repressor. The chromogenic substrate commonly known as X-gal (which gives a blue colour after hydrolysis by β-galactosidase) is an example of a substrate that does not act as an inducer. Therefore in order to obtain blue colonies it is necessary to add an inducing agent such as IPTG to the medium.

Regulatory mutants of the lac *operon*

Much of the evidence for the above model is derived from studies of regulatory mutants of the *lac* operon. These fall into several categories.

(1) *Constitutive mutants*. In these cells, the enzymes of the *lac* operon are produced at maximum level even in the absence of any inducer. Such mutants are of two types: (a) *lacI* mutants which are defective in the production of the repressor (or the repressor cannot bind to the operator), and (b) operator-constitutive (O^c) mutants in which the change is in the operator itself, preventing recognition by the repressor protein. Partial diploid strains can be constructed in which the regulatory mutation is located on the chromosome and the wild type regulatory gene on a F′ plasmid (see Chapter 5). With $LacI^-$ strains (Figure 3.13), the introduction of a plasmid carrying a functional *lacI* gene restores inducibility. In other words, the *lacI* mutation is recessive. This is in accord with the model, since the *lacI* gene carried on the plasmid will produce an active repressor that is capable of binding to the chromosomal *lac* operator (the repressor is said to be capable of acting *in trans*).

With operator constitutive mutants on the other hand, the introduction of the plasmid does not restore inducibility (Figure 3.14); the repressor coded for by the plasmid is equally unable to bind to the chromosomal operator. These mutations are described as *cis*-dominant: the operon of which they are part is

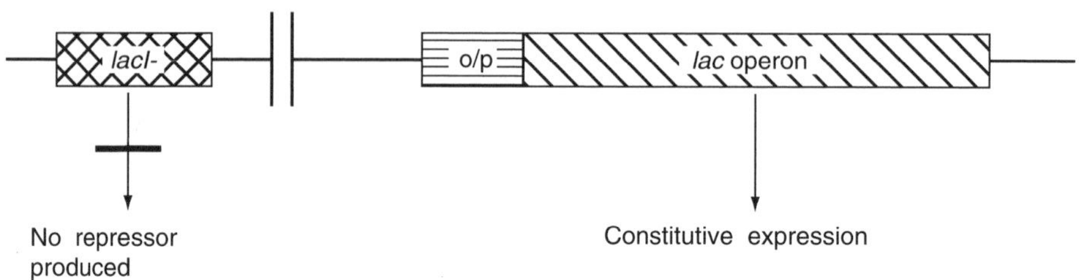

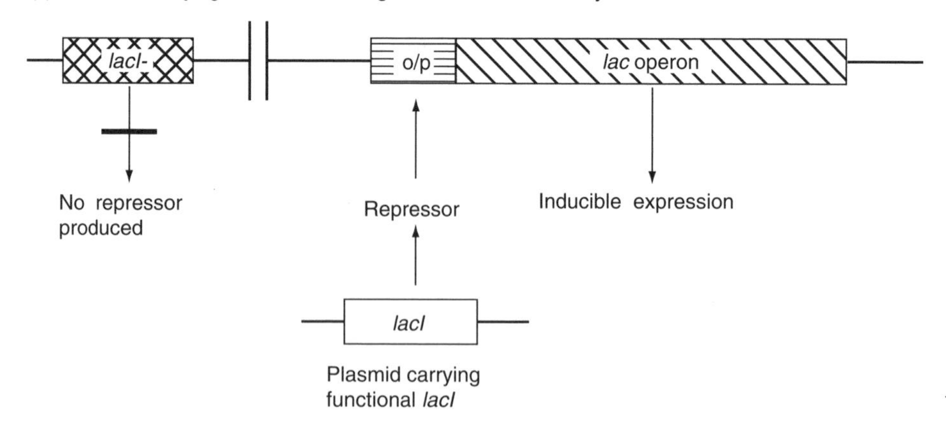

Figure 3.13 *lacI* mutation is recessive

constitutively expressed irrespective of any genes on an introduced plasmid. They are not completely dominant; if the plasmid in Figure 3.14 contained a functional *lac* operon (rather than just the *lacI* gene), that operon would be inducible as normal and would be unaffected by the O^c mutation on the chromosome. Thus the O^c mutation operates *in cis* but not *in trans*.

(2) *Non-inducible mutants*, which again are unaffected by the presence or absence of the inducer, but in this case the level of the enzymes is always low. There may be a variety of reasons for this behaviour, but the most significant one from our point of view is a different type of *lacI* mutation that abolishes the ability of the repressor protein to recognize and respond to the inducer.

(3) *Super-repressor (lacIq) mutants.* These cells are characterized by an over-production of the repressor, commonly due to a mutation in the promoter of the *lacI* gene (remember that the *lacI* gene is not part of the *lac* operon and is transcribed from a different promoter). These mutants are useful in genetic manipulation where the normal level of *lac* repressor is not sufficient to repress

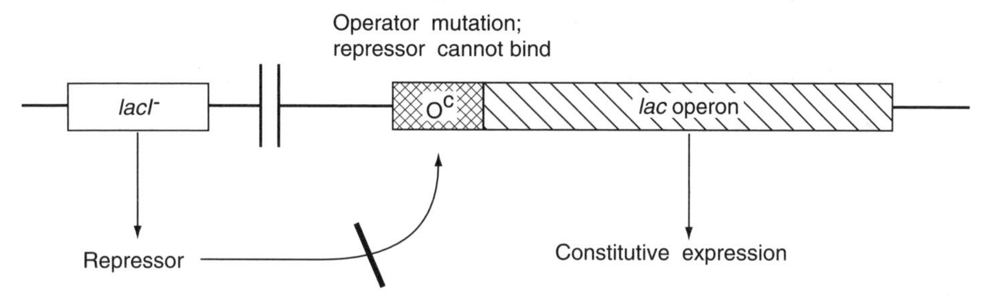

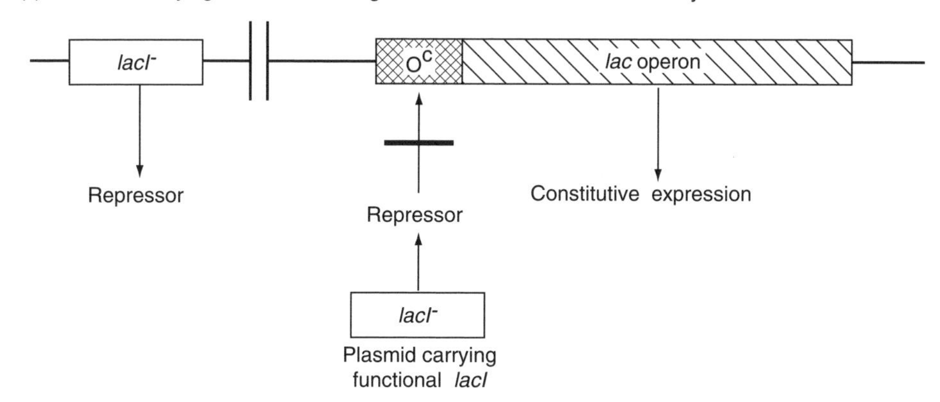

Figure 3.14 *lacO^c* mutation is cis-dominant

all the copies of the β-galactosidase gene on the multi-copy plasmids (see Chapter 8).

Catabolite repression

If *E. coli* is grown in a medium containing, as carbon and energy sources, both glucose and lactose, it will preferentially use glucose and the lactose will not be metabolized until the glucose has been used (Figure 3.15). The genes of the *lac* operon are repressed while there is glucose present. This is a specific example of a widespread effect – the repression of a set of genes in the presence of an easily metabolized substrate – which is known as catabolite repression.

The molecular basis of catabolite repression of the *lac* operon in *E. coli* is as follows. In the presence of glucose the level of ATP within the cell rises as the glucose is broken down to release energy; at the same time, the level of cyclic AMP (cAMP), a cellular alarm molecule, decreases due to activation of cAMP phosphodiesterase. In the absence of glucose, adenylate cyclase is activated and levels of cAMP rise. The activity of the promoter of the *lac* operon is dependent on stimulation by a combination of cAMP and a protein known as the cAMP receptor protein (CRP). Binding of cAMP to CRP causes a conformational change in the protein which allows it to recognize and bind to, specific sites on the DNA. If the level of cAMP in the cell is low, this stimulation cannot take place and expression of the *lac* operon will not occur.

The cAMP–CRP complex binds to a DNA site upstream from the promoter (-72 to -52 with respect to the transcription start point; see Figure 3.16) and

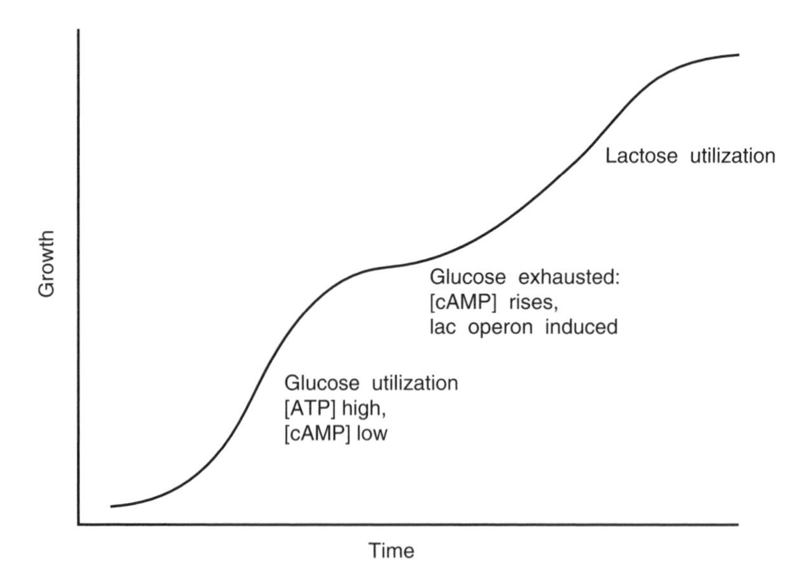

Figure 3.15 Diauxic growth and catabolite repression in *E. coli*

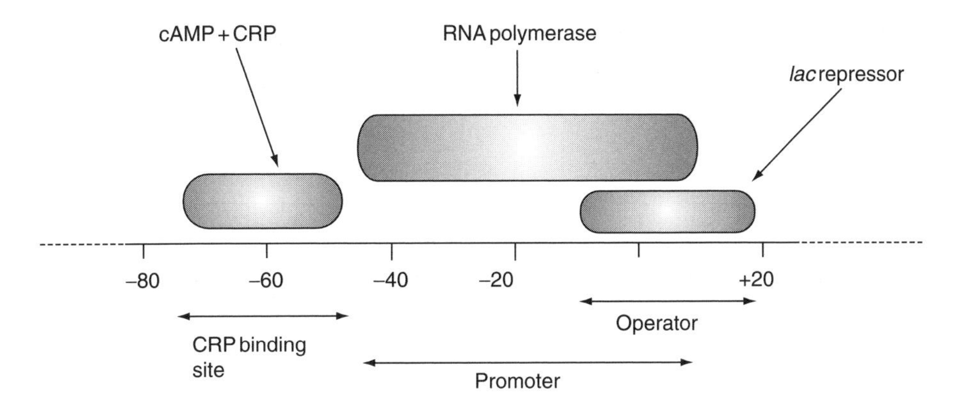

Figure 3.16 Protein binding sites in the regulatory region of the *lac* operon

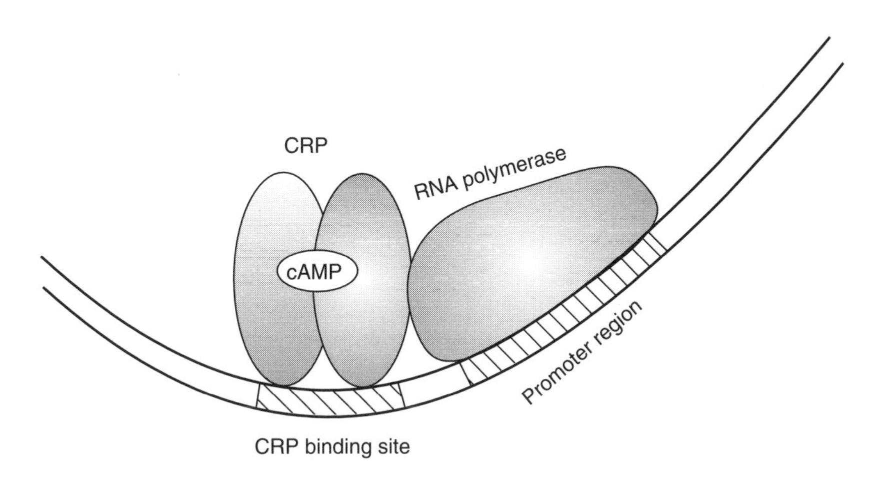

Figure 3.17 The cAMP–CRP complex causes DNA bending

interacts directly with RNA polymerase to promote binding of the latter enzyme to the promoter (Figure 3.17). In the absence of the cAMP–CRP complex, RNA polymerase binds very weakly to the *lac* promoter. Despite the term catabolite repression, it should be clear that the role of the CRP is a *positive* one; it is an activator (when bound to cAMP) not a repressor. Hence, some people prefer to call it catabolite activator protein (CAP).

An additional effect of the binding of the cAMP–CRP complex is that the DNA becomes bent at this point (Figure 3.17). Many regulatory proteins bend DNA which has local affects on the winding of the helix and may make the promoter more accessible to the RNA polymerase.

Many other operons are sensitive to catabolite regulation involving the cAMP–CRP complex, although details such as the position of the binding site vary between operons. Furthermore, binding of cAMP–CRP can have different consequences, being in some cases inhibitory to transcription rather than stimulating it.

Arabinose operon

The *lac* repressor is an example of a negative regulator, i.e. binding of the repressor to the operator prevents transcription, while CRP (in the presence of cAMP) is a positive regulator. Regulation of the arabinose operon in *E. coli* provides an example of a protein that can act as both a positive and a negative regulator.

In the absence of arabinose, a dimer of the regulatory protein AraC binds to two widely separated sites on the DNA, forming a loop and repressing transcription from the promoter of the *ara* operon (Figure 3.18). The formation of loops can be an important part of the regulation of gene expression and enables the involvement of DNA sites that are not close to the transcriptional start position. The AraC protein consists of two largely independent regions (domains): the

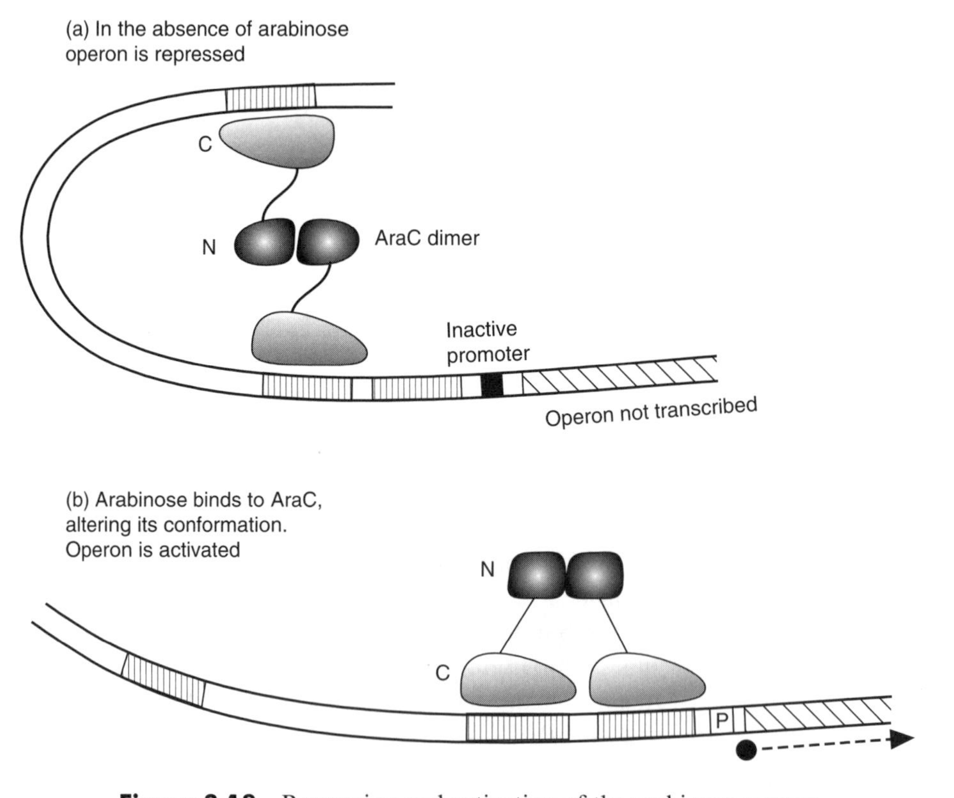

Figure 3.18 Repression and activation of the arabinose operon

NH_2-terminal domain binds arabinose and mediates dimerization, while the COOH-terminal domain contains the regions that bind to the DNA. When arabinose binds to the NH_2-terminal domain, it alters the way that the dimer forms and hence its ability to contact different sites on the DNA. As a consequence, AraC no longer binds to the upstream site (and therefore the DNA is no longer looped), but binds instead to two adjacent sites close to the promoter, resulting in activation of the promoter. Thus in the absence of arabinose AraC is a negative regulator while in the presence of the substrate it acts as a positive regulator.

Other transcriptional regulators

As mentioned earlier, the *lac* repressor belongs to the class of regulators containing helix-turn-helix motifs. This is the most common DNA-binding protein motif in prokaryotes. Many other transcriptional regulators with very diverse roles also contain this motif, which can be used to identify putative regulator genes in genome sequence data. In *E. coli* alone there are 314 transcriptional regulators with helix-turn-helix motifs. Generally, in this class of regulator the protein is guided to specific DNA sequences by the helix-turn-helix motif which is linked to a larger regulatory domain which regulates the DNA binding activity. In the case of CRP (see above), the signal molecule cAMP binds to the regulatory domain and allosterically controls the DNA binding activity of the regulator.

Amongst other examples, a protein known as FNR, which has a similar structure to CRP, controls the transcription of a wide range of genes whose functions are necessary for growth under oxygen limitation. Regulatory proteins such as FNR (and CRP) that influence the activity of a large number of genes are known as *global regulators*. The DNA-binding activity of FNR is regulated by oxygen availability with a cluster of iron and sulphur molecules $[4Fe\text{-}4S]^{2+}$ forming the actual oxygen sensor. OxyR is also a member of this class of regulators, and globally regulates the expression of genes in response to hydrogen peroxide stress. In this protein, the sensor for the oxidizing agent is a motif containing two cysteine residues. Oxidation of these residues by hydrogen peroxide causes the formation of an intramolecular disulphide bond and results in a conformational change which allows OxyR to bind to its target genes. Alternative mechanisms for global regulation are considered later in this chapter.

3.2.4 Attenuation: *trp* operon

The *lac* and *ara* operons are both examples of inducible systems, i.e. they are switched on in the presence of the relevant substrate. On the other hand, when dealing with biosynthetic pathways, such as the production of amino acids, the converse applies: expression of the genes concerned should be switched *off* if the end-product is present. This can also be achieved by the use of regulatory proteins, with only a minor modification of the model. For example, the *trp* operon, coding for the genes required for synthesis of tryptophan, is regulated (in *E. coli)* by a protein, TrpR, which is unable by itself to interact with the operator region of the *trp* operon; it requires binding of tryptophan to form the active repressor.

The *trp* operon is also regulated by a completely different process, known as attenuation, which involves transcriptional termination as described earlier in this chapter. The operon (Figure 3.19) contains a sequence (of 162 bases), known as the leader sequence, between the transcription start point and the start of the first structural gene. The leader sequence has several mutually complementary regions that can form alternative stem–loop structures under different conditions. Of these alternative secondary structures, only the 3:4 stem–loop structure will actually cause termination.

The leader sequence contains a region that can be translated into a short peptide (Figure 3.20). As soon as a long enough portion of the mRNA is made, ribosomes will bind to it and start translation. The sequence however contains two tryptophan codons, thus requiring the presence of tryptophan within the cell for incorporation into the peptide. In the absence of tryptophan, the ribosomes will therefore stall at that point. This is within the sequence designated 1; the presence of the ribosomes on this sequence will block its pairing with sequence 2, which is thus free to form a stem–loop structure with region 3 as soon as the latter region is produced. This structure (2:3) is not a terminator and does not block further transcription, but it sequesters region 3 and prevents it pairing to region 4 (see Figure 3.21b). This therefore prevents the formation of the 3:4 stem–loop structure which would otherwise terminate transcription. Thus in the absence of tryptophan, expression of the *trp* operon is permitted.

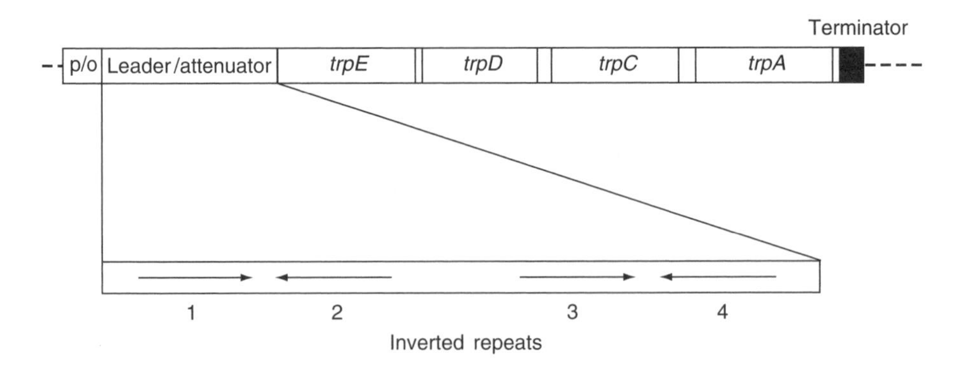

Figure 3.19 Structure of the *trp* operon

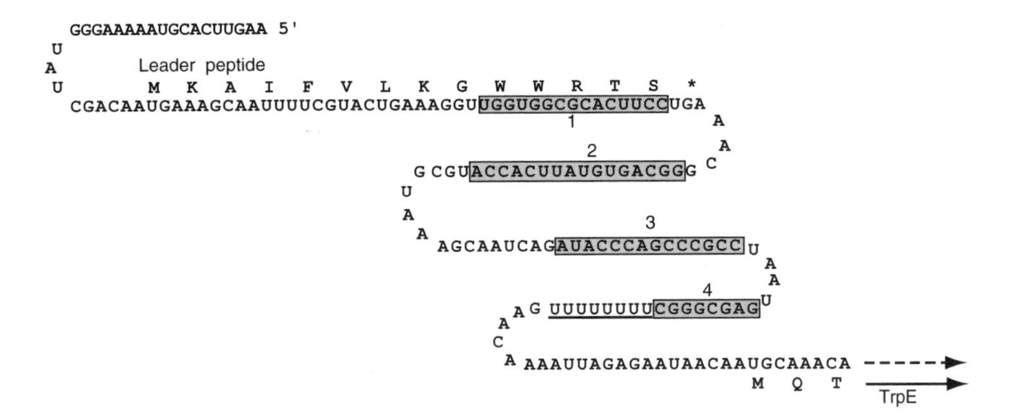

Figure 3.20 Structure of the leader–attenuator region of the *trp* operon mRNA. The boxed sequences 1–4 are the regions capable of forming alternative stem–loop structures

In the presence of sufficient tryptophan, the ribosomes can proceed unimpaired through the leader sequence until they reach the stop codon which is adjacent to the start of region 2. Although the ribosomes would eventually dissociate from the mRNA, they occupy this position for long enough to ensure that region 2 is prevented from forming a stem–loop structure. If 2:3 does not form rapidly, then as soon as region 4 is made, the more stable 3:4 structure will form (Figure 3.21c). Since 3:4 is a genuine terminator, transcription will stop. The presence of tryptophan therefore prevents expression of the *trp* operon. Similarly, if protein synthesis is prevented altogether, then 1 will pair with 2 and 3 with 4, also leading to repression of the operon (Figure 3.21a).

Similar attenuation methods control a number of other operons concerned with the biosynthesis of amino acids. In each case, the operon has a leader sequence that is translated into a peptide containing multiple codons for the amino acid concerned. For example, the *his* operon in *E. coli* has a leader sequence that codes for a run of seven histidines, while the *pheA* attenuator region has seven phenylalanine codons. Stalling of the ribosome at these codons prevents the formation of the terminator structure.

In *B. subtilis*, the *trp* operon is also controlled by attenuation at the transcriptional level (Figure 3.22), but the mechanism is an intriguing variation of that described above for *E. coli*. In this case stem–loop structures can occur in regions A:B or C:D of the leader sequence of the *trp* mRNA but only the latter structure will actually cause termination. In the absence of tryptophan the structure A:B forms; because of the overlap between B and C, this prevents the formation of C:D so transcription of the operon occurs. The key differences between the *E. coli* and *B. subtilis* system are the presence of 11 repeated DNA sequences (GAG or UAG) in the leader mRNA and an RNA binding protein called TRAP (*trp* RNA-binding Attenuation Protein) which, in the presence of tryptophan, is able to bind

(a) Absence of protein synthesis

mRNA

5′

1

2

3

4

Termination

Stem-loop causes transcriptional termination

(b) Protein synthesis with limiting tryptophan

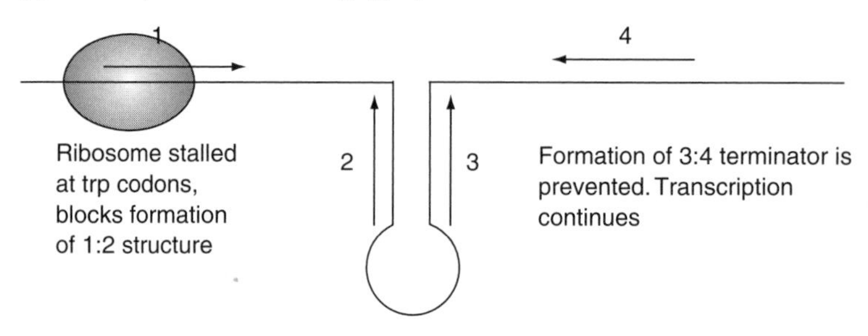

(c) Protein synthesis with sufficient tryptophan

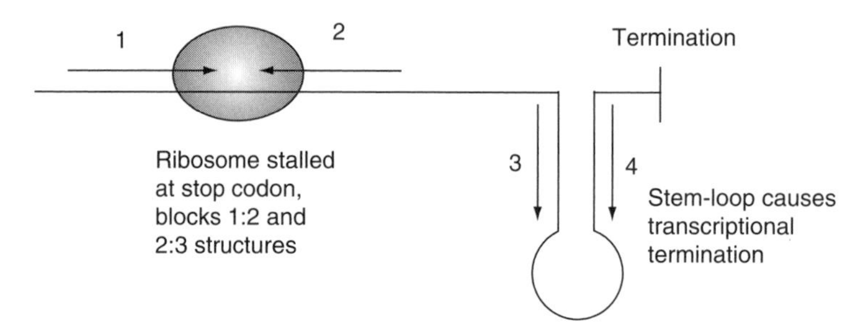

Figure 3.21 Attenuation control of the *trp* operon. (a) In the absence of protein synthesis, the terminator stem–loop 3:4 is able to form, and the operon is not transcribed. (b) If protein synthesis occurs in the presence of limiting amounts of tryptophan, ribosomes will stall at the tryptophan codons in the leader region, blocking formation of the 1:2 stem–loop. When the RNA polymerase transcribes region 3, it will pair with region 2. The 2:3 structure is not a terminator, but it sequesters region 3, thus preventing formation of the 3:4 terminator and allowing transcription of the operon. (c) In the presence of sufficient tryptophan, the ribosomes will proceed as far as the stop codon, thus blocking both regions 1 and 2. This allows the 3:4 termination structure to form, preventing transcription of the operon

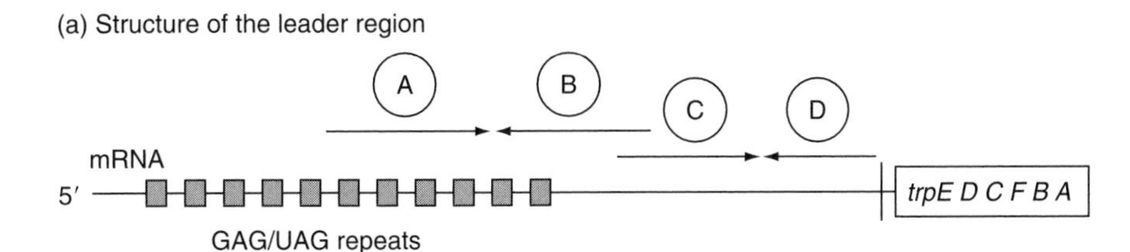

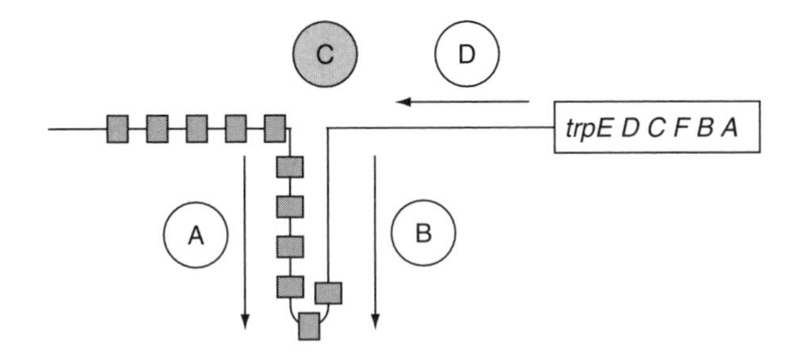

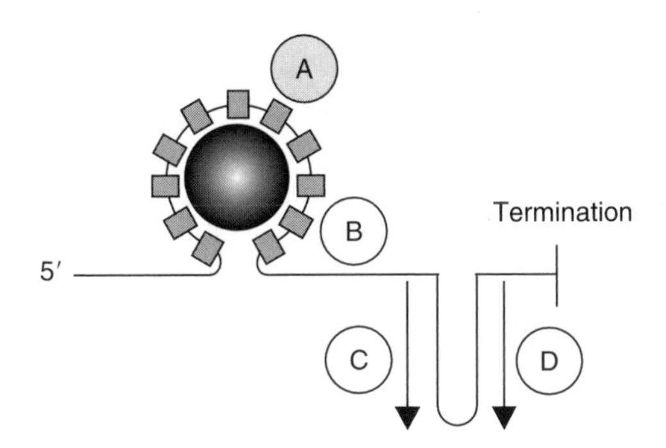

Figure 3.22 Attenuation control of the *trp* operon in *Bacillus subtilis*. (a) The leader region contains two pairs of complementary sequences, enabling two alternative stem–loop structures A:B and C:D. The partial overlap of B and C prevents both structures forming. The leader also contains 11 repeats of GAG or UAG, which can bind TRAP (*trp* RNA-binding Attenuation Protein) in the presence of tryptophan. (b) In the absence of tryptophan, TRAP does not bind, the A:B stem–loop forms and the terminator C:D cannot form. Thus transcription of the operon occurs. (c) In the presence of tryptophan, TRAP binds to the GAG/UAG repeats, which blocks region A and prevents the formation of the A:B stem–loop. Region C is free to pair with region D to form the terminator structure, preventing transcription of the operon. Shaded circles indicate regions that are blocked from forming stem–loop structures

to these. Since the GAG/UAG repeats cover the whole of A, and part of B, the A:B stem–loop structure cannot form, which allows the formation of the C:D stem–loop terminator. Consequently transcription of the *trp* operon is prevented.

3.2.5 Two-component regulatory systems

The systems described above respond to conditions *within* the cell. Bacteria also have mechanisms for sensing and responding to external conditions and other stimuli, without such conditions altering the internal state of the cell. The mechanism for transmitting external signals to the interior of the cell is known as *signal transduction*.

One of the most common of such mechanisms are the two-component regulatory systems (Figure 3.23). In general, these systems comprise an integral membrane protein called a histidine protein kinase (HPK) and a separate cytoplasmic protein called a response regulator (RR). The HPK has two domains. The input domain is usually found on the outside of the cell in an ideal position to detect environmental signals. In contrast, the transmitter domain is located on the cytoplasmic face of the cell membrane, positioned to interact with the RR. When a stimulus causes a conformational change in HPK, the HPK autophosphorylates at a conserved histidine residue and subsequently transfers this phospho group to the response regulator. In this form the RR is able to bind to DNA to regulate transcription of the target genes.

The RR also consists of two domains: a receiver domain containing an aspartate residue which accepts the phospho group, and an output domain which can bind to DNA. Dephosphorylation of the response regulator is also important to terminate the signal and can be carried out by the RR itself or by a specific phosphatase. The above description outlines the orthodox scheme for this type of sensor. However, since the domains of the two component proteins are modular, the scheme is highly adaptable and a single response pathway can often contain multiple RRs and other variations on this simple scheme (Figure 3.23b). Although these are still often termed two-component systems because of the historical context, it is important to note that they will contain more than two components and are sometimes termed *phosphorelay* systems.

Over 300 two-component systems are now known in bacteria. *E. coli* and *Ps. aeruginosa*, for example, possess 32 and 89 RRs, and 30 and 55 HPKs, respectively. Indeed, there is so far only one known example of a bacterium (*Mycoplasma genitalium*) that does not contain these systems. Mycoplasmas are believed to have evolved into a minimal life form, containing just enough genes for replication and growth and have lost much regulatory capacity in order to reduce genome size.

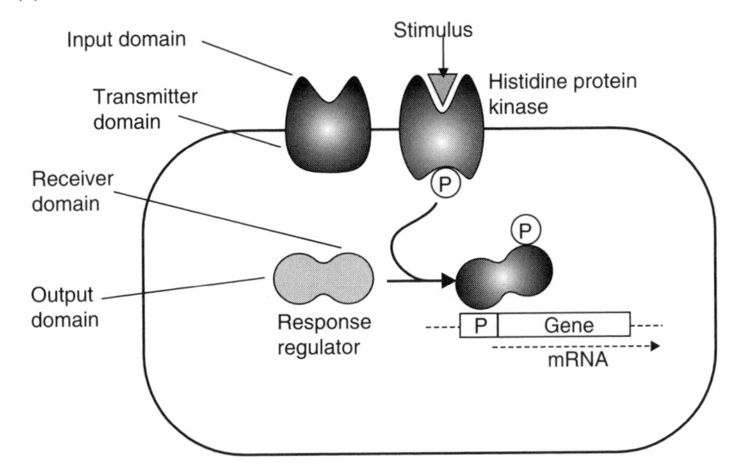

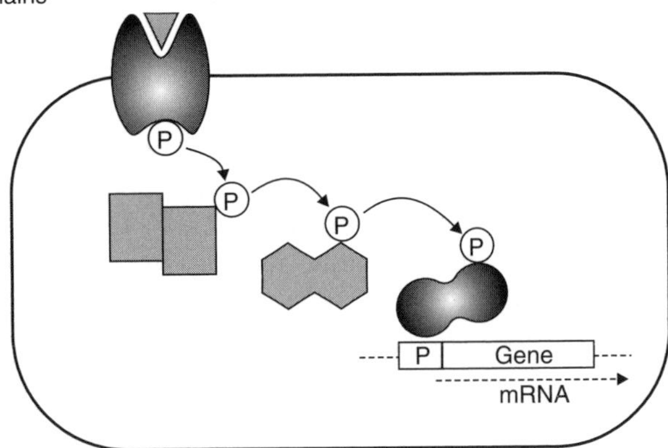

Figure 3.23 Two component regulatory systems. (a) In an orthodox system, an external stimulus interacts with a specific membrane protein, causing a conformational change in the protein which activates autophosphorylation on the cytoplasmic face of the protein. The phosphate is then transferred to the receiver domain of a response regulator, enabling the output domain of the regulator to bind to operator sites of the regulated gene(s). (b) In a phosphorelay system, there is a chain of phosphorylation reactions, leading to the activated regulator protein

Amongst the wide variety of two-component systems responding to different stimuli, the HPK output domains are similar but the input domains, where the actual sensing mechanism resides, are very diverse. For example, the input domain from FixL (an HPK from Rhizobium) contains a haem group which responds to oxygen whilst the input domain from VirA from the plant pathogen *Agrobacterium tumefaciens* senses phenolic compounds released from plant wounds. In *Salmonella* the pleiotropic PhoP-PhoQ system regulates virulence and acid tolerance but actually senses extracellular Mg^{2+} concentrations through the periplasmic input domain of the HPK, PhoQ, which contains several Mg^{2+}-binding acidic amino acids.

3.2.6 Global regulatory systems

The regulatory mechanisms described may control not just a single operon but a very large number of unrelated genes. Such systems are referred to as *global regulation*. For example, the cAMP–CRP system, described above, is not just a regulator of the *lac* operon but affects the expression of some 200 genes. Other examples of global regulatory systems include the heat-shock response, acid response, oxidative stress response, cold shock response and osmotic stress response.

Many pathogenic bacteria also use global regulatory systems to control the expression of the mechanisms that they need to survive in the host and avoid the immune response. *Bordetella* species for example sense different stimuli including temperature, magnesium and nicotinic acid and globally regulate virulence gene expression in response to these stimuli.

The mechanisms involved in these global regulatory systems are very diverse and sometimes overlap to form networks which can cross-talk. The global heat shock response, for example is controlled by σ^{32}. This alternative σ-factor regulates the expression of proteins known as molecular chaperones which can re-fold damaged proteins or degrade denatured proteins. Although these genes are often referred to as heat-shock genes, their expression can also be stimulated independently by a variety of other stress conditions, including bacteriophage infection. The heat shock response also overlaps substantially with the stringent response (see below) and with the SOS response referred to in Chapter 2 as the source of error-prone repair of DNA.

Another way in which cells can respond to changes in environmental conditions is by alteration in the degree of supercoiling of the DNA. In general, the overall level of supercoiling in the cell is controlled by a balance between the activities of DNA gyrase (which introduces negative supercoils) and DNA topoisomerase I (which removes supercoils). However, DNA topology is also influenced by the presence of histone-like proteins which bind to, and cause bending of, the DNA or even wrap it around them. Although the most abundant of these proteins, known as HU, is evenly distributed, others such as integration host factor (IHF), H-NS (histone-like nucleoid structuring protein) and FIS (Factor for Inversion Stimulation) are to some extent sequence specific and therefore affect the topology of DNA in defined regions.

If we combine this information with the knowledge that some promoters are affected by the degree of supercoiling of the DNA (some being stimulated by elevated supercoiling, while others prefer relaxed DNA), then it is clear that conditions, including anaerobiosis and changes in the osmolarity of the surrounding medium which affect either global or local supercoiling, may be expected to affect the expression of these genes.

3.2.7 Feast or famine and the RpoS regulon

Until quite recently, most of the fundamental concepts of molecular genetics and metabolism were obtained using *E. coli* cells that were growing rapidly under laboratory conditions. However, in nature enteric bacteria like *E. coli* live under very variable nutritional conditions. As a consequence, evolution has optimized *E. coli* to be able to respond efficiently to conditions of both 'feast and famine'. The transition from logarithmic growth to stationary phase, and starvation, is associated with metabolic and morphological changes that result in a remarkable resistance to a variety of stress conditions such as heat, acid, high salt and oxidative stress. The central regulator for many of these changes and for preparing the cell for famine and conditions of no growth is the alternative σ-factor RpoS which controls the expression of 30 or more genes.

3.2.8 Quorum sensing

For many years bacteria have been considered merely as individual cells, yet as long ago as 1905 Erwin F. Smith wrote 'The only explanation I can think of is that a multitude of bacteria are stronger than the few, and thus by union are able to overcome obstacles too great for the few'. This remarkably prophetic statement summarizes the phenomenon of *quorum sensing*, which only began to be unravelled in the 1980s, whereby bacteria measure and respond to their own population density. The principle, as illustrated by Figure 3.24, is that bacteria may both secrete and respond to a diffusible signal. At low cell density, the concentration of the signal in the surrounding medium is low, so the bacteria do not respond. When the cell density is high, the concentration of the signal is also high, and the bacteria respond by activating expression of a specific set of genes.

The phenomenon was originally uncovered following studies on the regulation of bioluminescence in marine Vibrio species. These species are able to live two separate life cycles, as free-living marine bacteria or as commensal bacteria occupying and providing bioluminescence in the light organs of fish and squid. Bioluminescence is an energetically expensive process and in these bacteria is tightly controlled by cell population density. So, when high numbers are present, as is the situation in the light organ, bioluminescence occurs, but when they are free living, isolated, cells it does not. In these species, the signal molecule, also called an *autoinducer*, is an acyl homoserine lactone which is synthesized by LuxI

(a) Low cell density
Low level of signal
Target genes off

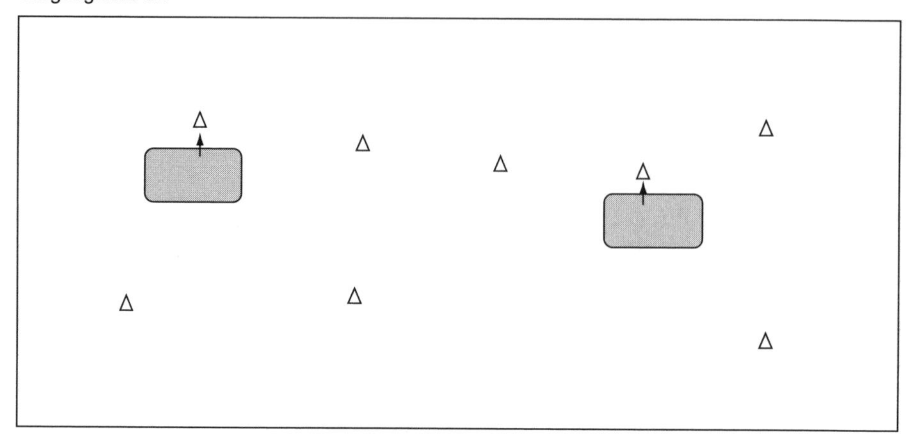

(b) High cell density
High level of signal
Target genes on

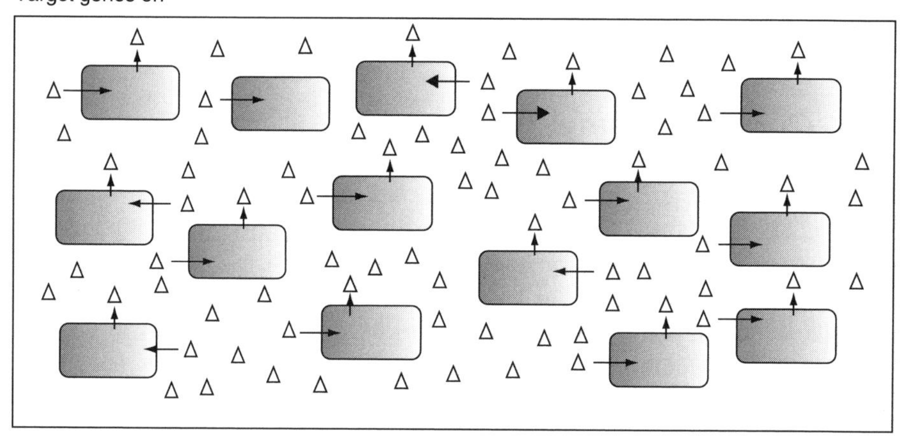

Figure 3.24 Quorum sensing in bacteria. Some aspects of bacterial physiology can be influenced by the concentration of bacteria. This is mediated by sensing the concentration of a secreted signal (shown as triangles). (a) At a low cell density, the level of the signal in the environment will be low and the cells will not respond. (b) When the cell concentration is higher, the concentration of the signal will rise. This will be detected by the cells and the expression of the relevant genes will be activated

(acyl homoserine lactone synthase) and freely diffuses across the cell membrane into the environment surrounding the cell which has made it (Figure 3.25). If few cells are present, then production of this compound is limited and consequently it is not recognized as a signal. However, when many cells of the same species are present, the amount of acyl homoserine lactone in the environment (and hence

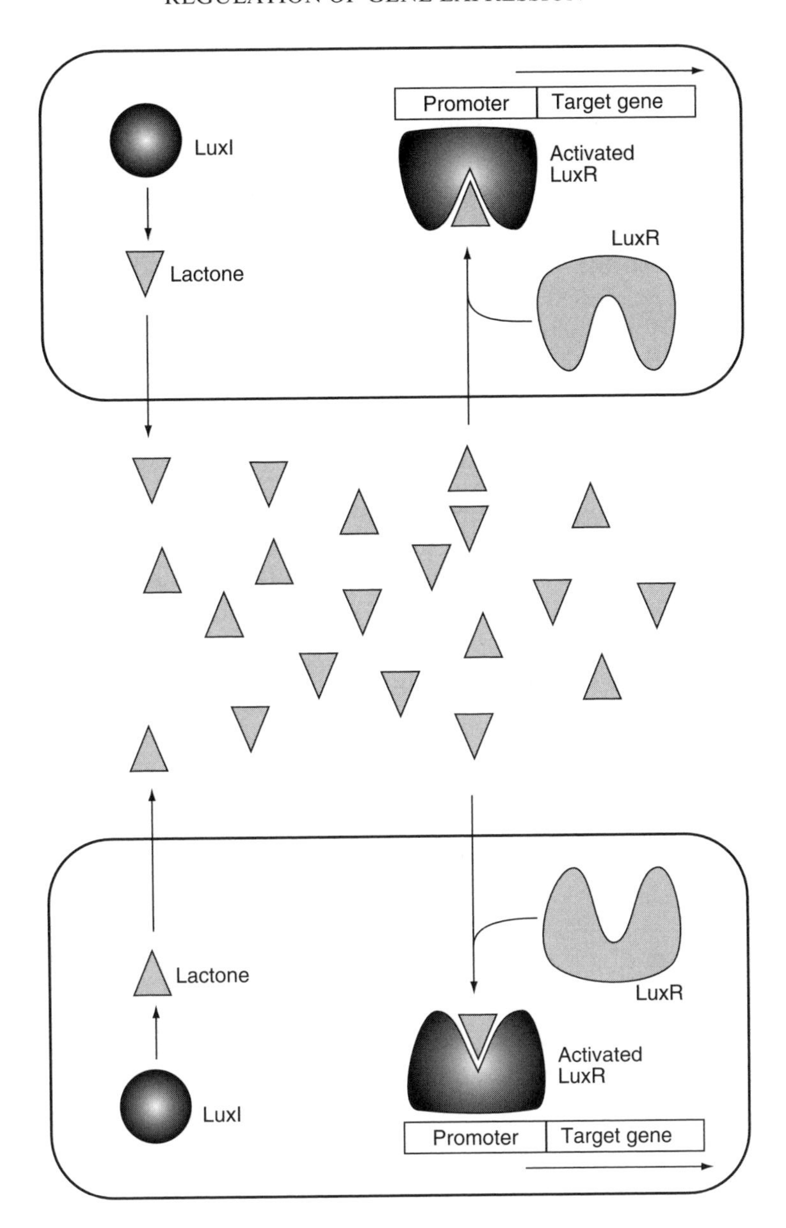

Figure 3.25 Quorum sensing: LuxI/LuxR system. In many Gram-negative bacteria, quorum sensing is mediated by acyl homoserine lactones produced by LuxI. The lactones can diffuse across the cell membrane and enter other cells where they interact with LuxR. If the lactone concentration is sufficient, the activated LuxR will switch on the genes concerned

taken up by other cells) is increased simply because more cells are producing it. When it reaches a threshold concentration, it stimulates a transcriptional regulator (LuxR) which controls the expression of genes necessary for bioluminescence.

Apart from marine vibrios, related systems have been identified in over 50 Gram-negative bacterial species.

Gram-positive bacteria also regulate a variety of processes in response to cell density, using small secreted peptides (Figure 3.26), rather than the acyl homoserine lactones used by Gram-negative bacteria. Initially, a polypeptide signal precursor is synthesized which is later cleaved to yield a small peptide (8–20 amino acids). When the external concentration of this peptide signal reaches a threshold concentration, an HPK of a two-component system (see above) detects the signal and activates a RR which then stimulates transcription of the target genes. A marked contrast with the Gram-negative system is that these peptides do not diffuse across the cell membrane – they are specifically secreted and the cells respond to the *extracellular* concentration via a signal transduction mechanism.

Quorum sensing facilitates multicellular cooperation in bacteria. Myxobacteria (such as *Myxococcus xanthus*) are predatory bacteria which obtain nutrients through the lysis of other bacteria, by secreting lytic extracellular enzymes. The attack of a single myxobacterial cell would be ineffective since the digestive enzymes and any released nutrients would be rapidly diluted. As a consequence, myxobacteria use a 'wolf-pack' hunting strategy, in which quorum sensing co-

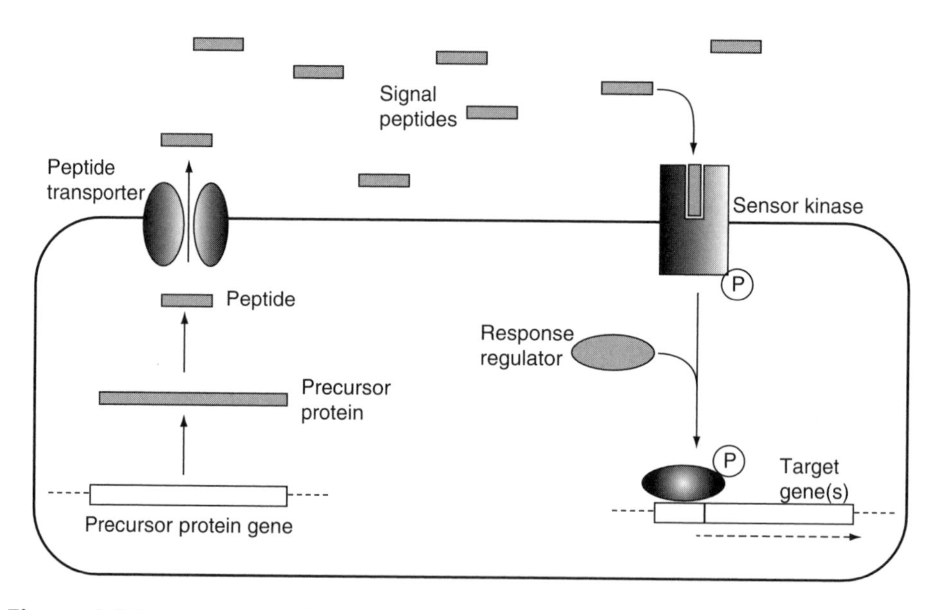

Figure 3.26 Quorum sensing: Gram-positive bacteria. Quorum sensing in Gram-positive bacteria is typically mediated by small peptides which are transported across the membrane by a specific peptide transporter. These peptides are not taken up by the target cells, but are recognized by a membrane receptor which transmits the signal to the interior of the cell (see also two-component regulation)

ordinates the formation of a cooperatively feeding swarm of individual cells. We will look at other aspects of communication in *M. xanthus* in Chapter 9.

Similarly, the plant pathogen *Erwinia carotovora* uses a homoserine lactone quorum sensing system to ensure that digestive enzymes, which attack plant structures, are only produced if there are sufficient cell numbers to mount a concerted and effective attack on the plant tissue. As another example, we will see in Chapter 6 that some mechanisms involved in the transfer of genetic information between bacteria are also dependent on quorum sensing and only work at high cell density.

3.3 Translational control

3.3.1 Ribosome binding

After the mRNA has been produced, the next step is the attachment of ribosomes and the translation of the sequence into protein. This stage seems to play a comparatively minor role in the natural control of gene expression in bacteria. This is not really surprising as it would be rather wasteful to produce large amounts of mRNA that are not required for translation. Translational control can however become important with genetically-engineered bacteria, when very high levels of transcription of a specific gene have been achieved.

Binding of the ribosomes to the mRNA occurs at specific ribosome binding sites (see Chapter 1) up to seven bases upstream from the AUG translation initiation codon. In bacteria, this sequence determines where the ribosomes will bind to the mRNA and thus determines which AUG codon is used for the initiation of translation.

The actual sequence of a ribosome binding site (RBS) and its distance from the start codon can vary somewhat, so it is to be expected that there will be weak and strong ribosome binding sites, just as there are weak and strong promoters. However, the sequence of an RBS does not seem to affect the *level* of translation – it either works or it does not. But the distance separating the ribosome binding site from the initiation codon can have a powerful effect on gene expression.

In polycistronic operons, the ribosome may, after translating the first gene, dissociate from the mRNA. The next gene then must have a site to which ribosomes can attach in order for it to be translated. The efficiency of this can vary which may result in the subsequent genes being translated less effectively. This effect, known as polarity, is an alternative to the attenuation mechanism referred to earlier for achieving different levels of expression of genes within an operon.

Polarity is particularly marked if there is a nonsense mutation causing premature termination of translation. Such a mutation may prevent translation of the subsequent cistrons altogether. The absence of ribosomes translating that stretch of RNA may also allow the formation of stem–loop structures that result in premature transcriptional termination; these attenuator sites would normally be masked by the presence of the ribosomes.

In many cases however the initiation codon for the second gene is very close (within a few base pairs) to the termination codon for the previous gene – or may even overlap with it (see Figure 3.27). In such a situation, the 30S subunit of the ribosome does not dissociate. After release of the first polypeptide, and of the 50S subunit, the 30S subunit can contact the initiation codon of the next gene to restart the translation process.

An extension of this process, known as ribosomal frameshifting, has been shown to occur in a limited number of cases without a stop codon. In these cases, when the ribosome reaches a specific site in the mRNA (a *slippery* sequence, usually containing several adenine residues), it may shift back one base (a -1 frameshift) and then continue polypeptide synthesis but reading the mRNA in a different frame. This can result in two proteins being produced from the same mRNA (see Figure 3.28), one of which is a *fusion protein* that contains portions read in two different reading frames. Since this may happen only occasionally, it represents a way of achieving further downregulation of gene expression, which may be significant when the product is only needed at extremely low levels. It can be difficult to achieve such low levels of expression merely by regulating mRNA synthesis. The

```
PROTEIN 1:---| Arg | Gly | Arg | Val | stop |
mRNA      ---AGAGGUCGCGUAUGACGCUGCCUUAUA-----
PROTEIN 2:              | Met | Thr | Leu | Pro | Tyr |----->
```

Figure 3.27 Translation initiation in overlapping genes. The stop codon (UGA) for protein 1 overlaps the start codon (AUG) for protein 2 in a different reading frame

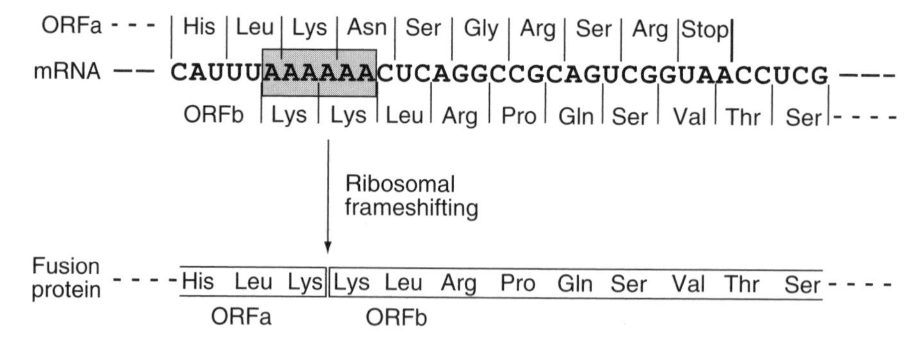

Figure 3.28 Ribosomal frameshifting. At the position indicated, the ribosome may shift back one base on the mRNA, giving rise to a fusion protein read from two different reading frames

best characterized examples come from the regulation of the mobility of insertion sequences (see Chapter 7).

3.3.2 Codon usage

The effectiveness of translation can also be influenced by the nature of the codons used throughout the gene. Most amino acids can be coded for by more than one codon. In some cases, the codons are effectively equivalent, since the same tRNA will recognize both equally well (see Chapter 1). But in many cases, a different tRNA species is responsible for recognition of the different codons and some of these tRNA species are known to be present in the cell at quite low levels. A gene that contains many codons that require these 'rare' tRNA molecules will then be expected to suffer delays in translation that may affect the amount of end-product formed.

Some indirect evidence of this *codon usage* effect comes from the study of gene sequences of *E. coli*, which shows that highly expressed genes have a high degree of 'codon bias', i.e. they have a marked preference for codons that can be recognized by the common tRNA species. Genes that are expressed at moderate or low levels do not on the whole show such a codon bias and they may contain several codons that require relatively rare tRNAs. Codon usage appears to be more important for highly expressed genes.

3.3.3 Stringent response

Ribosomal protein synthesis provides a specific example of translational control. The production of these proteins needs to be coordinated so that equal amounts of each are made. This is achieved, in part, by autogenous control of translation, i.e. the accumulation of a ribosomal protein will repress the translation of the corresponding mRNA. This also provides a way of relating ribosomal protein production to the amount of rRNA. If there is a shortage of rRNA, free ribosomal proteins will start to accumulate and will shut down further translation of the relevant mRNAs. In this way, the cell is able to respond to changes in growth conditions. When the cells are short of nutrients, they need fewer ribosomes. This can be achieved by reducing rRNA production, with a consequent cessation of ribosomal protein production. This is one aspect of what is known as the *stringent response*.

The stringent response is triggered by amino acid starvation, which leads to the presence of uncharged tRNA occupying the A site of the ribosome. A protein known as the stringent factor is involved in the conversion of GTP to two unusual nucleotides, ppGpp (guanosine-5′-diphosphate-3′-diphosphate) and pppGpp (guanosine-5′-triphosphate-3′-diphosphate). These nucleotides prevent transcription of the rRNA operons; the production of ribosomal proteins will then be reduced as a consequence of the reduced amount of rRNA available.

3.3.4 Regulatory RNA

In addition to proteins acting as transcriptional regulators, in recent years regulatory RNA molecules (often denoted riboregulators) that regulate gene expression at the post-transcriptional level have been discovered. In many cases this involves the production of 'antisense' RNA. If a region of a gene, particularly the region including the ribosome binding site and translation initiation point, is transcribed in the opposite direction, an RNA molecule will be produced that is complementary to the mRNA. This molecule can hybridize to the mRNA, and thus block the binding of ribosomes and the initiation of translation. Interactions of this kind are known to be involved in the regulation of some bacterial genes such as *ompF*, which codes for a porin in the outer membrane. Translation of *ompF* mRNA is regulated by a short antisense RNA called MicF. Under stress conditions, the amount of MicF increases, and consequently the production of OmpF is decreased.

Typically, antisense RNA binds to mRNA made from the opposite DNA strand and is therefore specific for a single gene. However, some regulatory RNAs regulate multiple targets, often in different ways and with different consequences. For example, DsrA is an 87-kb RNA which regulates translation of two different genes, *hns* (which codes for the H-NS protein, a silencer of expression of a wide range of genes) and *rpoS* (coding for the stationary-phase sigma factor σ^S, see above). DsrA binds to the two ends of the *hns* mRNA, preventing translation. In contrast, its binding to *rpoS* mRNA prevents formation of a stem–loop structure that otherwise obscures the ribosome binding site; DsrA thus *stimulates* translation of *rpoS*. We will encounter other facets of antisense RNA activity when considering the control of plasmid replication in Chapter 5 and for genetic manipulation (Chapter 9).

3.3.5 Phase variation

Whilst the expression of most bacterial genes is controlled at the level of transcription and the amplitude of the response can be graded, a number of genes are controlled in an 'all or nothing' manner. The high frequency switching of gene expression between an ON or OFF state is called *phase variation* and is another important mechanism for regulating gene expression in bacteria. This will be discussed in detail in Chapter 7.

4

Genetics of Bacteriophages

Bacteriophages (or phages for short) are simply viruses that infect bacteria. They played a central role in the development of molecular biology, especially in our understanding of gene structure and expression and are also important, in the laboratory and in nature, in providing a way in which genes can be transferred from one bacterium to another. With the development of gene cloning, phages took on an additional role – as *vectors* for cloned DNA (see Chapter 8).

In many of their basic properties, phages are similar to other viruses. They contain either RNA or DNA enclosed in a protein coat. The phage infects by attaching to a specific receptor on the surface of the bacterium and the nucleic acid enters the cell. Some of the phage genes (the *early* genes) are expressed almost immediately, using pre-existing host enzymes; in general, these code for proteins required for replication of the phage nucleic acid. A number of copies of the phage nucleic acid are then made, and the expression of the *late* genes starts: these are mainly those needed for production of the phage particle. The nature of the switch from early to late gene expression varies between different phages. Later in this chapter we will consider some specific examples. The phage particles are then assembled and the cell lyses, liberating a number of phage particles, each of which can then go on to infect another bacterial cell (see Figure 4.1).

If a bacteriophage infects a liquid culture of bacteria, it may result in complete clearing of the culture. On the other hand, if the bacteria are spread on the surface of an agar plate, the phage particles liberated from an individual infected cell will only be able to infect neighbouring bacteria which will result in localized clearing of the bacterial lawn, referred to as *plaques*. This allows us to determine the concentration of bacteriophage particles in a suspension (the phage *titre*), using the assumption that the number of plaques corresponds to the number of bacteriophage particles present in the sample. Since the assumption is not always valid, the results are usually expressed in terms of *plaque-forming units*, or pfu. In practice, such assays are usually performed with the bacterial culture (the

Molecular Genetics of Bacteria, 4th Edition by Jeremy Dale and Simon F. Park
 ISBN 0 470 85084 1 (cased) ISBN 0 470 85085 X (pbk)

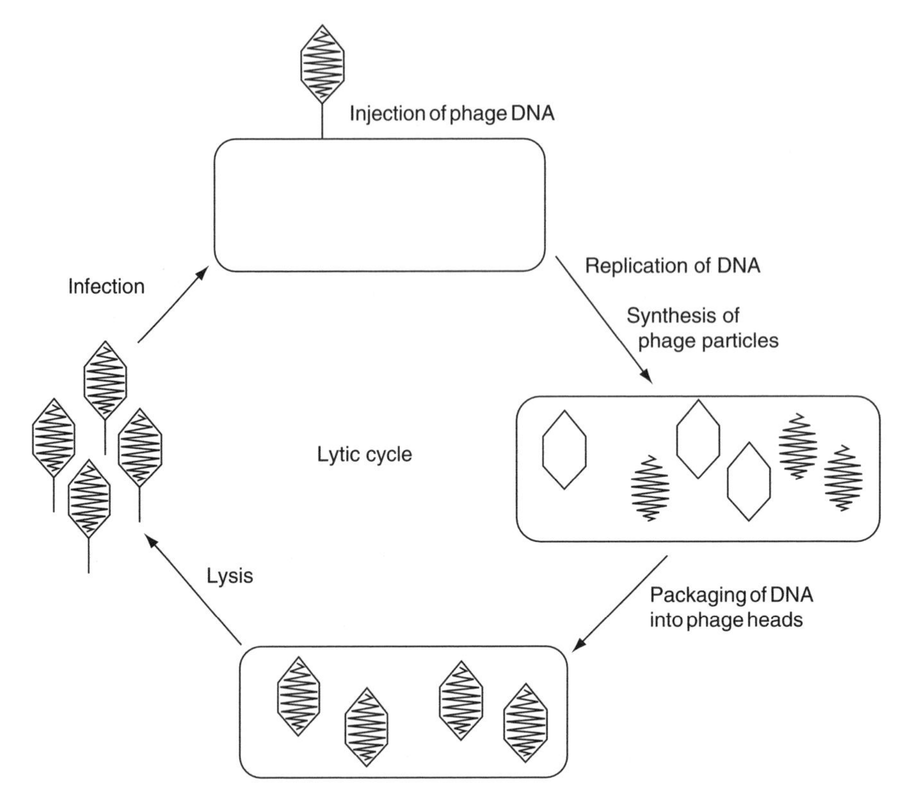

Figure 4.1 Lytic growth of bacteriophages

indicator cells) and the phage suspended in soft agar poured as an overlay on the top of an agar plate, rather than spread on the surface. This gives a three-dimensional plaque, which is easier to see.

Some bacteriophages, such as λ (lambda), do not always enter the lytic cycle described above. These are *temperate* phages and can establish a more or less stable relationship with the host cell, known as *lysogeny*. In this state, almost all the phage genes are completely repressed, as is replication of the nucleic acid, so no phage particles are produced until the lysogenic state breaks down again. Temperate bacteriophages will usually produce plaques in an agar overlay with a suitable bacterial indicator since many of the infected cells will lyse rather than become lysogenic. However, since those cells that enter the lysogenic state will continue to grow (and are resistant to further infection), the plaques will be turbid rather than the clear plaques seen with a virulent phage (i.e. a phage that is unable to establish lysogeny). Lysogeny is covered in more detail later in this chapter.

Many phages have been characterized to a greater or lesser extent; the main examples used in this chapter are all viruses that infect *E. coli*. These come in a variety of shapes and sizes as shown in Figure 4.2, including the small, simple

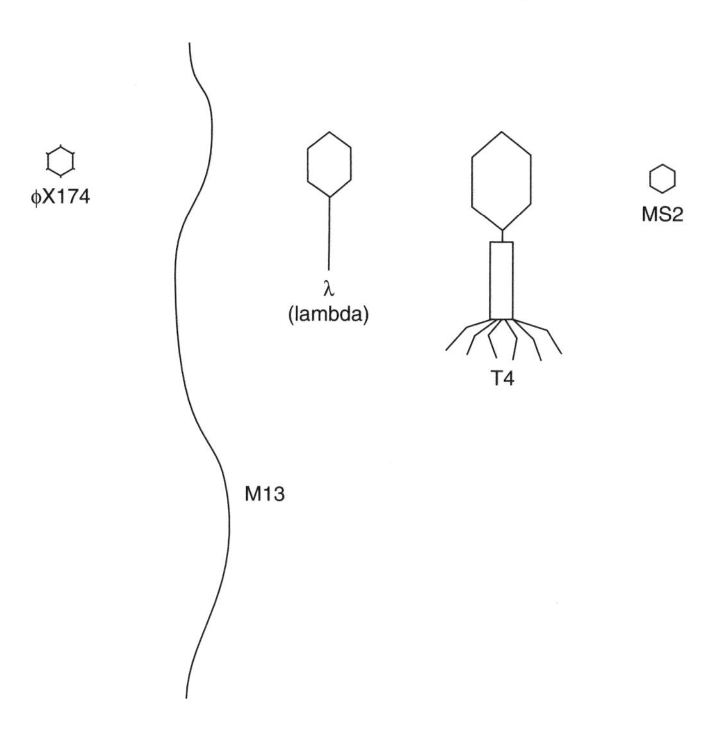

Figure 4.2 Morphology of selected bacteriophages. Note that the filamentous structure of M13 is longer than can be represented on this scale

structures of ϕX174 and MS2, more complex structures with identifiable heads and tails (λ, T4) and filamentous structures such as M13.

The phage particle consists of a nucleic acid molecule contained within a protein coat. In the simpler phages, such as MS2, the coat contains a number of copies of a single major protein which essentially polymerize spontaneously. The manner in which they associate defines both the shape and the size of the phage particle. In principle, filamentous structures can also be produced by spontaneous polymerization of protein subunits – but this is not sufficient to define the *length* of the filament. A common way of ensuring that filaments of a defined length are produced is to use a *template*, around which the protein polymerizes. As we will see later on, with filamentous phages such as M13 the phage DNA provides the template.

The assembly of M13, where the proteins making the phage coat polymerize around the DNA, is a marked contrast to the larger phages such as lambda and T4, where the DNA is packaged into pre-formed empty phage heads and therefore the size of the phage head is defined by the proteins of which it is composed. The size and complexity of these structures necessitates a different approach to their production – spontaneous polymerization would be too inefficient and the structures generally unstable until they are complete. Assembly of these phage

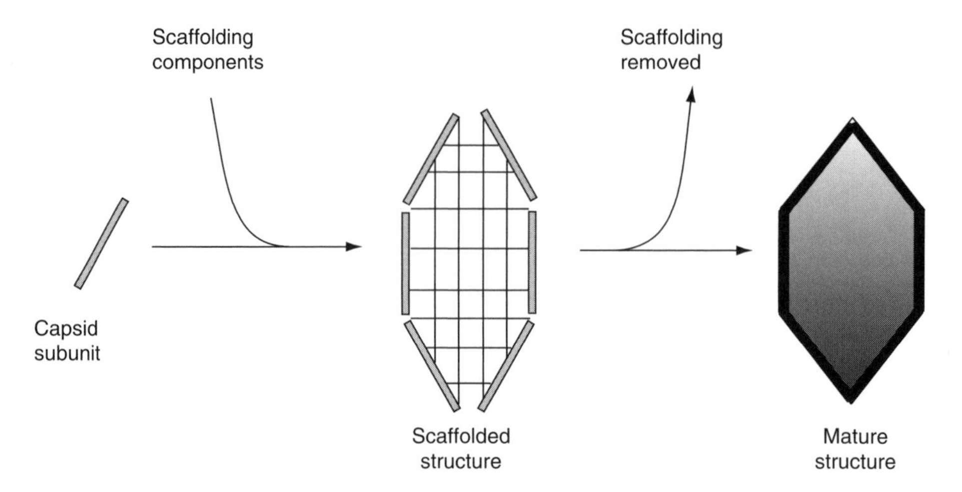

Figure 4.3 Scaffolded assembly. Assembly of complex phages is assisted by scaffolding proteins. These are removed once assembly is complete

heads therefore requires the temporary presence of a number of additional proteins, which are removed before the final maturation of the phage (see Figure 4.3). These components which assist with assembly but are not present in the final particle are referred to as *scaffolding* proteins.

Bacteriophages also differ in the nature of their nucleic acid content. Phages φX174 and M13 contain circular single-stranded DNA, MS2 has an RNA genome, while λ and T4 particles contain double-stranded DNA. Each of these phages is worth considering in more detail as they provide examples of fundamental molecular processes.

4.1 Single-stranded DNA bacteriophages

4.1.1 φX174

The phage φX174 is an icosahedral phage that contains a circular single-stranded DNA molecule of 5386 nucleotides. It codes for 11 proteins, each of which has been identified. Adding together the size of all those proteins comes to 2330 amino acids, which would require 6990 nucleotides (3×2330) – substantially more than the total length of the genome. How this is achieved becomes apparent from inspection of the genome sequence of the virus (Figure 4.4).

Firstly the genes are very tightly packed – there is very little non-coding sequence in the genome. In most cases, the end of one gene is directly adjacent to (or slightly overlaps with) the start of the next. Secondly, one of the proteins (A^*) corresponds to the C-terminal region of protein A, due to the use of an

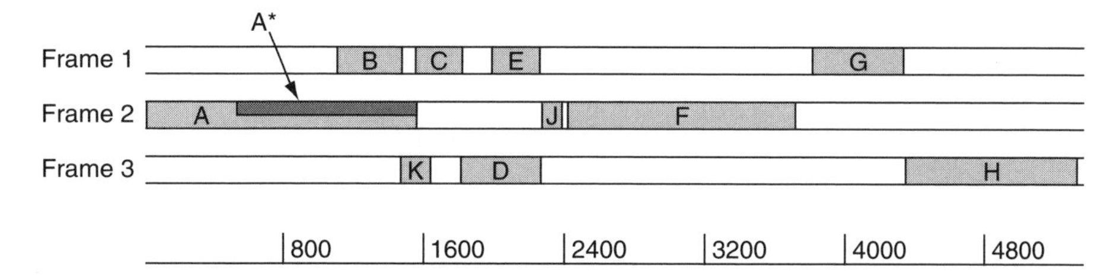

Figure 4.4 Genome organization of ϕX174. The DNA is actually circular, therefore it should be envisaged with the ends joined together

internal translation initiation site within the A gene. But even if we disregard protein A*, the coding capacity required still exceeds the length of the genome. This is achieved by the extensive use of overlapping genes. Since the genetic code is read in groups of three nucleotides, any DNA sequence is in theory capable of coding for three different proteins by use of each of the three possible reading frames. (In fact, double-stranded DNA could, theoretically, code for six proteins simultaneously, by transcription in both directions – but the ϕX174 genome is only transcribed in one direction). In ϕX174, protein B is encoded by part of the sequence that also codes for protein A in a different reading frame and so has a completely different amino acid sequence. Similarly gene E is entirely within gene D, and gene K overlaps with genes A and C in different reading frames. This economy of genetic coding is a feature of several small bacterial viruses, which take advantage of the ability of the bacterial protein synthesis machinery to initiate translation at multiple sites within a polycistronic message. Overlapping genes are used only to a very limited extent in larger viruses and the bacterial chromosome, where presumably the advantage of economy is more than offset by the constraints placed on the evolution of the genes involved.

Since only one strand of DNA is present in the phage particle, replication must involve features that are distinct from the replication of the bacterial chromosome, as outlined in Figure 4.5. After the single-stranded ϕX174 DNA enters the cell, it is converted into a double-stranded molecule (replicative form, RF) by synthesis of the complementary (or 'minus') strand (step A); it is this minus strand that is transcribed for the synthesis of phage proteins. This double-stranded form replicates in a way that is different from that of the chromosome, in that the two strands are copied separately. The minus strand provides a template for the production of further copies of the plus strand (step B); these are in turn converted to the double-stranded replicative form by synthesis of the complementary (minus) strand (step C). Although this mechanism ends up with the production of double-stranded DNA, it has the significant feature that the two strands are produced separately and by a different process. Some types of plasmids are replicated in a very similar manner and these will be discussed further in Chapter 5.

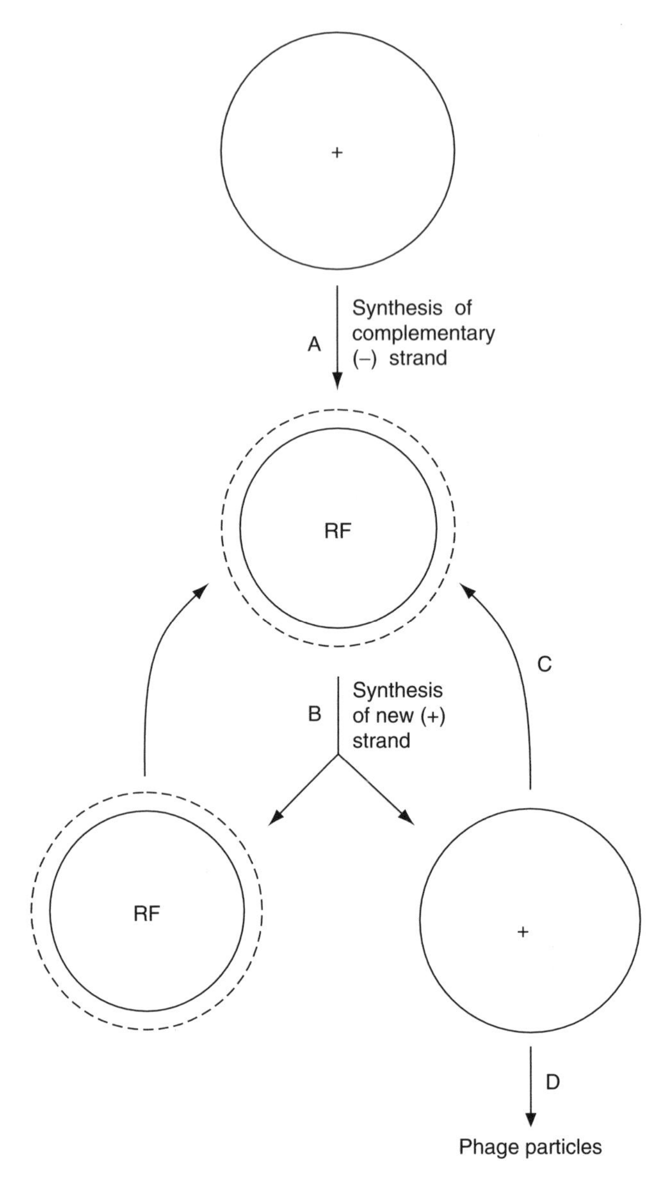

Figure 4.5 Replication of single-strand bacteriophages. The DNA entering the cell is single stranded and is converted (A) to a double-stranded replicative form. This generates (B) more single-stranded molecules which are converted (C) to the double-stranded RF form. Later (D) the single-stranded DNA is packaged into phage particles instead

Concurrently with this process, phage proteins are being produced and assembled into empty precursors of the phage particle. This results in a change in the DNA replication pathway. Instead of the new plus strands being converted to double-stranded RF, they are captured by the empty phage capsid precursors, leading to the production of mature phage particles (step D).

4.1.2 M13

The filamentous phages, such as M13, represent another type of single-stranded phage. These are 'male-specific' phages, since they infect the cell by attaching to the tips of the pili specified by the F plasmid (see Chapter 5). They are unusual amongst virulent bacterial viruses in that they are released from the cell without lysis. Although they produce plaque-like lesions in a bacterial lawn, these are areas of reduced growth rather than genuine plaques. After infection, a double-stranded replicative form is produced by synthesis of the complementary strand, in much the same way as ϕX174 (see Figure 4.5). Large numbers of this RF are found in infected cells and these can be isolated by standard plasmid techniques – a useful feature when these viruses are used as cloning vectors (see Chapter 8). Replication then proceeds via the synthesis of single-stranded DNA which is in turn converted to the double-stranded form, as shown in Figure 4.5.

A second unusual feature of these phages is that the DNA is not packaged into pre-formed empty phage heads. As the replication cycle proceeds, one of the phage proteins that is produced is able to bind to the single-stranded DNA and divert it into the production of phage particles by targeting it to the cell membrane where it is extruded from the cell, with phage coat proteins being polymerized around it during its passage through the membrane. The RF DNA remains within the cell, giving rise to continued production of more phage DNA, and hence more phage particles.

4.2 RNA-containing phages: MS2

Another type of male-specific phage is represented by MS2. This is an icosahedral RNA-containing phage which attaches to the sides of the F-pili, rather than to the tip as M13 does. It is an extremely simple phage, containing some 3600 nucleotides, coding for just three genes: a coat protein, a maturation protein and a replicase. The RNA in the phage particle is the coding strand, so it is both a replication template and a mRNA; regulation occurs at the translational level, mediated primarily by the extensive secondary structure of the RNA.

Replication of MS2 requires an RNA-directed RNA polymerase, an enzyme not normally present in bacterial cells. Replication of MS2 therefore cannot start until the MS2 replicase gene has been translated. The replicase synthesizes minus strands by copying the viral (plus) strand and then uses the minus strands for production of large amounts of the viral RNA. The coat protein aggregates around the plus strand RNA and the phage particles are released from the cell, possibly due to the mechanical damage caused by the very large numbers of particles produced (5–10 000 per cell).

4.3 Double-stranded DNA phages

The larger phages such as T4 and lambda are double-stranded DNA phages. A complication arises from the fact that the DNA within the phage particle is in a linear form. DNA in bacterial cells is usually circular. This is not accidental. Replication of linear DNA encounters the problem that the ends of a linear molecule cannot be copied by the usual mechanism. This problem does not occur with circular DNA. (Some bacteria have linear plasmids, or even linear chromosomes and have evolved alternative ways of ensuring that the ends are replicated; see Chapter 5). The linear phage DNA is therefore usually converted into a circular form before replication, although T4 adopts a different strategy (see below).

The general features of the lytic mode of replication of these viruses are similar in outline. DNA replication ultimately results in the production of a long DNA molecule, many times the length of a single phage genome. Concurrently, phage tails and empty phage heads are assembled; the multiple-length phage DNA molecule is cut into pieces of the correct length which are packaged into the empty phage heads, followed by attachment of the tails and lysis of the cell. Although this overall strategy is the same, the differences are significant and worth considering in more detail.

4.3.1 Bacteriophage T4

Replication and packaging

Bacteriophage T4, a virulent bacteriophage of *E. coli*, contains a linear DNA molecule of about 165 kb. This molecule is slightly longer than that needed to contain the complete phage genome, since it shows terminal redundancy: a sequence of about 1600 base pairs at one end of the molecule is repeated in the same orientation at the other end. The existence of this repeated sequence facilitates the production of long, multiple-length concatemers of the DNA as the product of replication, by recombination between the terminal repeats (Figure 4.6). This enables the complete molecule to be replicated as a linear structure, without circularization, and without loss of sequences at the ends.

The outcome is a number of linear DNA molecules that are many times longer than that needed to fill a phage head. By this time, the proteins necessary for formation of the phage particles have been produced and assembled into tails and empty heads. Packaging of the DNA into the heads is initiated and proceeds by

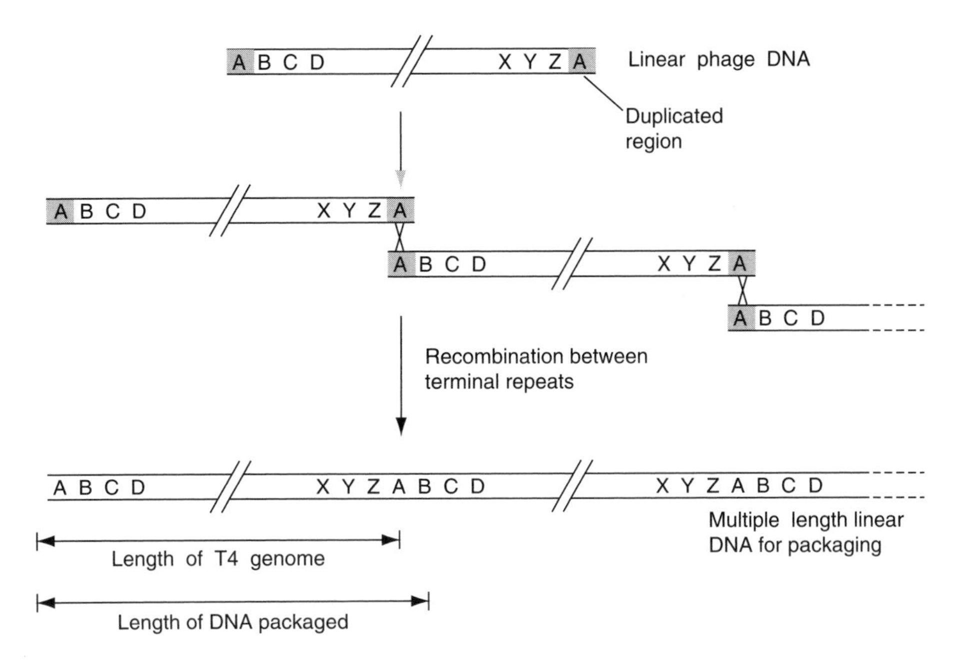

Figure 4.6 Replication of bacteriophage T4. Replication of the linear phage DNA, coupled with recombination between the terminal repeats, generates multiple length linear DNA for packaging into phage particles

coiling the DNA tightly inside the head structure until it is full. The DNA strand is then cut and the unincorporated DNA remains to be packaged into another phage head. This system is known as the 'head-full' mechanism (see Figure 4.7).

Terminal redundancy arises as a consequence of this packaging mechanism since the amount of DNA required to fill the head is greater than the complete length of the T4 genome. If packaging starts at a point A, it will continue past the far end of the genome and incorporate a second set of region A (Figure 4.7). A second consequence is that the population of phages produced shows circular permutation, that is, although the order of the genes is always the same when arranged on a circular map, the linear DNA from one particle starts and ends at a different point from that in another phage particle. Thus in the above example, the second phage starts and ends with the region B, the third with C, and so on.

Control of phage development

As outlined earlier in this chapter, a characteristic feature of most bacteriophages is that specific sets of genes are expressed at different stages after infection. The fundamental division is between the early genes which can be expressed immediately on infection using host enzymes and the late genes, the expression

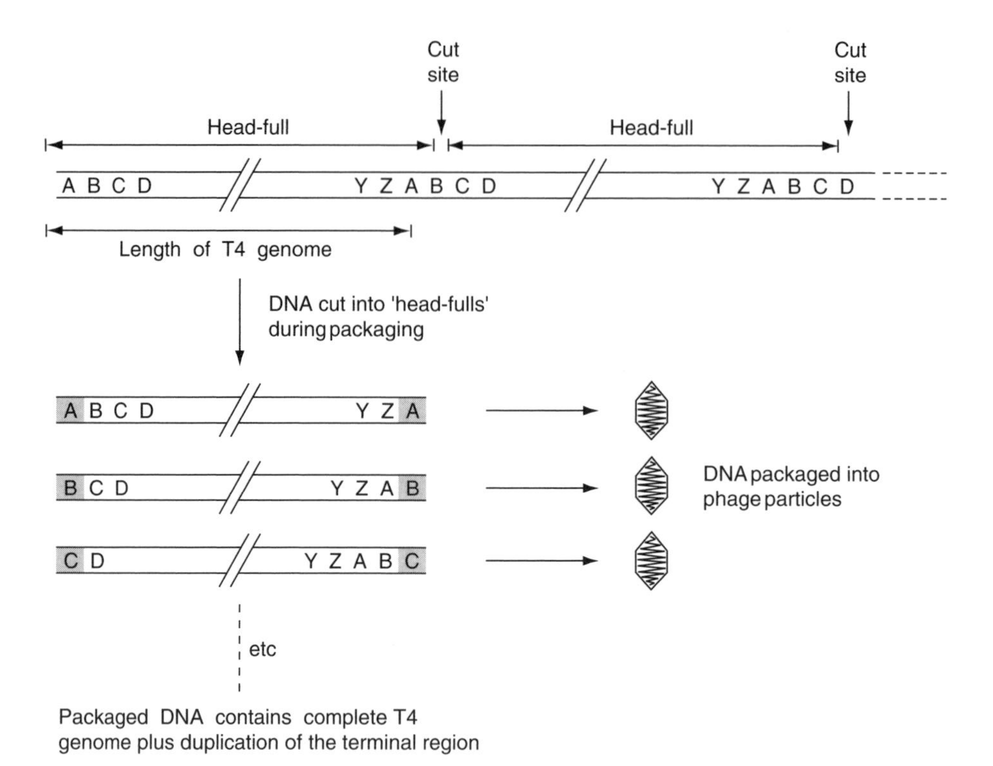

Figure 4.7 Terminal redundancy of bacteriophage T4. Multiple length linear DNA is the substrate for packaging into phage particles. The amount of DNA packaged in longer than the genome size, leading to terminal redundancy

of which requires a phage-encoded protein which is the product of one of the early genes.

With phage T4, the early genes are expressed from promoters that are recognized by the host RNA polymerase. A second group of genes, known as the quasi-late (or middle) genes, is expressed somewhat later in infection; these genes are also transcribed by host RNA polymerase but do not have a consensus promoter. Instead, two phage-encoded proteins are used to assist the RNA polymerase in binding to the DNA. Together these two groups of genes contain the information necessary for replication of phage DNA and regulating the synthesis of the late genes which carry the information for the production of the phage particles themselves. One of the ways in which the late genes are activated is by production of a new sigma (σ) factor that alters the specificity of the host RNA polymerase (see Chapter 3), but other regulatory mechanisms, including the presence of attenuators, are also involved in controlling the temporal expression of T4 genes.

Another phage with a similar, well-characterized, switch in gene expression is the *B. subtilis* phage SPO1, where the early genes are transcribed by the host RNA polymerase (using the primary sigma factor). One of these early genes (gene 28) codes for a sigma factor (gp28) that switches on a group of phage genes known as the middle genes. Two products of these middle genes (gp33 and gp34) in turn combine with the core RNA polymerase to activate the late phage genes.

4.3.2 Bacteriophage lambda

Replication and packaging

Bacteriophage lambda (λ) particles, like T4, contain linear double-stranded DNA. However, unlike T4, λ DNA does not show terminal redundancy, nor are the phages circularly permuted. The ends of the DNA are identical in every phage particle. This reflects differences in the replication and packaging processes between the two bacteriophages.

When λ infects *E. coli*, the linear DNA is injected into the cell, where it is converted into a circular form with the aid of a short length of unpaired bases (12 nucleotides) at each 5′ end of the DNA (Figure 4.8). These sequences (*cos* sites) are complementary to one another and are therefore cohesive, i.e. they tend to form base pairs with one another thus creating a (non-covalently linked) circular molecule. The action of DNA ligase within the cell rapidly seals the nicks in the circle to form a covalently closed circular DNA molecule.

This molecule replicates initially in what is known as the theta mode, so-called because of the similarity of the replicating structure to the Greek letter theta (θ). This is essentially similar to the replication of other circular molecules such as plasmids. At a later stage in infection there is a switch from the theta mode to rolling circle replication which yields a multiple length linear DNA molecule (Figure 4.8). As with T4, by this stage phage tails and empty heads have been produced. However the packaging mechanism is different.

With λ, the extent of the DNA to be packaged is determined by the position of specific sequences, the *cos* sites. A protein attached to the phage head recognizes a *cos* site in the multiple length DNA molecule and initiates packaging of the DNA. This proceeds until the next *cos* site is reached, when the protein cuts the DNA at each *cos* site. These cuts are asymmetric, that is, the two strands are cut at positions that are not opposite one another. There is a distance of 12 bases between the two cuts, leading to a sequence of 12 unpaired nucleotides at each end of the packaged linear DNA (see Figure 4.9). The head is then sealed, the pre-formed tail is added and the mature phage particles are released by lysis of the cell.

This contrasting packaging mechanism is of practical importance in the use of λ for gene cloning. With T4, the head-full mechanism ensures that all particles have just enough DNA to maintain the integrity of the particle: deletion of genes will increase the extent of terminal redundancy; addition of extra sequence will reduce it, or in extreme cases not all the T4 information will get into the phage. With λ on

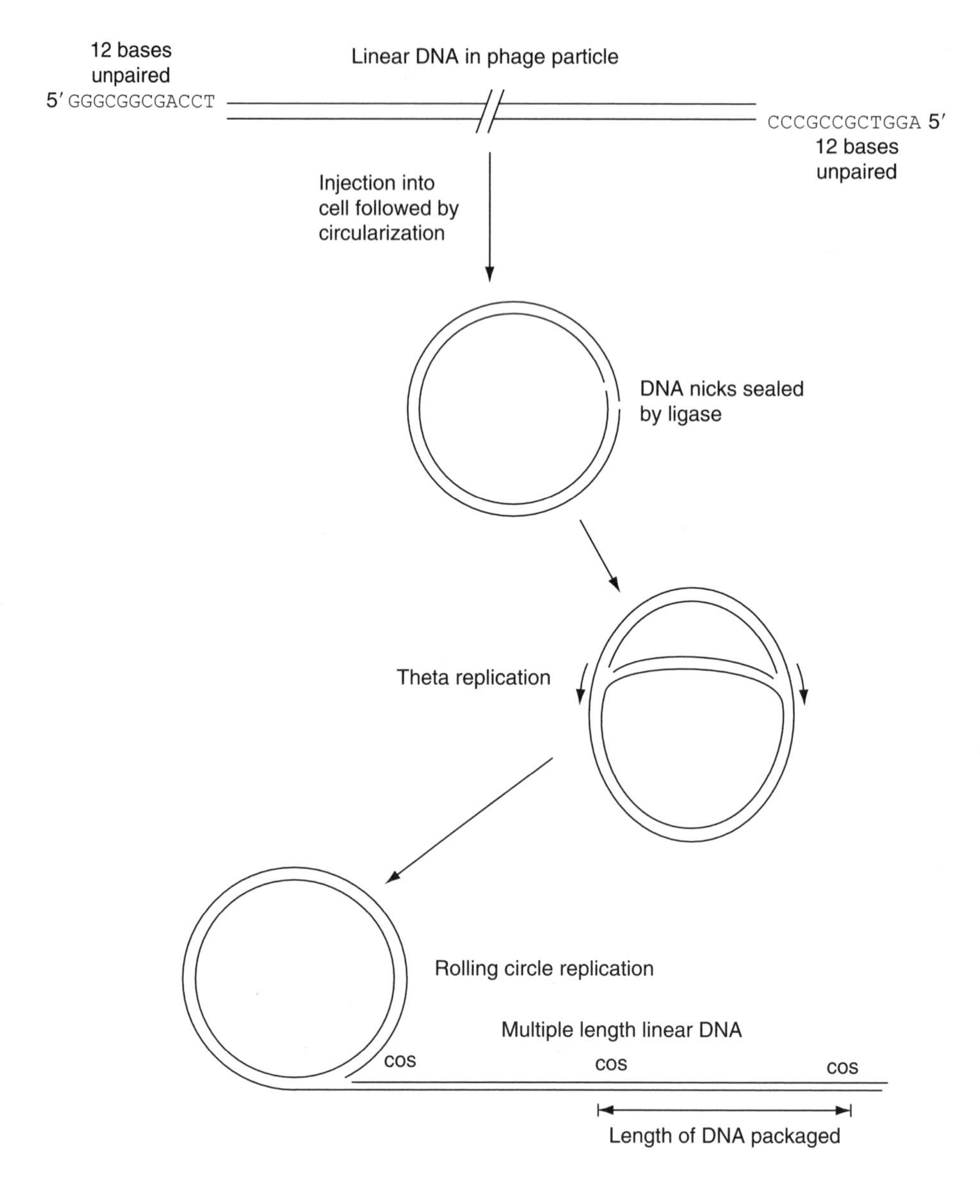

Figure 4.8 Replication of bacteriophage lambda DNA. After infection, the cohesive ends are ligated to form a circular molecule which replicates (theta mode) to generate more circular DNA. Later, replication switches to the rolling circle mode, generating multiple length linear DNA for packaging into phage particles

the other hand, since the distance between two *cos* sites determines the amounts to be packaged, deletion of DNA may mean that insufficient DNA is inserted into the phage head to maintain its physical integrity. On the other hand, addition of DNA fragments into the λ DNA molecule may have the result that the distance

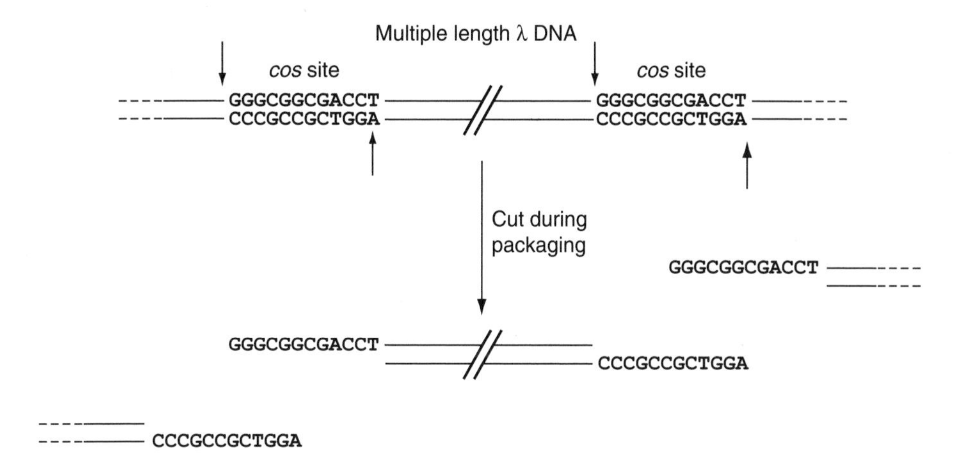

Figure 4.9 Cutting λ DNA at the cos sites

between two *cos* sites is greater than the capacity of the phage head; such DNA cannot be packaged. Phage λ therefore has *packaging limits*; the distance between two *cos* sites must be between 75 and 107 per cent of the wild-type sequence (i.e. between about 37 and 52 kb; wild-type lambda DNA has 48.5 kb of DNA; see Chapter 8 for a further discussion on the use of λ as a cloning vector).

Lysogeny

The bacteriophage λ is a temperate phage, which means that after infecting a bacterial cell it can establish a lysogenic relationship with its host, rather than entering the lytic cycle (Figure 4.10). When a bacterial culture is infected with lambda, or any other temperate phage, some cells will lyse and some will become lysogenic. In the lysogenic state the phage DNA is maintained as a *prophage*, and the lytic cycle is prevented by the action of a specific repressor.

If the lysogenic pathway is followed, then instead of replicating in the theta mode, the λ DNA will integrate into the host chromosome by recombination (Figure 4.11) involving specific sequences on the phage (*attP*) and the bacterial chromosome (*attB*). These two sequences (*attP* and *attB*) have a short (15 bp) identical region, the core sequence (Figure 4.12), although additional sequences either side of the core are necessary for integration. Recombination between these sequences requires the product of the λ gene *int*, which specifically recognizes the *attP* sequence. The mechanism is distinct from that of general recombination (see Chapter 6), especially in not requiring an extended region of homology. Since this recombination occurs only at the specific attachment sites, it is an example of *site-specific recombination.*

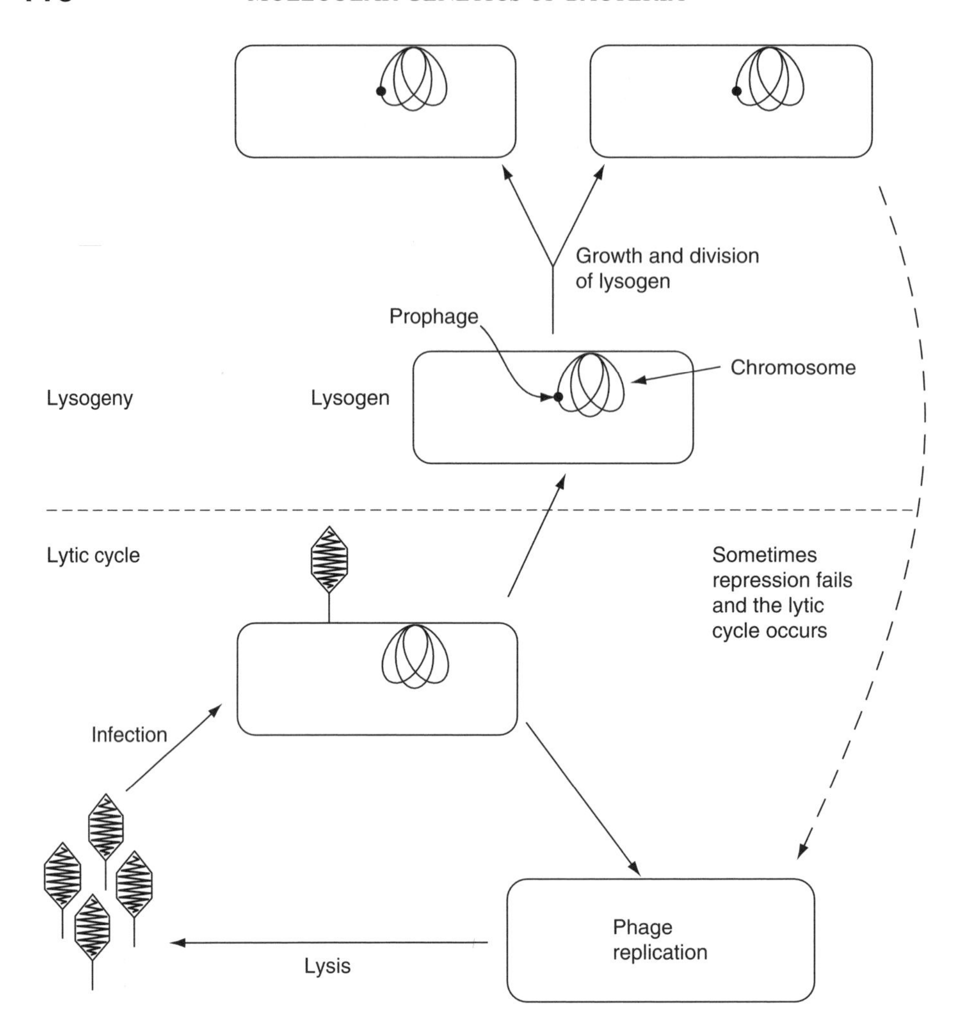

Figure 4.10 Lytic cycle and lysogeny: options for a temperate bacteriophage. After infection, a temperate phage may establish lysogeny as an alternative to the lytic cycle. In this diagram, the prophage is shown integrated into the bacterial chromosome, but with some phages the prophage exists as a plasmid

Since the *attP* and *attB* sequences are different (apart from the central core region), recombination between them produces two sequences which are different from either *attP* or *attB*, on either side of the core region. These sites are now called *attL* and *attR* since they are at each end of the integrated prophage. The Int protein, which carries out the site-specific recombination, is specific for *attP* and *attB* and will not catalyse recombination between *attL* and *attR*. The Int protein is therefore unable to promote the reverse recombination event that would lead to excision of the prophage. This ensures that integration is not reversible during the establishment of lysogeny. When the lysogenic state breaks down leading to entry into the lytic cycle (induction of the lysogen), the product of another gene (*xis*)

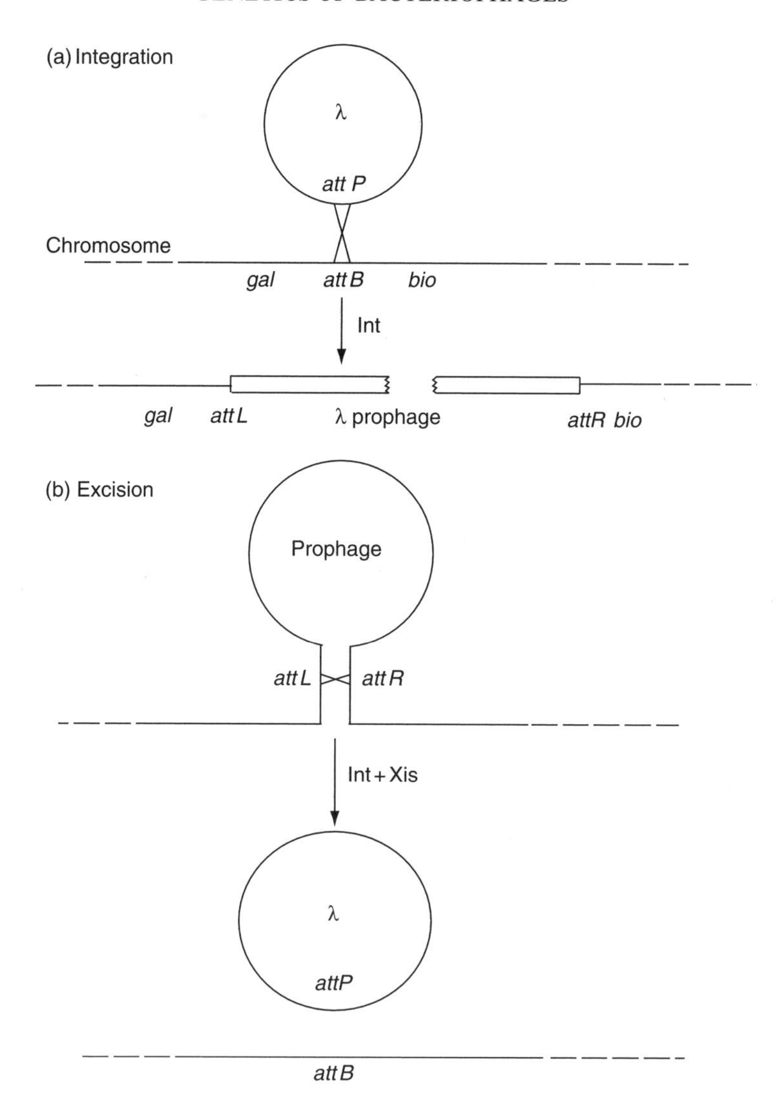

Figure 4.11 Integration and excision of λ DNA. (a) Integration occurs by site-specific recombination, mediated by the Int protein. (b) Excision requires both Int and Xis

interacts with Int to modify its specificity so that it can carry out excision leading to the formation of an intact chromosome and a circular phage molecule.

It should be emphasized that integration is not an essential feature of lysogeny. Some lambda mutants that are unable to integrate can still form lysogens, in which the prophage is maintained as an extrachromosomal circular DNA molecule – in effect it is a plasmid. Such lambda lysogens are normally unstable, but some temperate phages normally form stable lysogens in which the prophage is a plasmid. The bacteriophage P1 (which we will encounter again later on as a transducing phage and as a cloning vector) is a good example.

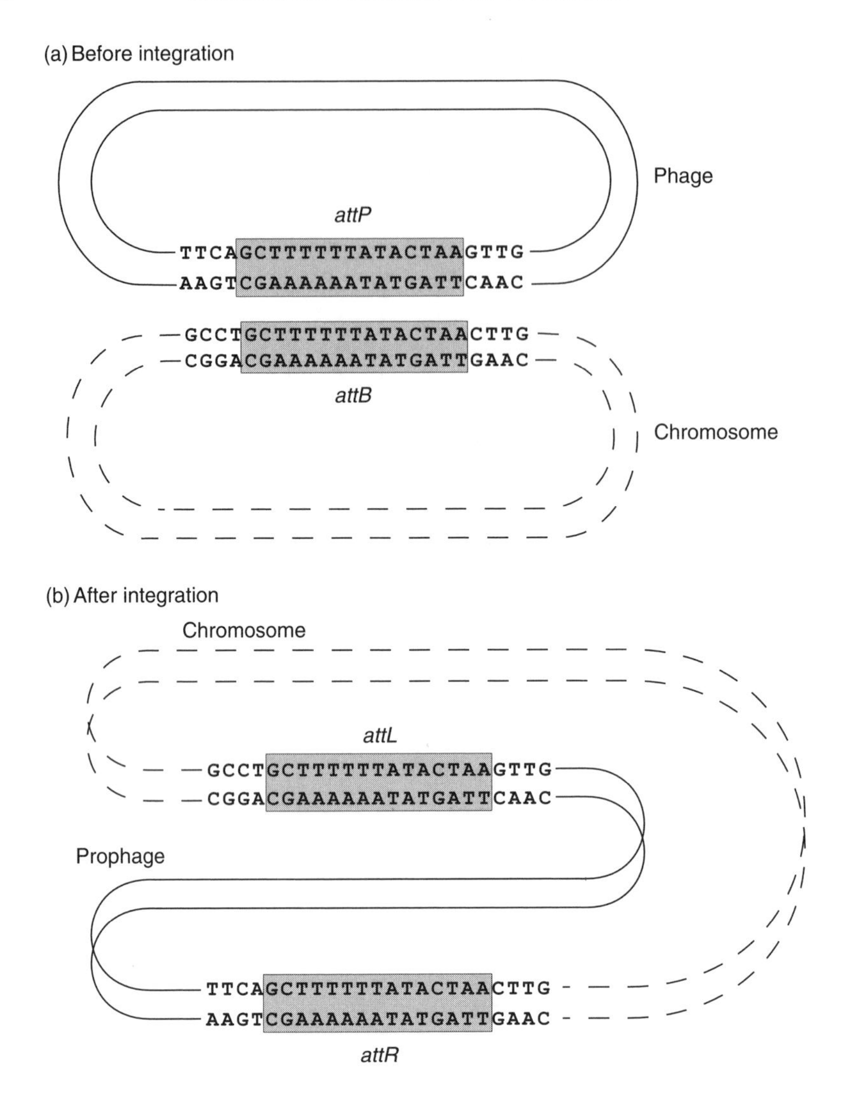

Figure 4.12 Structure of the attachment sites of λ. (a) Integration requires site-specific recombination between *attP* (on the phage) and *attB* (on the chromosome). (b) After integration, the sites at either end of the prophage (*attL* and *attR*) have the same core sequence (shaded) as *attP* and *attB* but different flanking sequences

4.3.3 Lytic and lysogenic regulation of bacteriophage lambda

Regulation of bacteriophage λ is worth considering in more detail as it provides well understood examples of different regulatory mechanisms. In particular, we can look at the temporal control of the lytic cycle (i.e. how the sequential expression of different sets of genes is achieved) and the control of lysogeny. The latter includes the nature of the switch that sends an individual phage infection down the lytic or lysogenic routes (the lytic/lysogenic switch).

Temporal control of the lytic cycle

An approximate map of the major features of the λ genome is provided in Figure 4.13 which shows that the genes are arranged in functional groups. The early

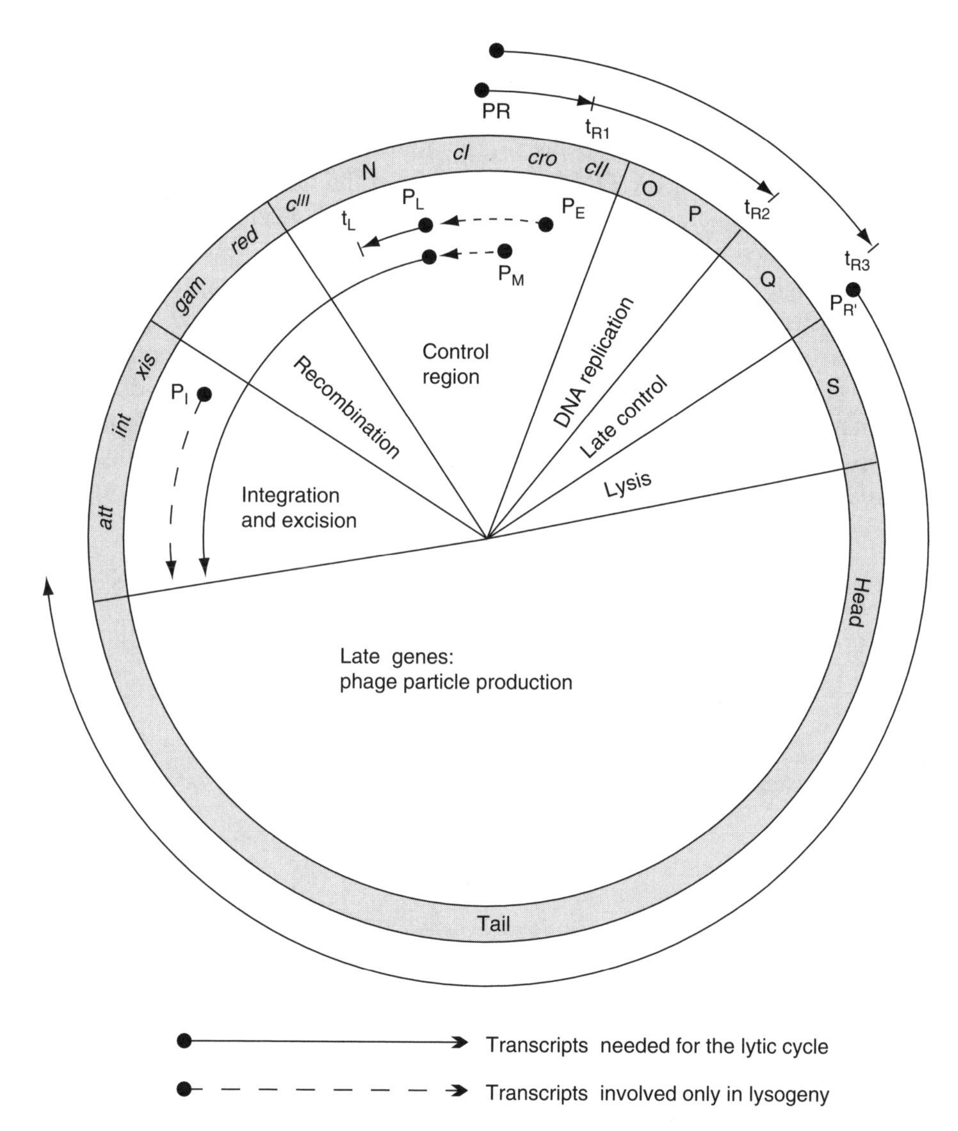

Figure 4.13 Bacteriophage lambda: genetic map and organization of transcripts

genes, concerned mainly with replication, are nearer to the control region, from where the major early transcripts originate. Later in infection, we find expression of the late genes responsible for synthesis of the head and tail proteins and production of the mature phage particle. An expanded representation of the control region is shown in Figure 4.14.

Soon after infection, transcription is initiated at two major promoters which are known as the major leftwards and rightwards promoters, P_L and P_R . These promoters are recognized very efficiently by *E. coli* RNA polymerase. Rightwards transcription from P_R however ends at the transcriptional terminators marked t_{R1} and t_{R2} (t_{R1} is a weak terminator and some of the transcripts ignore it); the leftwards transcript ends at t_{L1}. Thus only a comparatively small number of genes are expressed immediately, amongst which are the two genes *N* and *cro*.

The product of gene *N* has specific anti-terminator activity, i.e. it allows transcription to proceed through the t_{R1}, t_{R2} and t_{L1} terminators. The N protein binds to specific sites on the DNA upstream from the terminators and interacts with the RNA polymerase to allow it to ignore these transcriptional termination sites. The extended transcripts include a further set of genes (the 'delayed early' genes). The rightwards transcript then extends as far as a strong terminator t_{R3} that is unaffected by the action of the N protein. The delayed early genes expressed from the rightwards transcript are involved in DNA replication; those from the leftwards transcript, although not essential for lytic growth, do include genes such as *red* and *gam* that make phage production more efficient. (In order to keep things reasonably simple, we will ignore anything else that happens to the leftwards transcript).

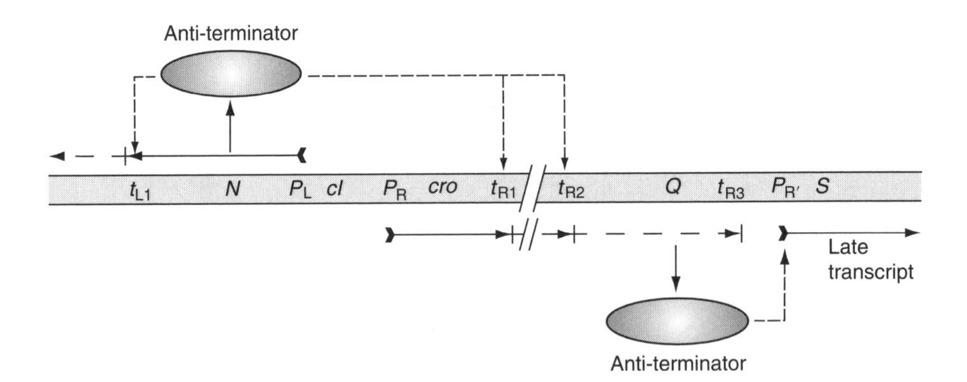

Figure 4.14 Regulation of the major lytic transcripts of bacteriophage lambda. Early transcription initiated at P_L and P_R terminates at t_{L1} and t_{R1}/t_{R2}. The product of gene *N* allows anti-termination and expression of delayed early genes, with the rightwards transcript terminating at t_{R3}. The gene *Q* product is also an anti-terminator and allows expression of the late genes from $P_{R'}$

The product of one of the delayed early genes (*Q*), which is also an anti-terminator, enables transcription of the late genes from a promoter $P_{R'}$ that lies between genes *Q* and *S*. This results in the production of a very large mRNA molecule which contains the information for all of the proteins of the phage head and tail and other products necessary for packaging and maturation of the phage (see Figure 4.13).

At the same time as the above events, the level of the product of one of the early genes (*cro*) has been building up. This acts as a repressor of the early promoters P_L and P_R, and thus prevents further synthesis of the early and delayed early transcripts which are no longer needed. The repression of P_L and P_R by Cro also plays a key role in the lytic/lysogenic switch (see below).

Control of lysogeny

The above description of the control of the lytic cycle ignores the alternative mode of replication of λ, which is via the establishment of lysogeny. The key gene in this respect is *c*I (the *c* stands for clear, since mutants that are unable to form lysogens give rise to clear plaques, as opposed to the turbid plaques that are characteristic of temperate phages). The *c*I gene codes for a repressor protein that switches off the lytic pathway, thus allowing a stable relationship to develop with the host cell. The way in which this repression operates is considered later.

The *c*I gene is expressed from a promoter known as P_E (for Establishment of repression), located to the right of gene *cro* (Figure 4.15). However, this promoter can only function in the presence of the product of the *c*II gene which is a positive regulator of the P_E promoter. The product of another gene, *c*III, acts to stabilize the *c*II gene product. (Both *c*II and *c*III gene products are highly susceptible to proteolytic degradation and are thus short lived). These genes are expressed from

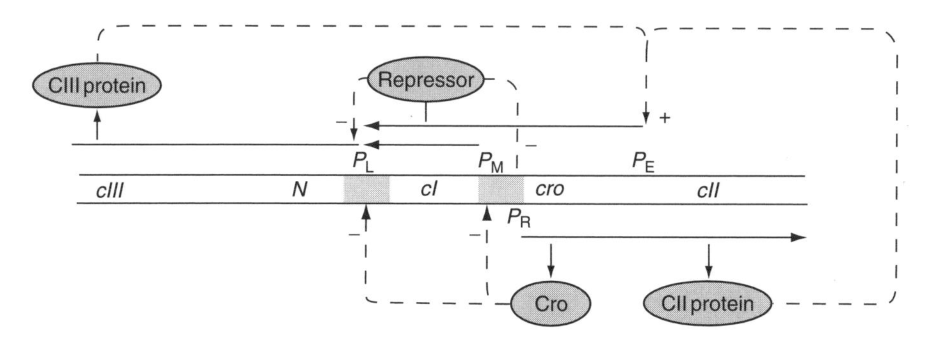

Figure 4.15 Control of the lytic/lysogenic switch in λ. CII stimulates transcription of the *cI* repressor gene from P_E; CIII stabilizes CII. CI represses P_L and P_R and stimulates P_M allowing further synthesis of repressor. Cro represses P_L, P_R and P_M

the rightwards and leftwards promoters, P_R and P_L respectively, but are beyond the initial termination sites, so require the presence of gene *N* and are therefore not expressed immediately on infection. Once these gene products have been made, the P_E promoter is stimulated.

The repressor protein (CI) then binds to two operator sites on the λ DNA; these are O_L and O_R, which are adjacent to the main early promoters P_L and P_R respectively. One consequence of this binding is that both of these promoters are turned off. This prevents further synthesis of several key products, notably those of genes *N* and *cro*. However, the action of the repressor protein in turning off promoters P_L and P_R also switches off genes *c*II and *c*III which are necessary for transcription from P_E. But if the lysogenic state is to be maintained, continued synthesis of the repressor is essential. This is carried out, at a lower level, from another, weaker, promoter P_M (the Maintenance promoter) which is adjacent to the O_R/P_R region. The binding of the repressor to O_R not only inhibits P_R activity but stimulates the activity of P_M. The CI protein thus has both a negative and a positive regulatory function.

As mentioned earlier, the Cro protein also represses P_R and it does this by binding to the O_R site. In contrast to the binding of the repressor however, the binding of Cro inhibits the activity of the P_M promoter. The Cro product therefore prevents synthesis of the CI repressor in two ways: by switching off the *c*II gene it indirectly switches off the establishment promoter P_E, and by binding to the O_R operator it directly switches off the maintenance promoter P_M. The decision between the lytic cycle and lysogeny can therefore be regarded (albeit simplistically) as a competition between the *cro* and *c*I gene products for binding to the O_R site. These interactions are summarized in Figure 4.16. These interactions may become easier to understand by reference to the more detailed consideration of the binding of CI and Cro to the O_R operator region in the next section.

Why should lambda have this complicated manner of controlling its life cycle? The benefits are simple and very elegant. Decisions, once taken, are self-reinforcing and irrevocable. Early after infection, if lysogeny is to be established, there has to be rapid production of substantial amounts of repressor, so that

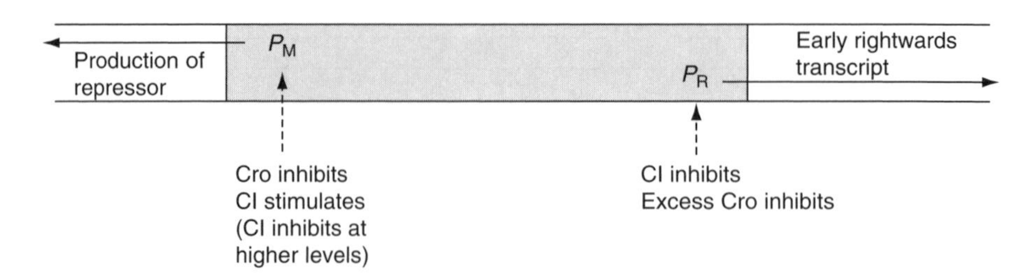

Figure 4.16 Summary of the effects of Cro and CI on P_M and P_R

further phage transcription can be prevented before it has gone past the point of no return. This is achieved by the use of P_E, which is a high level promoter. On the other hand, once lysogeny is established, continued synthesis of repressor is required at only a comparatively low level, which occurs by the use of P_M. The lysogenic decision is thus self-reinforcing, with the presence of the CI repressor on the O_R site preventing transcription of other genes but also ensuring continued synthesis of the repressor.

If however the repressor is removed from the O_R site (which occurs sometimes spontaneously and also as a consequence of various stresses that cause induction of the prophage), this not only relieves the repression of P_R, but also switches off P_M, thus preventing further synthesis of repressor to replace that which has been destroyed. This is therefore also a self-reinforcing switch mechanism.

Mechanism of binding of repressor and Cro proteins to the operator site

In the above description, the operator site O_R was treated as though it were a simple DNA sequence to which either the repressor or Cro could bind. The nature of this operator site and of the binding of these proteins is worth considering further because of the information this system has supplied regarding the binding to DNA of regulatory proteins in general. Only the O_R site will be considered, but there is considerable similarity in the binding to the O_L site.

The O_R operator in fact consists of three similar adjacent regions of DNA, O_{R1}, O_{R2} and O_{R3} (Figure 4.17). The rightwards promoter P_R overlaps with O_{R1} and the maintenance promoter P_M overlaps with O_{R3}. Both the CI repressor and the Cro protein contain a characteristic structure with a helix-turn-helix motif that is typical of many regulatory proteins (see Chapter 3). This is capable of interacting with each of these three sites although with different affinities and different consequences.

The affinity of the repressor is greatest for O_{R1}, but binding is cooperative; the repressor bound to O_{R1} stimulates binding of a second molecule to O_{R2}. Binding of RNA polymerase to P_R is thus prevented, while stimulation of transcription from P_M occurs by protein–protein interactions between the repressor bound to O_{R2} and the RNA polymerase. In the normal lysogenic state, O_{R3} is unoccupied, but very high concentrations of repressor can result in binding to O_{R3} as well. This will switch off P_M until the level of repressor falls below that needed to saturate O_{R3}. Thus not only is the expression of all the genes other than the repressor, turned off, but this also ensures that expression of *c*I is maintained at a low level which is an advantage for the stable maintenance of the phage as there is very little burden on the cell.

Although Cro can bind to all three sites, its relative affinity for these sites is different from that of the repressor: it will bind first of all to O_{R3}. This prevents RNA polymerase binding to the maintenance promoter P_M, thus preventing the

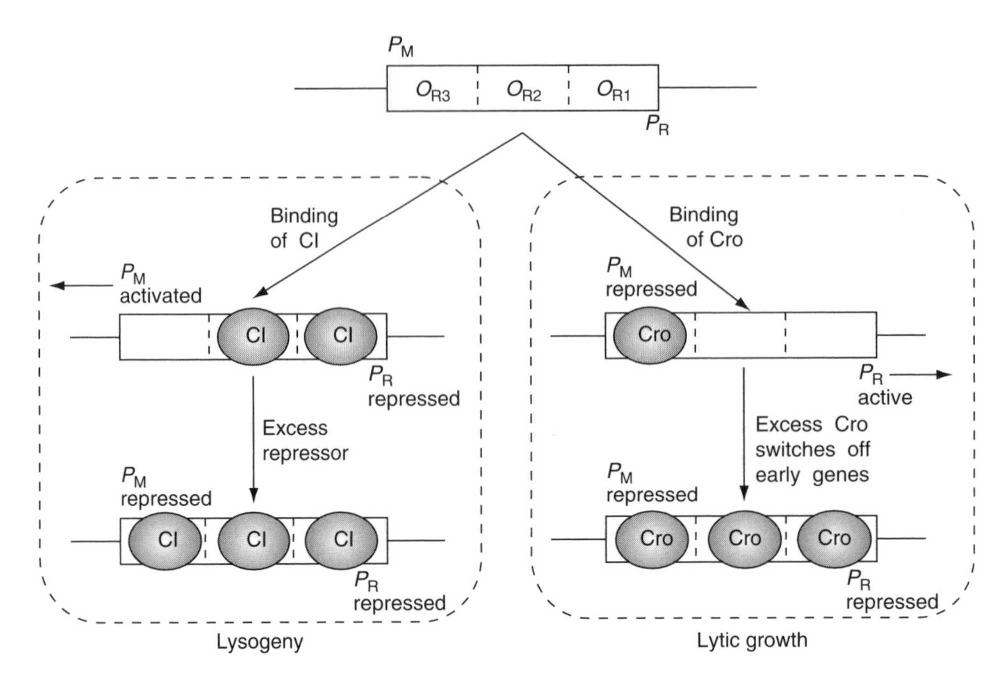

Figure 4.17 Interaction of Cro and CI with the operator site O_R

phage entering the lysogenic state. Only subsequently, as the level of Cro builds up, will it bind to sites 1 and 2; by this time the phage is committed to the lytic mode and sufficient of the transcripts from P_L have been produced. The role of Cro now is to turn off the expression of the early genes that are no longer needed and thus to direct gene expression towards the late genes that are needed for production of phage particles.

In the light of this knowledge, we can refine the model for the control of lambda and the lytic–lysogenic decision. The key events are the binding of the CI repressor to O_{R1}, which will prevent the lytic cycle, versus the binding of Cro to O_{R3}, which will prevent the establishment of lysogeny by repressing P_M. The binding of Cro to the operator site does not prevent synthesis of repressor from P_E, as long as active CII protein remains. Mutations that increase the stability of CII and host mutations (*hfl*, for high frequency of lysogenization) that reduce the activity of proteolytic enzymes (and hence prolong the activity of CII), will therefore result in an increased frequency of lysogenization.

Superinfection immunity and zygotic induction

A lysogenic cell is resistant to infection by further λ particles (or related phages). This is called *superinfection immunity*. The reason for this should now be clear:

such a cell already contains pre-formed repressor protein which is therefore able to bind immediately to the O_L and O_R sites on the incoming DNA. Consequently, the incoming phage DNA will be repressed before transcription can even start.

The converse phenomenon is seen when a lysogen is used as a donor to transfer DNA to another cell by conjugation (see Chapter 6). At a certain time after the start of conjugation, the prophage (which is inserted into the chromosome) is transferred just like any other part of the chromosome. The recipient cell however, if it is not lysogenic, contains no repressor. The transferred prophage is therefore able to escape from repression and initiate a lytic cycle in the recipient, leading to a sudden fall in the number of viable recipients. This phenomenon is known as *zygotic induction*.

4.4 Restriction and modification

As described earlier in this chapter, the number of bacteriophages in a preparation is usually assayed by counting the number of phage plaques produced using a sensitive bacterium as an indicator. The assumption behind this procedure is that there is a direct correspondence between the number of phage particles and the number of plaques, i.e. that every phage particle gives rise to a lytic infection. In technical terms, it is said that the *efficiency of plating* (e.o.p) is 1. Sometimes that is not true and the e.o.p. is much less than 1. This could happen when the indicator strain is different from the one used to grow the phage. This may of course mean that the new indicator is not sensitive to the bacteriophage. The few plaques obtained may be due to spontaneous mutation in the bacteriophage, yielding host range mutants that are now able to infect the previously resistant organism.

However, a reduction in the number of phage plaques may be due to a different phenomenon known as *host-controlled restriction and modification*. Consider the situation shown in Figure 4.18. A suspension of λ bacteriophage particles has been obtained by growth on a host *E. coli* strain known as *E. coli* C; these are referred to as λ.C. If this preparation is assayed using *E. coli* C and a different strain (*E. coli* K) as the indicator organisms, for every 10 000 plaques obtained on *E. coli* C (which is assumed to have an e.o.p. of 1), there is only one plaque on *E. coli* K (e.o.p. = 10^{-4}). If a plaque from the *E. coli* K plate is picked by stabbing it with a toothpick and resuspending the material recovered in a small amount of buffer, the resulting phage preparation (now referred to as λ.K) has an e.o.p of 1 using either strain as the indicator. All the phage are now capable of growing on both strains.

This might be taken to indicate that a mutant phage which was altered in its host range had been selected, i.e. the original population contained a small number of mutants that were able to grow on strain K. However, if these apparent mutants are used to infect strain C again, the phage will again show its original

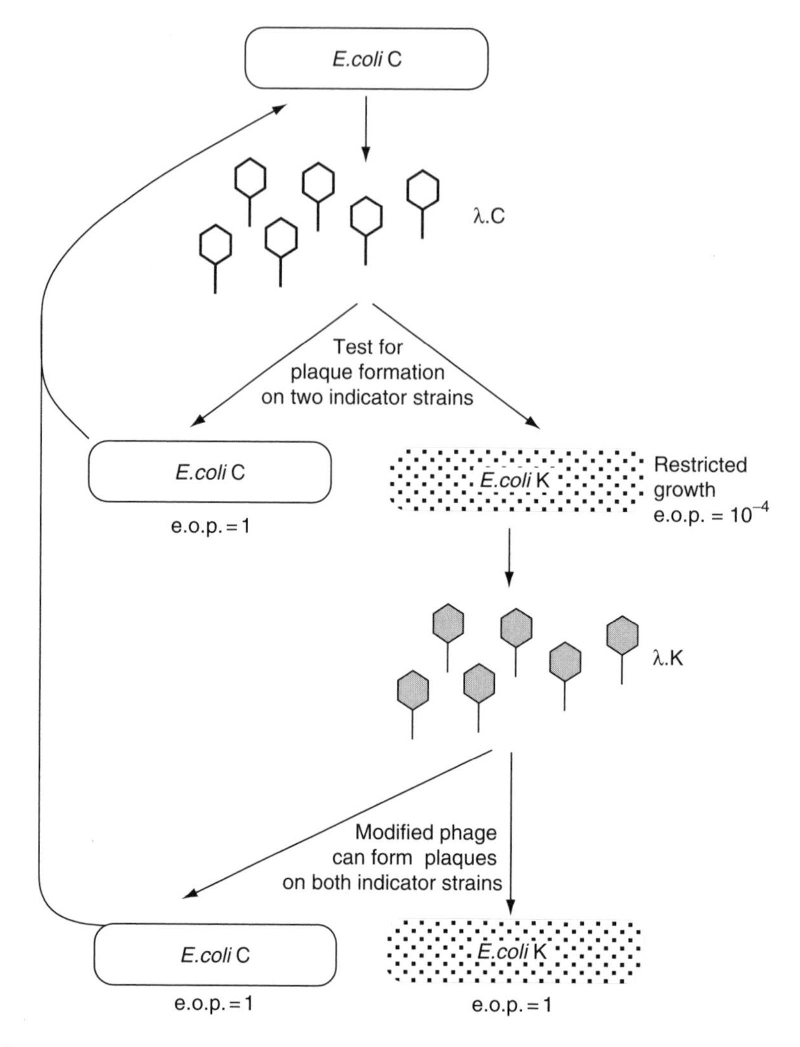

Figure 4.18 Host-controlled restriction and modification of bacteriophages

e.o.p. of 10^{-4} on strain K. It behaves in the same way as the original λ.C population. It is highly unlikely that this behaviour could be caused by mutation in the normal sense of the word.

All this becomes clearer once the true explanation is known. Strain K produces an enzyme (endonuclease) that is capable of recognizing foreign DNA and degrading it. Such an enzyme is referred to as a *restriction endonuclease*, since it results in a restriction of the growth of the foreign bacteriophage. There are a large number of such restriction enzymes now known. Most of those commonly encountered are Type II restriction endonucleases which recognize specific DNA sequences and cut the DNA strands at that point (see Box 4.1). They have a

Box 4.1 Restriction endonucleases

Restriction endonucleases (restriction enzymes) cut DNA molecules at specific positions. Some examples are shown below. The designation of these enzymes comes from the source organism, using the first letter of the genus name plus the first two letters of the species with additional letters/numbers to indicate the specific enzyme (since one species may produce several different restriction enzymes).

If each of the bases occurs equally often in the DNA (i.e. the G+C content is 50 per cent, which is approximately true for *E. coli* but not always for other organisms) and if the distribution of the bases is random, then a six base sequence would be expected to occur, on average, every 4^6 bases (= 4096 bases). So an enzyme like *Eco*RI would be expected to cut the DNA into fragments with an average size of 4 kb. For enzymes recognizing a four-base site, the expected average fragment size is 4^4 (=256) bases.

Note that the recognition sites are usually palindromic; in other words, if the double-stranded sequence is turned round (maintaining the polarity of the strands), it is exactly the same.

*Eco*RI cuts DNA asymmetrically between the G and A residues, as shown by the slash (/). This results in each fragment having four unpaired bases (in this case 5′AATT) at both ends. These single-stranded regions are complementary and will therefore tend to pair with similar regions on other fragments (or at the other end of the same fragment); they are thus referred to as 'sticky' ends.

Some enzymes, such as *Pst*1, leave a four-base sticky end region at the 3′ end of each fragment, while others such as *Sma*I cut in the centre of the recognition sequence, generating blunt-ended fragments.

Examples of restriction endonucleases

Enzyme	Recognition site	Number of bases	Ends generated	Source of enzyme
*Eco*RI	G/AATTC	6	5′ sticky	*Escherichia coli*
*Hin*dIII	A/AGCTT	6	5′ sticky	*Haemophilus influenzae*
*Bam*HI	G/GATCC	6	5′ sticky	*Bacillus amyloliquefaciens*
Pst I	CTGCA/G	6	3′ sticky	*Providencia stuartii*
*Sma*I	CCC/GGG	6	Blunt	*Serratia marcescens*
*Sau*3A	/GATC	4	5′ sticky	*Staphylococcus aureus*
*Alu*I	AG/CT	4	Blunt	*Arthrobacter luteus*

central role in genetic manipulation (see Chapter 8). The *Eco*K restriction endonuclease is a Type I enzyme; it recognizes a specific set of sequences (AACNNNNNNGTGC, where N stands for any nucleotide), but cuts the DNA at a variable distance away from the recognition site. The effect is the same: incoming DNA is degraded by the restriction enzyme, which thus protects the cell against phage infection.

Cells that produce a restriction enzyme must have some mechanism that prevents destruction of the host's own DNA. This is achieved by a second enzyme that recognizes the same specific base sequence as the restriction enzyme, but instead of cutting the DNA it *modifies* it (usually by methylation of one of the bases) so that the DNA is no longer a substrate for the restriction endonuclease.

In the example above, phage grown on strain C (which does not possess the restriction-modification system) will have non-methylated DNA. As soon as it is injected into a cell of strain K, the restriction endonuclease will degrade it; only occasionally will the DNA evade this effect and survive to form a phage plaque. The virus particles within this plaque, having grown in the presence of the modifying enzyme, will have methylated DNA and therefore be capable of infecting further cells of strain K. However, recycling them through strain C will again yield non-methylated DNA that will show a low efficiency of plating on strain K.

4.5 Complementation and recombination

With bacteria, many of the mutants commonly studied are those that affect the ability to degrade or to synthesize various compounds. The nature of bacteriophage growth however, means that some ingenuity has to be employed to discover suitable genetic markers. Since a large proportion of the genes are essential for phage growth, considerable reliance has to be placed on conditional mutants. Amongst the commonest types of conditional mutants are those which make the strain temperature sensitive. Another type of conditional mutant contains a translation-terminating mutation (e.g. an amber codon) within a specific gene. These mutants are unable to grow on a wild-type host strain but will grow normally on an amber-suppressor host (Chapter 2).

Other phage characteristics that are useful as genetic markers are the appearance of the plaques and the range of host strains that can be infected. The *r*II mutants of the *E. coli* phage T4 are characterized by the production of large plaques (*r* stands for rapid lysis), and by being able to form plaques on one *E. coli* strain (*E. coli* B) but not another (which we will call *E. coli* K). The mechanism of this strain dependence need not concern us here. The study of these mutants, by Benzer and co-workers, provided much information about gene structure.

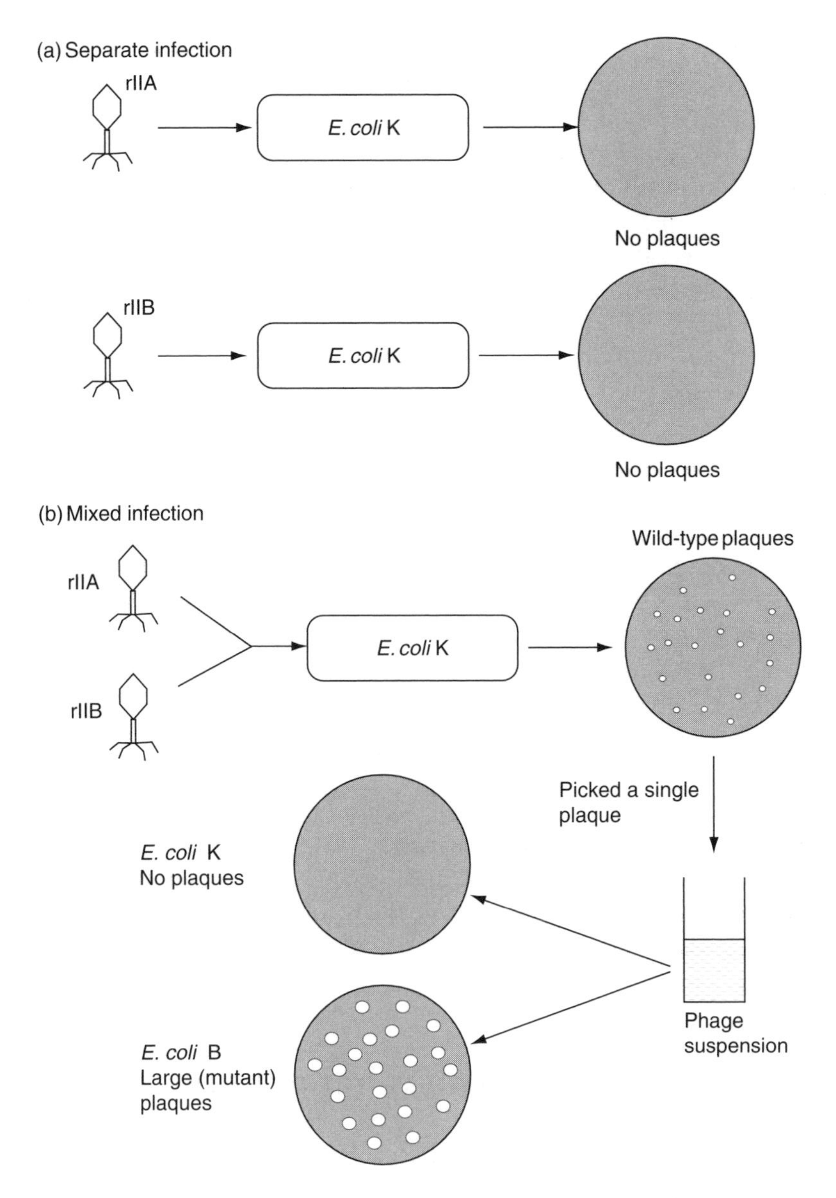

Figure 4.19 Complementation of *r*II mutants of T4

The mutants could be divided into two groups, *r*IIA and *r*IIB, since a mixed infection by one member of each group resulted in the formation of wild type plaques on *E. coli* K (see Figure 4.19), but a mixture of two mutants of the *r*IIA type, or of two *r*IIB mutants, was unable to form plaques on *E. coli* K. This effect is not due to recombination between the phages, since recovery and testing of the progeny phage shows that they are still mutants, i.e. they are unable to form plaques on *E. coli* K and form large plaques on *E. coli* B.

This phenomenon is in fact an example of *complementation*. The rII protein consists of two separate polypeptide chains, coded for by the *r*IIA and *r*IIB genes respectively. In the mixed infection, the phage that is defective in rIIA still codes for a functional rIIB polypeptide and vice versa. The two polypeptides can combine to produce an intact rII protein and thus enable the production of wild type plaques. The DNA chains are still independent, and each still contains the original mutation: the progeny phage are therefore still mutant. The complementation test allowed a refinement of the concept of a gene. The term *cistron* was coined for this purpose. If the two mutations are in different cistrons, complementation can occur; if they are in the same cistron, there will be no complementation. A cistron is the region of DNA that encodes a single polypeptide chain. Nowadays we simply refer to such a region as a gene, using the term cistron only to emphasize a particular point (as in *polycistronic mRNA*). Further examples of complementation and its uses will be encountered in Chapter 9.

Although complementation is distinct from recombination, the latter can also occur when bacterial cells are infected with a mixture of bacteriophage strains. The frequency of recombination is low, compared to complementation (which will occur in every cell that has undergone a mixed infection). However, if recombination yields progeny phage that can grow on a host that neither parent can infect, the presence of very rare recombinants can be detected simply by plating on the selective host. Furthermore, extremely large numbers of phage can be used. This enabled the mapping of mutations at sites very close to one another, in some cases at adjacent nucleotides. At the time, this was a revolutionary development since no other system of genetic analysis had approached that level of resolution. Nowadays, of course, the combination of gene cloning, DNA probes and nucleic acid sequencing has made it a routine matter to identify mutations at the molecular level.

4.6 Why are bacteriophages important?

This chapter has so far focused on the molecular biology of bacteriophages. Investigation of these properties has not only enabled molecular biologists to understand bacteriophages themselves, but has also contributed very extensively to the development of molecular biology in a wider sense. In later chapters it will become evident how bacteriophages are useful tools for genetic modification, as they provide some of the vectors that are used for the transmission of cloned DNA. But bacteriophages have further significance in the laboratory, in commercial processes and in nature.

4.6.1 Phage typing

All bacteriophages are, to a greater or lesser extent, specific in their host range. Generally, they are specific for members of a certain species (or sometimes species that are very closely related). But in some cases the specificity may be more restricted than that – they may be able to infect some strains but not others. We can use this to distinguish between strains of the same species. This is known as *phage typing*.

The best example of the use of phage typing is in distinguishing strains of *Staphylococcus aureus*. This bacterium, which is a common cause of infection in hospitals, is a common skin organism and is also able to persist for many weeks in dust. So if a ward has several cases of *Staphylococcus aureus* infection over a short period of time, it could be just an unfortunate coincidence. On the other hand, it might indicate that there is a source of infection within the ward (such as a carrier who is shedding large numbers of the pathogen and thus repeatedly infecting patients). If this is the case, action must be taken to identify and control the source of infection. Phage typing provides a means of distinguishing between these possibilities by testing each of the isolates for sensitivity to a battery of bacteriophages. If there is a single source of infection, all strains from infected patients will be the same and will show the same pattern of phage sensitivity.

Bacteriophages also have potential in a diagnostic laboratory for the detection and identification of pathogenic bacteria. This makes use of the amplification that arises from the multiplication of the bacteriophage – commonly each infected bacterium will yield 100 or more phage particles, so giving rise to a 100-fold increase in the sensitivity of detection – and also the specificity of the bacteriophage, thus enabling a specific pathogenic species and not closely related non-pathogens to be detected. The sensitivity and ease of detection can be enhanced by using a *reporter* phage, i.e. one that has been engineered to carry a gene coding for an easily detected enzyme. The firefly luciferase gene is often used for this purpose as luciferase can be easily and specifically detected by means of the light emitted. (The use of reporter genes for various purposes is considered further in Chapter 9). The potential of this technique is greatest for those bacteria that are not easily detected in clinical samples – such as *Mycobacterium tuberculosis* which takes several weeks to form colonies – but is yet to achieve widespread routine use.

4.6.2 Phage therapy

The increasing problem of bacterial resistance to antibiotics has encouraged searches for alternative ways of treating bacterial infections. One of these is the use of bacteriophages to kill the invading bacteria. This has been advocated on and off ever since phages were first discovered. However, although the concept is superficially attractive, it suffers from several problems. The main difficulty is that the phages themselves are antigenic and will be eliminated from the blood by the body's own immune system (or even worse would set up an adverse immunological reaction). Furthermore, phages are often strain specific and even an initially sensitive bacterial strain can develop resistance to a bacteriophage, so success is uncertain. Claims that since bacteriophages are a 'natural' way of killing bacteria and that the bacteria will therefore not develop resistance, are misguided.

However, more subtle approaches may be successful. Bacteriophages commonly lyse bacteria through the action of enzymes that attack the bacterial cell wall, so it may be possible to use the enzyme itself, rather than the whole phage, as an antibacterial agent. In one example, a lytic enzyme derived from a phage that attacks *Bacillus anthracis* (the causative agent of anthrax) was able to cure a high proportion of experimentally-infected mice.

4.6.3 Phage display

An important characteristic of many proteins is their ability to bind specifically to other molecules (*ligands*). This includes not only enzymes binding to their substrates, but also for example, hormone–receptor interactions, antigen–antibody binding and drug target recognition. In the past, the identification of a protein able to bind to a particular ligand was a very labour intensive procedure. Bacteriophages have provided an alternative known as *phage display*.

The principle of this technique, as illustrated in Figure 4.20, is that random fragments of DNA are incorporated into the genome of a bacteriophage (commonly the filamentous phage M13) so that the protein encoded by each DNA fragment is fused to one of the proteins that coats the phage particle. This produces a mixture of phages (known as a *library*) which display different proteins on their surface. Within this mixture, a small number of phages will display the protein that binds to the ligand of interest. If the surface of a well in a microtitre plate is coated with the chosen ligand, the required phage particles will bind to it and the rest can be washed away. The bound phage, enriched for particles displaying the correct protein can then be eluted and amplified by infecting a suitable *E. coli* strain. This process, known as *panning*, can be repeated several times to obtain a pure phage preparation that provides a source of the gene required.

In this way, genes of interest can be enriched from vast libraries simply as a function of the affinity of the corresponding polypeptide for a specific target molecule, be it an antigen, antibody, cell-surface receptor, hormone or drug. This has revolutionized the study of protein–ligand interactions and drug discovery.

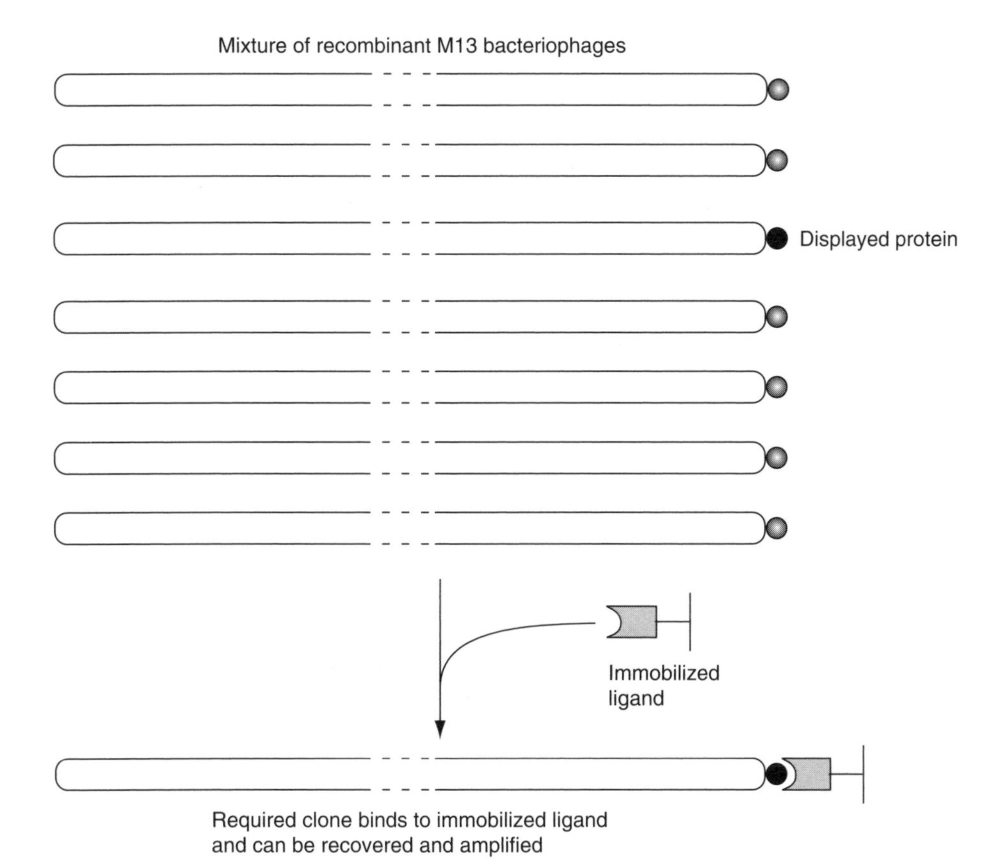

Figure 4.20 Phage display. Using gene cloning techniques, fragments of DNA are incorporated into M13 phage DNA yielding a mixture of recombinant phages some of which will display the required protein on the surface as a fusion with a coat protein. The required phage can be recovered from the mixture because of the ability of the displayed protein to adhere to an immobilized ligand

4.6.4 Bacterial virulence and phage conversion

Another important characteristic of bacteriophages is their association with determinants of pathogenicity in many medically important bacteria. For example, scarlet fever occurs concurrently with throat infections caused by certain strains of *Streptococcus pyogenes*. The symptoms are due to a protein toxin made by the bacterium. However, the gene that codes for the toxin is not a bacterial gene but is carried by a bacteriophage. Only strains of *S. pyogenes* that are lysogenic for this phage are able to produce the toxin and hence cause scarlet fever.

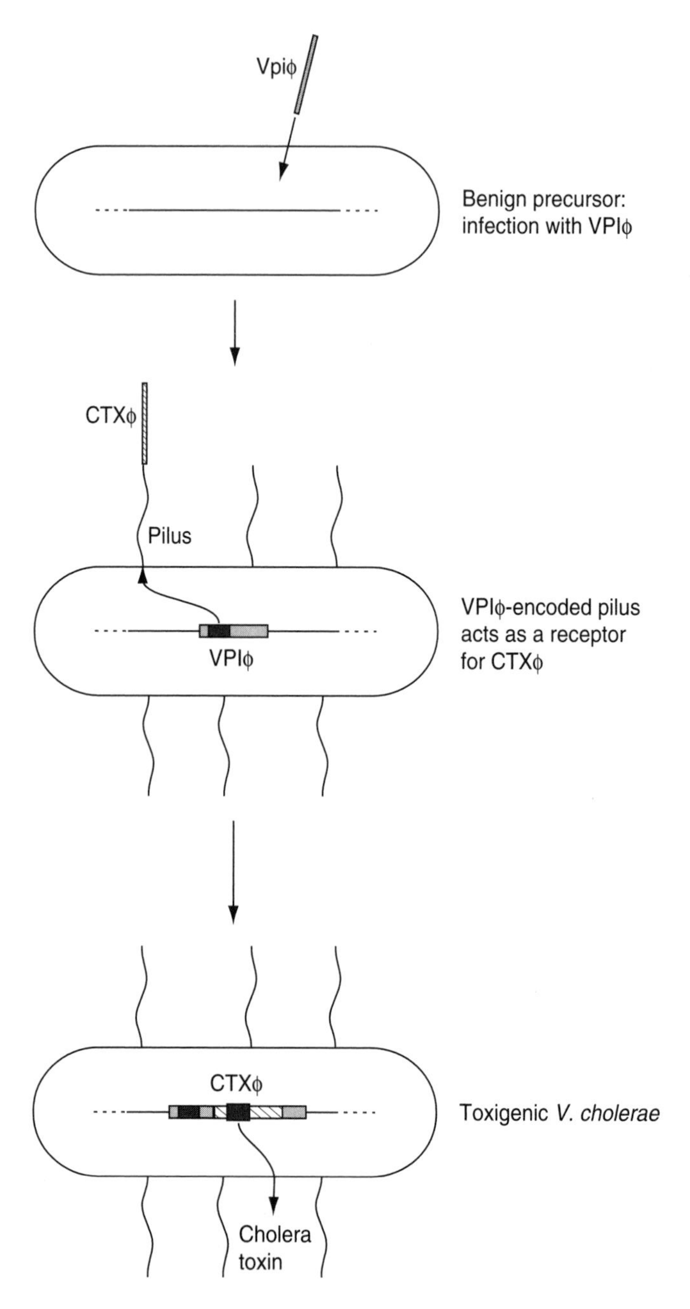

Figure 4.21 Bacteriophages and virulence of *Vibrio cholerae*. A putative, non-pathogenic, precursor is infected with the VPI phage. The VPI-encoded pilus provides an attachment site for the CTX phage which includes the toxin genes. Insertion of the CTXϕ DNA into the VPI prophage results in a toxigenic strain of *V. cholerae*

A similar phenomenon is shown by the bacterium that causes diphtheria (*Corynebacterium diphtheriae*). Again, the production of a toxin is required for the disease to occur and the toxin gene is carried by a specific bacteriophage.

In both of these examples it will be seen that initially non-pathogenic strains can be converted to pathogenicity by infection with a specific phage: this process is referred to as *phage conversion*. A further example, with an intriguing twist, is found in *Vibrio cholerae*. The genes encoding the subunits of the cholera toxin (*ctxA* and *ctxB*) are carried by the filamentous phage CTXϕ (CTXphi), which is related to M13 (see above). In toxigenic strains, CTXϕ is integrated in a specific chromosomal site which contains a cluster of virulence-related genes. Such groups of virulence genes are not uncommon in pathogenic bacteria and are referred to as *pathogenicity islands* (see Chapter 9). It has been suggested that, the *V. cholerae* pathogenicity island (VPI), is actually another integrated bacteriophage, VPIϕ, and the receptor for CTXϕ infection is a pilus encoded by the VPIϕ phage. Consequently, prior infection by VPIϕ may be a prerequisite for infection by CTXϕ. Therefore a model (Figure 4.21) for the evolution of virulent *V. cholerae* can be produced, starting with a harmless organism that becomes infected with VPIϕ. In consequence, it becomes susceptible to CTXϕ infection which results in the acquisition of the toxin genes leading to a bacterium capable of causing cholera.

Genome sequencing of bacteria has disclosed the presence of many previously unsuspected bacteriophages inserted into the chromosome. In most cases, it is not known whether these have any effect on the phenotype of the bacterium. Indeed it is likely that most of these bacteriophages have no effect at all. A remarkable exception is the case of the dangerous food-borne pathogen *E. coli* O157:H7. The genome of this strain contains an additional 1.4 Mb of O157:H7-specific DNA that is not present in benign laboratory strains of *E. coli* K-12. It is now known that this additional sequence contains 24 prophages and prophage-like elements and that these occupy nearly half of this additional genetic material. In this case, it is clear that bacteriophages have played a predominant role in the emergence of O157:H7 as a severe pathogen.

So the genetic composition of a bacterial species does not just consist of the chromosome with minor variations in sequence due to mutations. In most species, the chromosome also carries a variable number of bacteriophages inserted into it, which may or may not influence the characteristics of the organism. In the next chapter, another major source of variation due to the presence of additional, independent, DNA molecules known as *plasmids* will be discussed and in Chapter 6 the role of bacteriophages in the transfer of genes between bacteria will be described.

5 Plasmids

As described in the previous chapter, the overall genetic composition of a bacterial cell includes bacteriophages integrated into the chromosome (prophages). Even more important in terms of the effect on the phenotype of the cell are extrachromosomal DNA elements known as *plasmids*. Although these are considered to be a separate phenomenon from bacteriophages, it is not always possible to draw a firm line between them. Some bacteriophages (such as P1) do not integrate into the chromosome but in the prophage state exist as separate DNA molecules which are essentially plasmids. Bacteriophages such as M13 also replicate as plasmids. Conversely, some plasmids can integrate into the chromosome quite efficiently, as will be described in Chapter 6.

Plasmids and phages provide an important extra dimension to the flexibility of the organism's response to changes in its environment, whether those changes are hostile (e.g. the presence of antibiotics) or potentially favourable (the availability of a new substrate). This extra dimension therefore consists of characteristics that are peripheral to the replication and production of the basic structure of the cell – they are the optional extras. Their role in contributing these additional characteristics is particularly significant because of the relative ease with which they can be transferred between strains or between different species (see Chapter 6).

5.1 Some bacterial characteristics are determined by plasmids

5.1.1 Antibiotic resistance

The most widely studied plasmid-borne characteristic is that of drug resistance. Many bacteria can become resistant to antibiotics by acquisition of a plasmid, although plasmid-borne resistance to some drugs such as nalidixic acid and

Molecular Genetics of Bacteria, 4th Edition by Jeremy Dale and Simon F. Park
 ISBN 0 470 85084 1 (cased) ISBN 0 470 85085 X (pbk)

rifampicin does not seem to occur. (In those cases, resistance usually occurs by mutation of the gene that codes for the target protein). The antibiotic resistance genes themselves are many and varied, ranging from plasmid-encoded beta-lactamases which destroy penicillins to membrane proteins which reduce the intracellular accumulation of tetracycline. The ability of plasmids to be transferred from one bacterium to another, even sometimes between very different species (Chapter 6), has contributed greatly to the widespread dissemination of antibiotic resistance genes. Bacteria can become resistant to a number of separate antibiotics, either by the acquisition of several independent plasmids or through acquiring a single plasmid with many resistance determinants on it. Some examples will be discussed later in this chapter. Transposons (Chapter 7) are thought to have played a major role in the development of drug resistance plasmids, by promoting the movement of the genes responsible between different plasmids or from the chromosome of a naturally resistant organism onto a plasmid.

It should be appreciated that other mechanisms of antibiotic resistance also occur and that such resistance is not always due to plasmids: indeed many of the bacteria that are currently causing problems of hospital cross-infection are either inherently resistant or owe their antibiotic resistance to chromosomal genes.

5.1.2 Colicins and bacteriocins

Another property conferred by some plasmids that has been widely studied is the ability to produce a protein which has an antimicrobial action, usually against only closely-related organisms. One group of such proteins, produced by strains of *E. coli*, are capable of killing other *E. coli* strains, and are hence referred to as *colicins*, and the strains that produce them are *colicinogenic*. (These terms are more familiar than the general ones, bacteriocin and bacteriocinogenic and will therefore be used in this chapter). The colicin gene is carried on a plasmid (known as a Col plasmid), together with a second gene that confers immunity to the action of the colicin, thus protecting the cell against the lethal effects of its own product. One particular Col plasmid, ColE1, is of special importance because of the detailed information that is available concerning its replication and control (see later in this chapter) and also because most of the commonly used *E. coli* cloning vectors are based on ColE1 or a close relative.

5.1.3 Virulence determinants

The previous chapter discussed how bacteriophages can carry genes that code for toxins and that the presence of the phage is necessary for pathogenicity. In some bacterial species toxin genes are carried on plasmids rather than phages. For example, some strains of *E. coli* are capable of causing a disease that resembles cholera (although milder). These strains produce a toxin known as LT (labile toxin – to distinguish it from a different, heat-stable, toxin known as ST). The LT toxin is closely related to the cholera toxin, but whereas the gene in *V. cholerae* is carried by a prophage, the LT gene in *E. coli* is found on a plasmid.

Plasmids can also carry other types of genes that are necessary for (or enhance) virulence. One of the most dramatic examples of this is the 70-kb virulence plasmid of *Yersinia* species. This plasmid which is found in species of *Yersinia* (including *Yersinia pestis*, the causative organism of plague) has been aptly described as a mobile arsenal since it encodes an integrated system which allows these bacteria to inject effector proteins into cells of the immune response to disarm them, to disrupt their communications or even to kill them.

Box 5.1 provides examples of virulence factors which are carried by bacteriophages and plasmids in various pathogenic bacteria. This is by no means an exhaustive list.

5.1.4 Plasmids in plant-associated bacteria

A different type of pathogenicity is seen with the plant pathogen *Agrobacterium tumefaciens*, which causes a tumour-like growth known as a crown gall in some plants. Again, it is only strains that carry a particular type of plasmid (known as a Ti plasmid, for Tumour Inducing) that are pathogenic; in this case however, pathogenicity is associated with the transfer of a specific part of the plasmid DNA itself into the plant cells. This phenomenon has additional importance because of its application to the genetic manipulation of plant cells (see Chapter 8).

Members of the genus *Rhizobium* also 'infect' plants, although in this case the relationship is symbiotic rather than pathogenic. These bacteria form nodules on the roots of leguminous plants. Under these conditions the bacteria are able to fix nitrogen and supply the plant with a usable source of reduced nitrogen, a process of considerable ecological and agricultural importance. The genes necessary for both nodulation and nitrogen fixation are carried by plasmids.

5.1.5 Metabolic activities

Plasmids are capable of expanding the host cell's range of metabolic activities in a variety of other ways. For example, a plasmid that carries genes for the fermentation of lactose, if introduced into a lactose non-fermenting strain, will convert it to one that is able to utilize lactose. Such plasmids can cause problems in diagnostic laboratories where organisms are often identified on the basis of a limited set of biochemical characteristics. Commonly the potentially pathogenic *Salmonella* genus is differentiated from the (usually) non-pathogenic *E. coli* species primarily

Box 5.1 Plasmids and phages can carry virulence genes

Plasmids are best known for their ability to confer resistance to antibiotics. Amongst the many other characteristics that can be mediated by plasmids (or bacteriophages), the influence on virulence is especially significant.

The table shows a selection of some of the better characterized virulence genes carried by plasmids and phages. There are many other virulence determinants for which there is no evidence of either plasmids or bacteriophages being involved. With some of the examples shown, the genes involved may be chromosomally located in some strains.

Bacterial species	Disease	Virulence gene(s)	Location
Corynebacterium diphtheriae	Diphtheria	Toxin	Phage
Streptococcus pyogenes	Scarlet fever	Toxin	Phage
Vibrio cholerae	Cholera	Toxin	Phage
Shigella spp.	Dysentery	Invasion/adhesion	Plasmid
Yersinia enterocolitica	Gastroenteritis	Yops (outer membrane proteins)	Plasmid
Clostridium botulinum	Botulism	Toxin	Phage
Clostridium tetani	Tetanus	Toxin	Plasmid
Escherichia coli	Gastroenteritis	Enterotoxins	Plasmids
Escherichia coli	Gastroenteritis	Adhesion	Plasmids

because of the inability of *Salmonella* to ferment lactose. In some cases, the detection of serious epidemics of Salmonella infections has been delayed because the causative agent had acquired a lactose-fermenting plasmid.

A large number of other genes have also been found on plasmids, including those for fermentation of other sugars such as sucrose, hydrolysis of urea, or production of H_2S. Many of these were initially identified because of the confusion they caused in biochemical identification tests.

Biodegradation and bioremediation

Another type of plasmid-mediated metabolic activity is the ability to degrade potentially toxic chemicals. One such plasmid, pWWO, obtained from *Pseudomonas putida*, encodes a series of enzymes that convert the cyclic hydrocarbons toluene and xylene to benzoate (upper pathway in Figure 5.1) and a second

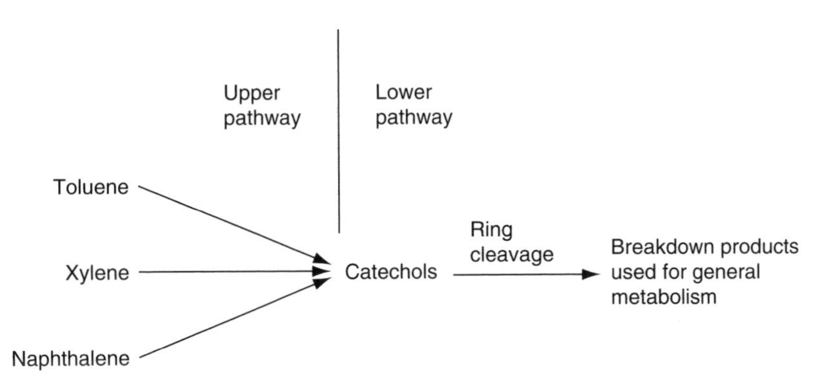

Figure 5.1 Degradation of cyclic hydrocarbons

operon responsible for the degradation of benzoate, via ring cleavage of a catechol intermediate, into metabolic intermediates that can be used for energy production and biosynthesis (lower pathway – see Figure 5.1). This organism can therefore grow using toluene as a sole carbon source. The enzymes of the upper pathway are specialized; other plasmids code for upper pathway enzymes with different specificities, enabling the organism to convert other chemicals into benzoate and catechol derivatives which can be degraded by the lower pathway enzymes. Plasmid-mediated degradation includes naphthalene and camphor, as well as chlorinated aromatic compounds such as 3-chlorobenzoate and the herbicide 2,4-D (dichlorophenoxyacetic acid).

The ability to degrade environmentally damaging chemicals is potentially useful in clearing up polluted sites (bioremediation). There is therefore considerable interest in extending the range of chemicals that can be degraded by microorganisms, both by modification of existing pathways and also by screening bacteria isolated from contaminated sites for novel activities. The usefulness of such strains is also potentiated by plasmids which confer resistance to toxic metal ions, notably copper and mercury.

5.2 Molecular properties of plasmids

Bacterial plasmids in general exist within the cell as circular DNA molecules with a very compact conformation, due to supercoiling of the DNA. In many cases, they are quite small molecules, just a few kilobases in length, but in some organisms, notably members of the genus *Pseudomonas*, plasmids up to several hundred kilobases are common. However, it is worth noting that the standard methods for isolating plasmids (see below) are geared to the separation of small covalently closed circular DNA, and the occurrence of large plasmids, or alternative forms such as linear plasmids, may be underestimated.

It is convenient to regard plasmids from *E. coli* as consisting of two types. The first group, of which ColE1 is the prototype, are relatively small (usually less than 10 kb), and are present in multiple copies within the cell. Their replication is not linked to the processes of chromosomal replication and cell division (hence the high copy number), although there are some controls on plasmid replication (as discussed later in this chapter). Replication of these plasmids can continue under certain conditions (such as inhibition of protein synthesis) that prevent chromosome replication, giving rise to a considerable increase in the number of copies of the plasmid per cell. This phenomenon, known as *plasmid amplification*, is very useful for isolating the plasmid concerned.

The second group of plasmids, exemplified by the F plasmid, are larger (typically greater than 30 kb; F itself is about 100 kb) and are present in only one or two copies per cell. This is because their replication is controlled in essentially the same manner as that of the chromosome; hence when a round of chromosome replication is initiated, replication of the plasmid will occur as well. It follows therefore that plasmids of this type cannot be amplified. In general, these large plasmids are able to promote their own transfer by conjugation (they are known as conjugative plasmids: see Chapter 6).

The existence of these two groups can be rationalized on the basis of their different survival strategy. Members of the first group rely on their high copy number to ensure that, at cell division, if the plasmid molecules partition randomly between the two cells, then each daughter cell is virtually certain to contain at least one copy of the plasmid (Figure 5.2a). For example, with a plasmid that is present in 50 copies per cell, the chance of one daughter cell not receiving any copies of the plasmid is as low as 1 in 10^{15}.

However, high copy number imposes a size constraint. Replication of a plasmid imposes a metabolic burden that is related to the size and copy number of the plasmid. The greater the burden, the greater the selective pressure in favour of those cells that do not possess the plasmid. Hence it is logical that high copy number plasmids will also be small. ColE1, for example, is 6.4 kb in size. If there are 30 copies per cell, this represents about 4 per cent of the total DNA of the cell. The F plasmid on the other hand (c. 100 kb), if it were to be present at a similar copy number, would add nearly 70 per cent to the total DNA content which would inevitably make the cell grow much more slowly and any cells that had lost the plasmid would have a marked selective advantage.

But the information required to establish conjugation in *E. coli* is quite extensive (see Chapter 6). With the F plasmid for example about 30 kb (out of 100 kb) consists of genes required for plasmid transfer. It follows therefore that a small plasmid will not be able to carry all the information needed for conjugative transfer.

The second type of plasmid has evolved a different strategy (Figure 5.2b). Firstly, linking replication of the plasmid to that of the chromosome ensures that there are at least two copies of the plasmid available when the cell divides.

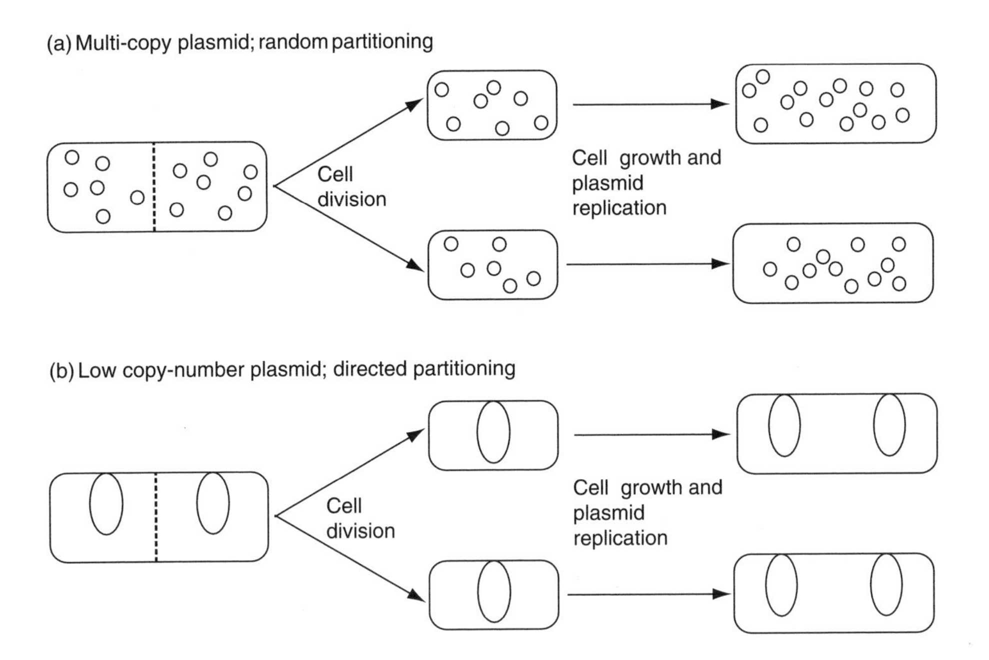

Figure 5.2 Partitioning of plasmids at cell division

Secondly, random partitioning will not be sufficient to ensure that each of the daughter cells receives a copy; so the plasmid must be distributed between the progeny in a directed manner, in much the same way as the copies of the chromosome are distributed. The ability to transfer by conjugation provides a back-up mechanism since any plasmid-free cells that arise in the population by failure of the partitioning mechanism will then be able to act as recipients for transfer of the plasmid.

It is necessary to be aware that this picture, although useful, is a highly simplified one, and there are many exceptions, even in *E. coli*. There are numerous examples of small plasmids that have a low copy number, although none of them are conjugative, and some examples of larger plasmids that exist in multiple copies. In addition, in other organisms the picture is less clear; for example in *Streptomyces* it seems that quite small plasmids are able to promote their own transfer by conjugation.

5.2.1 Plasmid replication and control

In order to understand the reasons for the different behaviour of plasmids as described above, we need to look at the mechanisms of plasmid replication and how it is controlled. This should be compared with the description of chromosome replication in Chapter 1. Many plasmids are replicated as double-stranded circular molecules. The overall picture with such plasmids is basically similar to that of the chromosome, in that replication starts at a fixed point known as *oriV* (the vegetative origin, to distinguish it from the point at which conjugative transfer is initiated, *oriT*), and proceeds from this point, either in one direction or in both directions simultaneously until the whole circle is copied. However there are some aspects of replication that differ from that of the chromosome, especially for the multicopy plasmids. Two examples that have been studied intensively are ColE1 and R100. Other plasmids with quite different modes of replication are dealt with later on.

Replication of ColE1

The colicinogenic plasmid ColE1 (Figure 5.3) is a comparatively small (6.4 kb) plasmid that carries just the genes for production of colicin E1, and immunity to it, together with functions involved in plasmid maintenance. This is probably the best understood of all plasmids. Replication starts with the production of an RNA primer (RNA II), starting from a site 555 bp upstream from *oriV* (see Figure 5.4). Transcription occurs through the origin (*oriV*), and RNA II is cut at a specific site by RNase H (which cuts RNA molecules when they are present as

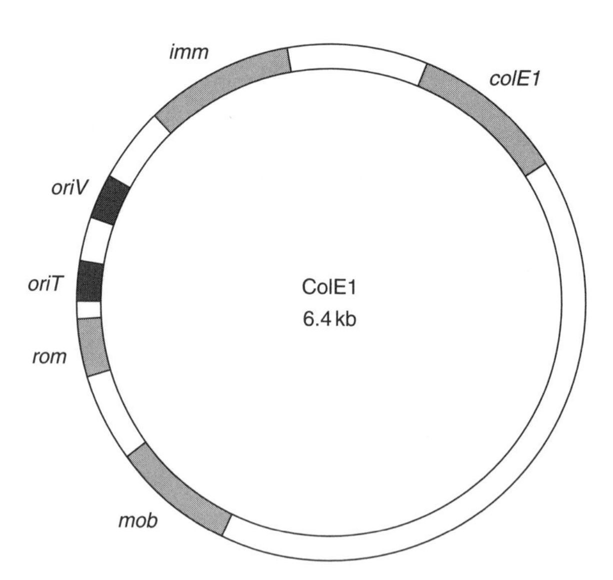

Figure 5.3 Genetic map of the plasmid ColE1. *colE*1, *imm*: genes for production of, and immunity to, colicin E1; *mob* codes for a nuclease required for mobilization; *rom* codes for a protein required for effective control of copy number; *oriT*: origin of conjugal transfer; *oriV*: origin of replication

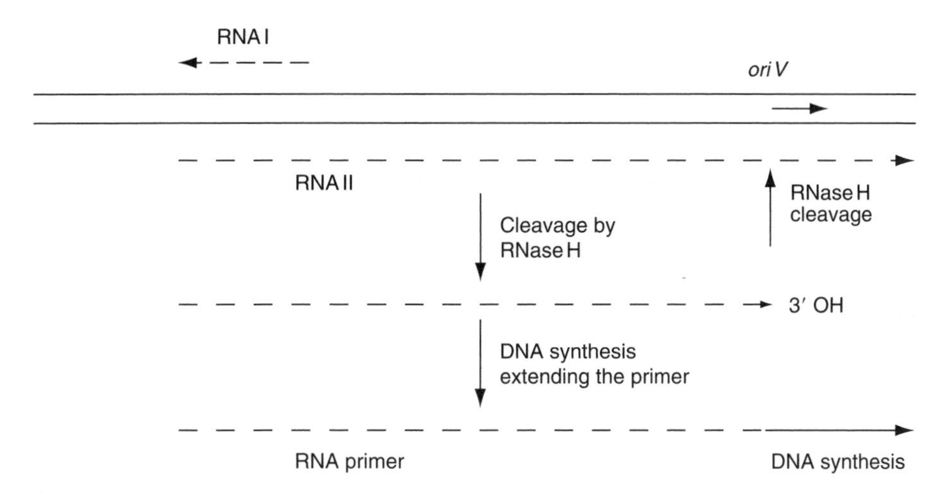

Figure 5.4 Structure and control of the origin of replication of the ColE1 plasmid. RNA II, after cleavage by RNaseH, acts as a primer for DNA synthesis. RNA I binds to RNA II and prevent RNase cleavage and hence prevents initiation of replication

an RNA–DNA hybrid). DNA synthesis then occurs by addition of deoxynucleotides to the 3′ OH end of the RNA primer.

The RNA primer is known as RNA II because there is another RNA molecule produced from the same region, which is called RNA I. This is transcribed from the opposite strand to RNA II, and is complementary to the first 108 bases of RNA II. The presence of RNA I is inhibitory to replication, because binding of RNA I to RNA II prevents cleavage of RNA II by RNaseH, due to interference with the secondary structure of RNA II. So, although the copy number is high, replication is still controlled to some extent. An additional gene that controls replication is the *rom* (or *rop*) gene, which codes for a protein that facilitates the interaction of RNA I and RNA II. Derivatives of ColEI in which the *rom* gene has been deleted have a higher copy number.

The ColE1 plasmid is non-conjugative, that is, it is not able to transfer itself from one cell to another. However, in common with many other non-conjugative plasmids, it can be transferred by conjugation if the cell carries a compatible conjugative plasmid. This effect, which involves the *mob* and *oriT* sites (Figure 5.3), is known as *mobilization* and is described in Chapter 6.

Replication of R100

R100 is a low copy number, conjugative, resistance plasmid, which contains about 89 kb of DNA; it confers resistance to four different antibiotics (tetracycline, chloramphenicol, streptomycin and sulphonamides), as well as to mercury salts.

The structure of R100 is shown in Figure 5.5. It is immediately apparent that all the genes required for conjugative transfer, comprising nearly half of the plasmid, are clustered together, adjacent to the origin of replication (*oriV*), while the antibiotic resistance determinants are all found on the right hand half of the diagram. These large antibiotic resistance plasmids are commonly organized in this way, which is thought to reflect their evolution by sequential addition of resistance genes to a basic replicon, i.e. it started as a cryptic plasmid comprising just the transfer region and origin of replication, to which the various resistance genes have been added (individually or in blocks).

One mechanism for acquiring extra resistance genes is shown by the tetracycline resistance determinant (*tet*). This is flanked by two copies of an insertion sequence (IS*10*); this combination results in a mobile structure known as a *transposon* (Tn*10* in this case) that can move from one DNA site to another. Tn*10* has therefore presumably transposed into R100 from another plasmid. R100 also contains two copies of a different insertion sequence, IS*1*. Transposons and

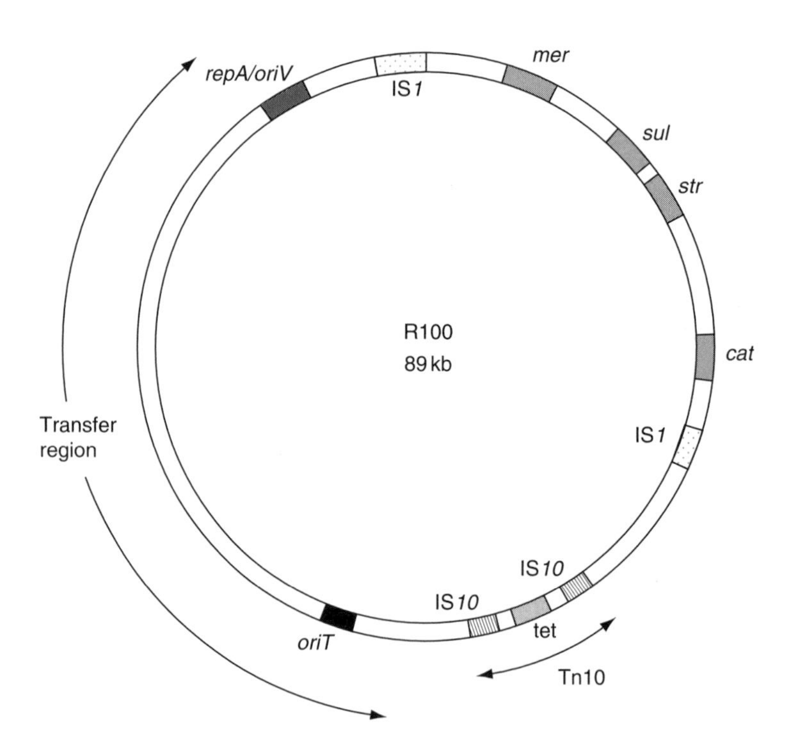

Figure 5.5 Genetic map of the conjugative *E. coli* plasmid R100. Resistance genes: *cat*, chloramphenicol (chloramphenicol acetyltransferase); *mer*, mercuric ions; *str*, streptomycin; *sul*, sulphonamides; *tet*, tetracycline. Other sites: *oriT*, origin of conjugative transfer; *repA/oriV*, replication functions and origin of replication. IS*1*, IS*10* are insertion sequences, Tn*10* is a transposon

insertion sequences play a key role in the evolution of plasmids, and are discussed further in Chapter 7.

Replication of R100 occurs from a single origin. In addition to the origin of replication, a gene known as *repA* (adjacent to *oriV*) is required for the initiation of replication from *oriV* (Figure 5.6). Plasmid copy number is controlled by two genes that regulate the production of the RepA protein. One of these, *copB*, codes for a protein that represses transcription of the *repA* gene. When the plasmid first enters a bacterial cell, the absence of CopB allows expression of RepA and so there is a burst of replication, until the level of CopB builds up to repress this promoter. From then on, expression of RepA occurs at a low level from the *copB* promoter. The second regulatory gene, *copA*, then regulates expression of RepA. This gene codes for an 80–90-nucleotide untranslated RNA molecule. The *copA* gene is within the region of DNA that is transcribed for production of RepA, but is transcribed in the opposite direction (it is an *antisense* RNA). The *copA* RNA is therefore complementary to a short region of the *repA* transcript, and will bind to it, interfering with translation of the RepA protein. When the plasmid replicates, the number of copies of the *copA* gene is doubled and the amount of the *copA* RNA will therefore increase; this causes a marked reduction in further replication initiation, until cell division restores the original copy number.

R100 is unable to co-exist with other related plasmids such as R1; this *incompatibility* is one way of classifying plasmids (see below). R100 and R1 belong to the IncFII group of plasmids. The *copA* gene is responsible for the incompatibility of R100 and R1. The sequence of the *copA* gene is very similar in these two plasmids and the products are interchangeable. The R100 *copA* RNA will therefore inhibit replication of R1 and vice versa, which results in one or the other plasmid being lost at cell division. It should be noted however that with other plasmids there are different causes of incompatibility.

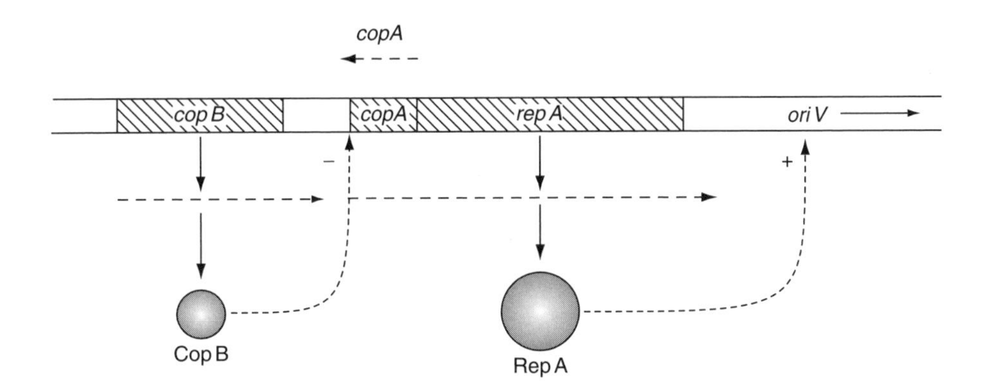

Figure 5.6 Replication control of the plasmid R100. The RepA protein is needed for initiation of replication. Transcription of *repA* is repressed by CopB and translation of the *repA* mRNA is inhibited by the antisense *copA* RNA

Plasmids like R100 also contain a region known as the *par* locus (for partitioning) that is necessary for accurate partitioning of plasmid copies at cell division. This sequence can only act in *cis*, i.e. it must be present on the plasmid itself and not on some other plasmid. (Genes that are able to have an effect on other DNA molecules are said to be active *in trans*, while those that can only affect the same DNA molecule are *cis*-acting). It is likely that all low copy number plasmids have a *par* sequence (or some other mechanism for ensuring accurate partitioning), while plasmids such as ColE1 rely primarily on high copy number to ensure that each daughter cell receives a copy of the plasmid.

Control of plasmid replication by DNA repeats (iterons)

Regulation of plasmid replication through control of the expression of RepA, as described above, is adequate for many low-copy plasmids. But the absence of a direct connection between the number of plasmids and their ability to replicate means that we would expect a distribution of copy number in a narrow range either side of a mean value. For example, most cells might have five copies, but some would have six and others would have four. In the presence of an active and effective partitioning system, this is adequate for stable inheritance of such a plasmid. But some plasmids, such as F, are present as only a single copy, after cell division, replicating once per cycle so that at cell division there are two copies. There is no room for a statistical distribution of plasmid copy number. Stable maintenance of such a plasmid requires a much tighter control of replication that is directly linked to the number of copies in the cell.

This is provided by the presence of repeated DNA sequences, 17–22 bp long, known as *iterons*, which are found in the replication initiation regions. For example, the F plasmid has nine repeats of a 17-bp sequence. The RepA protein binds to these iteron sequences. When there is more than one copy of the plasmid, the RepA protein can bind to iterons on both copies, coupling them together (Figure 5.7). This prevents further replication of the plasmid. This 'coupling' or 'handcuffing' model provides a mechanism for extremely tight control of the number of copies of the plasmid.

It also provides another mechanism for plasmid incompatibility. If two different plasmids carry related iteron sequences (so that RepA can bind to both), the plasmids will be coupled together and replication will be prevented. The two plasmids cannot therefore be stably maintained in the same cell.

Plasmid replication via single-stranded forms

Replication of plasmids in *E. coli* usually seems to follow the route described earlier in this chapter, that is by copying both strands as part of the same process.

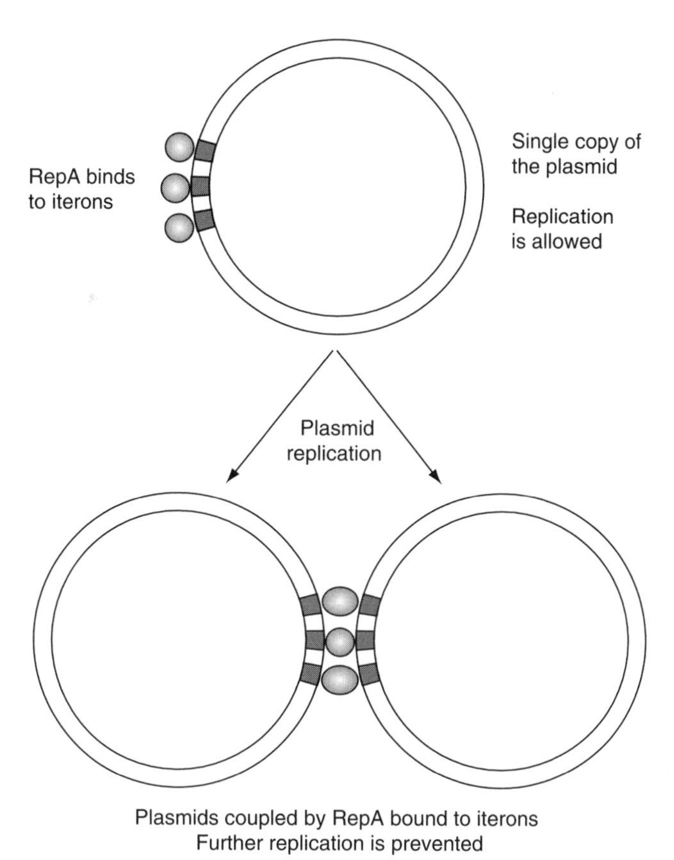

Figure 5.7 Coupling model for the control of iteron-containing plasmids

This model is not universally applicable. Many plasmids, especially in Gram-positive bacteria, replicate via a single-stranded intermediate (Figure 5.8). Although such plasmids are often referred to as single-stranded plasmids, it should be noted that the single-stranded form is only a replication intermediate and that normally the majority of plasmid molecules are double stranded.

A similar mode of replication has already been described, namely that of the single-stranded bacteriophages ϕX174 and M13 (Chapter 4). There is indeed a considerable similarity between these systems, not only in the general concept but also in the mechanisms involved. The general mechanism of replication of all of these elements, although differing in some details, follows the outline shown in Figure 5.8. A specific site (the plus origin) on the plus strand of the plasmid is first cut by a plasmid-encoded protein (Rep). The nicked DNA provides a site for initiation of DNA synthesis using host enzymes which displaces the old plus strand. While this is happening, the Rep protein remains attached to the free 5′ end of the nicked plus strand. When this process has travelled right round the

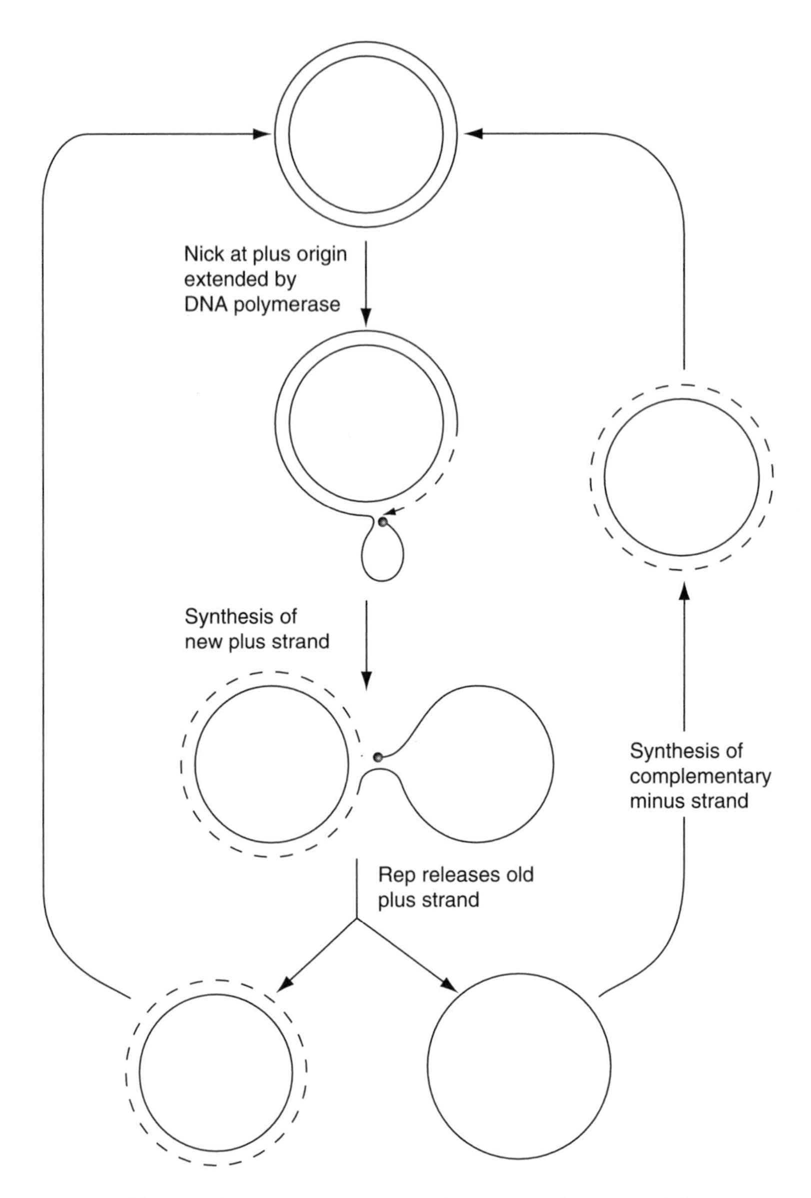

Figure 5.8 Replication of single-stranded plasmids

plasmid and returned to the plus origin, the Rep protein makes another nick to release the old plus strand and ligates the ends of this molecule to produce an intact single-stranded circular structure. This is then converted to the double-stranded form by synthesis of the complementary (minus) strand starting from a separate origin (the minus origin).

The same model can be applied to the single-stranded phages (see Figure 4.5). The important difference is, that with these phages, the single-stranded form is not merely a replicative intermediate but is incorporated into the phage particle. In the early stages of replication, the displaced single-stranded form is stabilized by coating with a host protein known as SSB (single-stranded DNA binding protein). Later in the cycle, the progeny viral strands are packaged into phage coats rather than producing more RF DNA. In the case of ϕX174, complementary strand synthesis is prevented because the viral strand is packaged directly into phage pro-heads as it is produced. With M13, a specific phage protein (gene V protein) accumulates and replaces SSB on the viral DNA, thus preventing complementary strand synthesis. The gene V protein is replaced by the proteins of the phage particle during the assembly of the phage at the cell membrane.

The sequences of the plus origins of a number of these elements have been determined and fall into several groups within which there is a high degree of similarity. For example, bacteriophage ϕX174 and the *S. aureus* plasmids pC194 and pUB110 all have the same short sequence at the origin. This is the recognition site for the Rep proteins of these elements, which also show sequence similarity. This indicates an evolutionary connection between bacteriophage ϕX174 and plasmids from Gram-positive bacteria.

The minus origins, on the other hand, are more diverse. Since this origin requires the action of host proteins, it must have evolved to fit in with the specificity of the current host. One consequence is that if such a plasmid is put into a different bacterial species, the plus origin may function quite well, but the minus origin, which requires host proteins, may be comparatively ineffective. This will result not only in frequent loss of the plasmid by segregational instability, but also in the generation of a variety of DNA rearrangements, since the single-stranded DNA forms that will accumulate will stimulate recombination.

Replication of linear plasmids

Another type of plasmid represents even more of a challenge to common ideas of bacterial DNA structure and replication. Linear DNA plasmids have been characterized from several bacterial genera including *Borrelia* and *Streptomyces*. In these bacterial species (and some others), the chromosome is also linear. The replication of these linear molecules poses something of a problem. As previously described (Chapter 1), one new DNA strand (the leading strand) is made continuously, while the other strand (the lagging strand) is made discontinuously – that is, it is produced as short fragments that are subsequently joined together. This is necessary as nucleic acids are always made in the 5′ to 3′ direction. Remember also that DNA synthesis requires a primer, which is normally generated by an RNA polymerase. This priming sequence is removed and replaced as the DNA polymerase reaches it when synthesizing the next fragment.

Now consider what happens as the replication fork approaches the end of a linear DNA molecule. The leading strand will continue right to the end, but what happens to the lagging strand? It is not possible to produce an RNA primer right at the 3′ end of the template strand, and in any case there is no mechanism for replacing the last bit of RNA with DNA. In consequence, the template strand cannot be copied right up to the 3′ end (see Figure 5.9).

Bacteria that have linear DNA (whether plasmids or chromosomes) have adopted different strategies to overcome this problem. In *Borrelia*, the ends of the two strands are joined together in a covalently closed hairpin structure (Figure 5.10). Similar structures are found at the ends of some animal viruses (e.g. poxviruses), as well as the *E. coli* bacteriophage N15 where the prophage is a linear DNA plasmid. There are several models that would account for the complete replication of such a structure, one of which is illustrated in Figure 5.11. Bidirectional replication initiating at a central origin of replication (*oriC*) would lead to a dimeric double-stranded circular molecule in which two copies of the genome are linked by copies of the hairpin loop sequence. This intermediate structure would then be processed by cutting and rejoining the DNA strands at each end to reform the covalently closed hairpin loops.

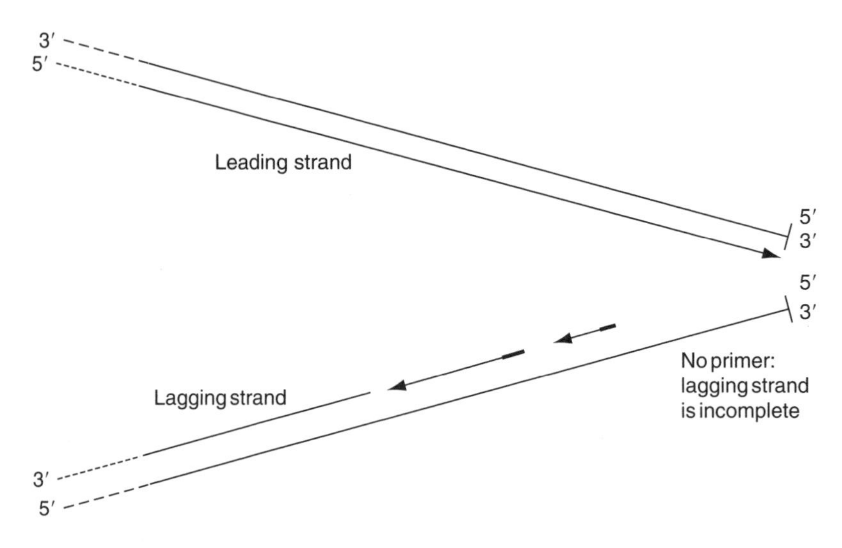

Figure 5.9 Incomplete replication of linear DNA. Normal modes of replication are unable to replicate the ends of a linear molecule

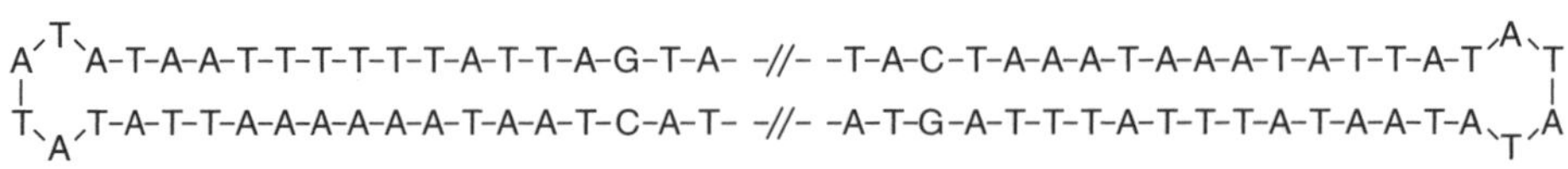

```
 T                                                                              A
A  A-T-A-A-T-T-T-T-T-T-A-T-T-A-G-T-A- -//- -T-A-C-T-A-A-A-T-A-A-A-T-A-T-T-A-T   T
|                                                                                |
T  T-A-T-T-A-A-A-A-A-A-T-A-A-T-C-A-T- -//- -A-T-G-A-T-T-T-A-T-T-T-A-T-A-A-T-A   A
 A                                                                              T
```

Figure 5.10 The ends of a linear plasmid from *Borrelia* are covalently joined

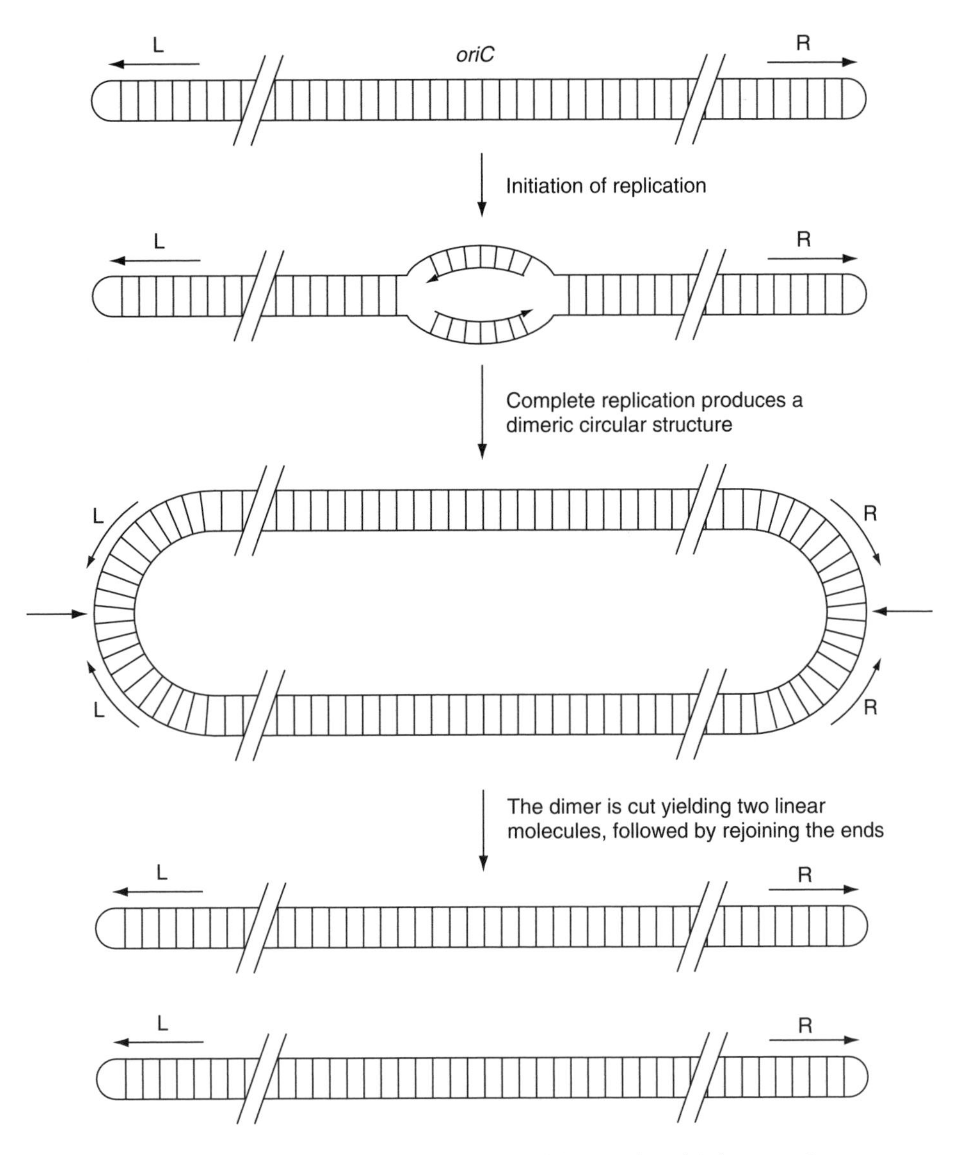

Figure 5.11 Model of replication of linear plasmids in *Borrelia*

In *Streptomyces*, the problem of replicating the ends of the linear DNA is solved in a different way. A key feature is the presence of a protein (terminal protein, TP), covalently attached to the 5′ ends of the DNA. The simplest model (Figure 5.12) is that this protein acts as a primer for DNA synthesis, allowing replication of the ends of the linear DNA – and indeed the replication of some linear DNA molecules (such as that of the *Bacillus subtilis* bacteriophage ϕ29) occurs in this way. But it is not an adequate explanation for the replication of

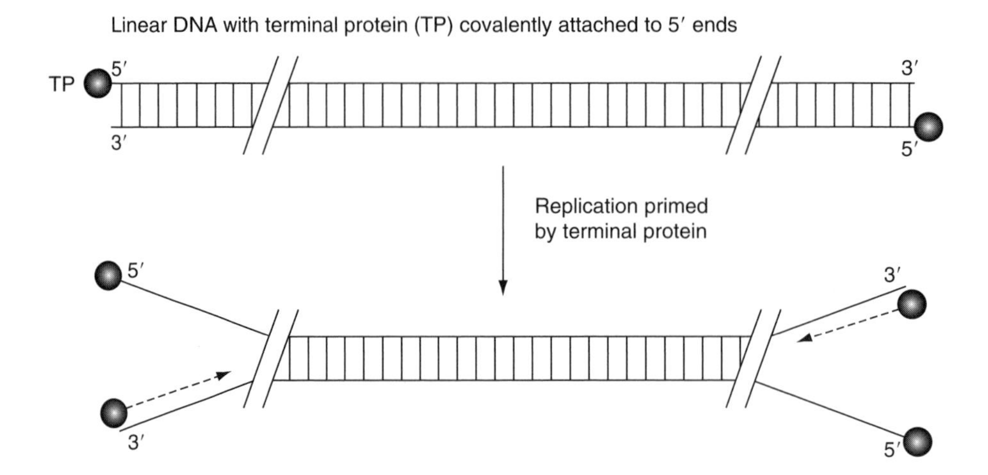

Figure 5.12 Replication of linear DNA primed by covalently attached protein

Streptomyces plasmids (and even less so for the replication of the linear chromosome). These have an origin of replication in an internal position in the molecule, from which replication occurs bidirectionally in a conventional manner. While the leading strand (made in the $5' \rightarrow 3'$ direction) can be synthesized right to the end of the template, the lagging strand will stop short, leaving a short portion of DNA unreplicated (Figure 5.13). The role of the terminal protein in this case lies in patching this unreplicated portion, assisted by the formation of secondary structures in the unreplicated strand due to the presence of inverted repeat sequences.

5.3 Plasmid stability

One of the characteristic features of plasmids is their instability. Plasmid-borne features are often lost from a population at a higher frequency than would be expected for the normal processes of mutation. The extent of this instability varies enormously from one plasmid to another. Naturally-occurring plasmids are usually (but not always) reasonably stable: selection will tend to operate in that way, and in addition in isolating the strain and looking for the plasmid the more stable plasmids will tend to be selected. Unstable ones will be harder to find.

Artificially constructed plasmids on the other hand are often markedly unstable. This is usually merely a nuisance on the laboratory scale, but can become a very expensive problem in the industrial use of strains carrying such plasmids.

There are three quite distinct phenomena associated with the concept of plasmid stability: (1) plasmid integrity, (2) partitioning at cell division and (3) differential growth rates.

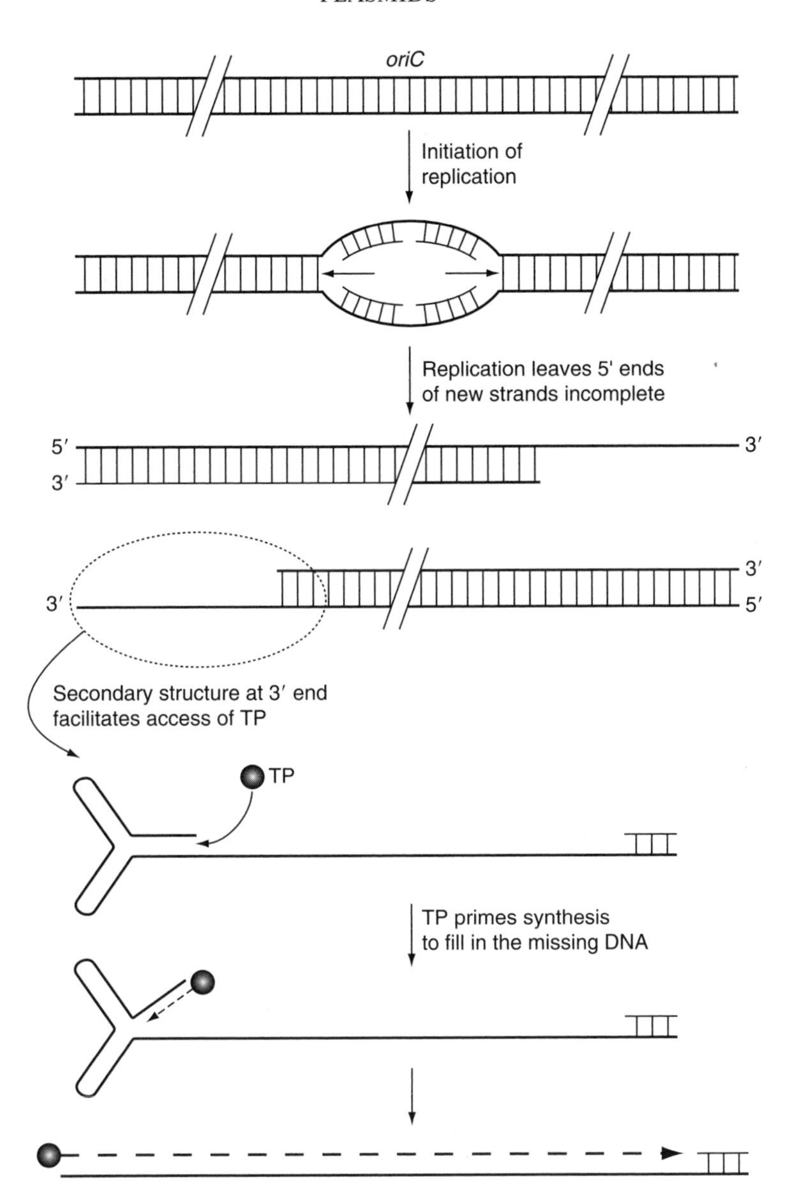

Figure 5.13 Model of replication of linear plasmids in *Streptomyces*. TP, terminal protein

5.3.1 Plasmid integrity

Integrity refers to maintenance of the *structure* of the plasmid. Even naturally-occurring plasmids can be 'unstable' in this respect, showing a tendency to lose genes due to the presence of recombination hot-spots. In particular, the presence of repeated sequences, due to the presence of transposons or insertion sequences (see Chapter 7) can lead to deletions or inversions due to recombination between the repeats. In Figure 5.14, plasmid B has suffered a deletion of the region containing the kanamycin and chloramphenicol resistance genes. Testing for ampicillin resistance might give the impression that the plasmid is quite stable. However, the deletion can be identified, in this case by testing for resistance to other antibiotics or (more generally) by determining the size of the plasmid. Identification of plasmids that have undergone a rearrangement (such as the inversion in plasmid C) is more difficult, since the plasmid remains the same size and may retain all of the original phenotypic characteristics. In many cases the only evidence (short of sequence data) may be a change in the restriction map of the plasmid, since the relative position of some restriction sites will have changed (see Chapter 7 for further discussion of the possible effects of inversions on plasmids or on the chromosome).

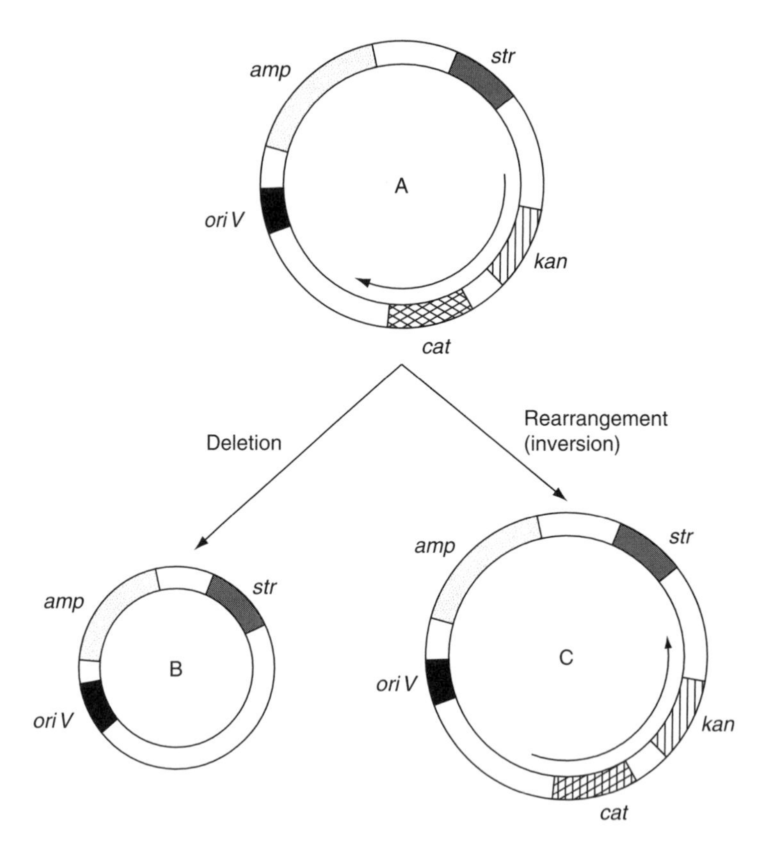

Figure 5.14 Plasmid integrity. Examples of two of the ways in which the structure of a plasmid can change. Plasmid B shows a deletion of the *cat* and *kan* genes, while plasmid C shows an inversion of this region. *amp, str, kan, cat*, resistance to ampicillin, streptomycin, kanamycin and chloramphenicol respectively. *oriV*, origin of replication

5.3.2 Partitioning

As mentioned previously, correct partitioning at cell division is essential if the plasmid is to be maintained in the culture (Figure 5.15). Although high copy number plasmids can rely principally on random distribution between the two daughter cells, this can be compromised by a tendency for plasmids to form multimeric structures during replication and also by recombination between monomers. Furthermore, since a dimer contains two origins of replication, it will be expected to replicate more efficiently than a monomer; multimers would replicate even more efficiently. This could potentially lead to what is known as a 'dimer catastrophe', in which the proportion of dimers, and higher multimers, increases to an extent that threatens the sustained maintenance of the plasmid.

It is worth pausing here to examine why it should matter whether the plasmid is present as monomers, dimers or multimers. The models for the control of plasmid copy number work essentially by counting the number of replication origins,

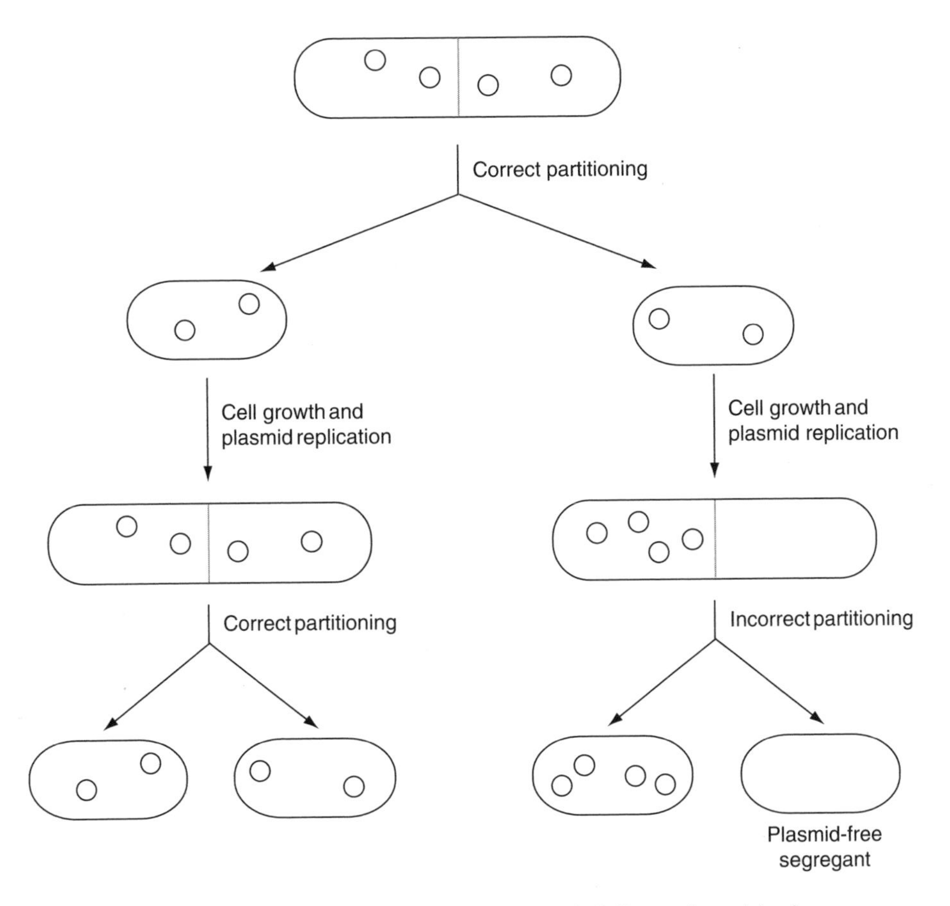

Figure 5.15 Plasmid segregation through failure of partitioning

rather than the number of molecules. So, a plasmid with a copy number of 30 would be present as 30 monomers, or 15 dimers or 10 trimers and so on. (In reality there would be a heterogeneous mixture of different sizes). So the existence of a substantial proportion of these plasmids as multimers will reduce the effective copy number when it comes to cell division and hence increase the likelihood that one of the daughter cells will not receive a copy of the plasmid.

The main mechanism for countering this effect involves site-specific recombination. For example, ColE1 contains a site known as *cer* which is a target for the action of the proteins XerC and XerD. In a dimer, there are two copies of the *cer* sequence and the Xer proteins will catalyse recombination between them – it breaks both DNA molecules, crosses them over, and rejoins them – thus resolving the dimeric structure into two monomers (see Figure 5.16). (XerC and XerD are actually host proteins, which carry out a similar function in resolving any chromosome dimers produced accidentally during replication). The importance of this system is demonstrated by the marked instability that results from deletion of the *cer* locus of ColE1.

Low copy-number plasmids cannot rely on random partitioning. As well as an active partitioning mechanism, some plasmids supplement their partitioning system with an ability to kill any cells that have lost the plasmid (post-segregational killing). These systems consist of two components: a stable, long-lived toxin and an unstable factor that either prevents expression of the toxin or acts as an antidote to it. For example the F plasmid contains an operon called *ccd* which consists of two genes, *ccdA* and *ccdB* (Figure 5.17). The CcdB protein is toxic because it interferes with DNA replication (through its action on DNA gyrase-

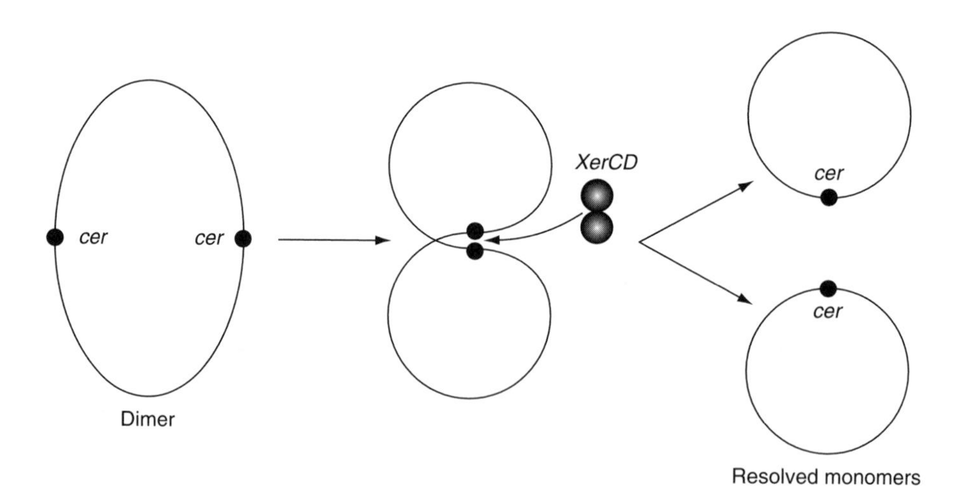

Figure 5.16 Resolution of plasmid dimers by site-specific recombination. XerCD causes site-specific recombination between two *cer* sites on a dimeric plasmid, leading to resolution into monomers

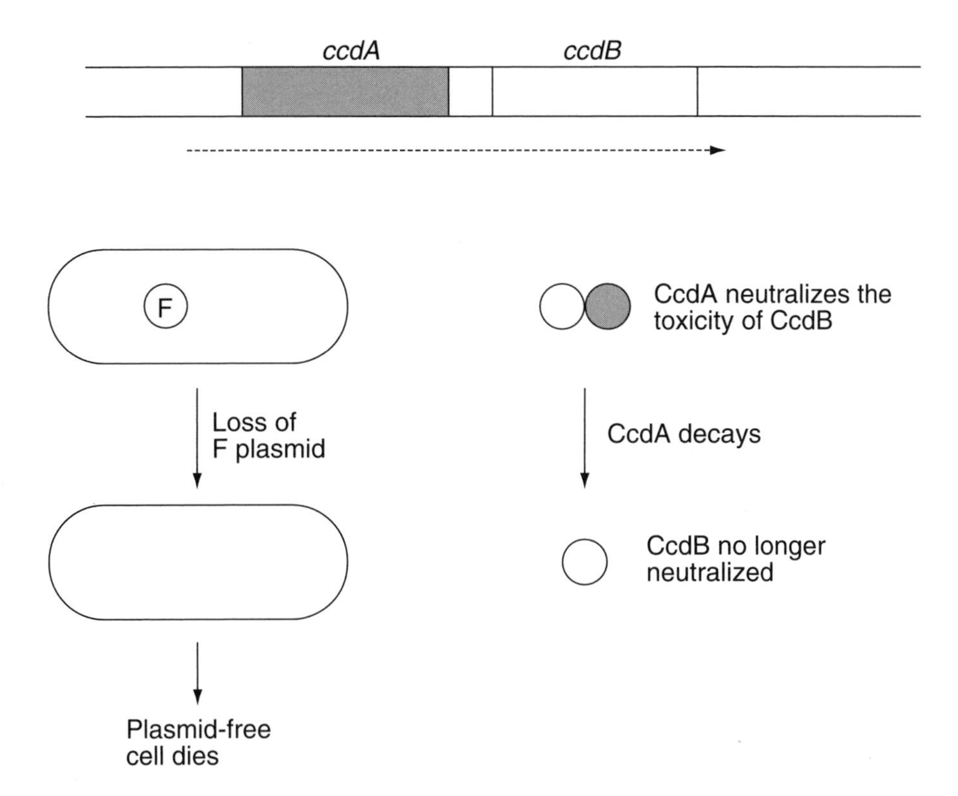

Figure 5.17 Role of *ccdAB* genes in maintenance of the F plasmid. The F plasmid-encoded CcdB protein is toxic, but is neutralized by the presence of CcdA. If the plasmid is lost, CcdA is rapidly destroyed exposing the cell to the toxic effects of the more stable CcdB protein

mediated supercoiling), while the (less stable) CcdA protein antagonizes the effect of CcdB. In a plasmid-free segregant, the CcdA product is destroyed by proteolytic enzymes, while the stable CcdB protein persists and kills the cell. Similar systems are found on a wide variety of unrelated plasmids.

Some plasmids possess an alternative mechanism for post-segregational killing, which works by regulating the expression of a toxin. For example, the plasmid R1 carries a gene, *hok* (*ho*st *k*illing), that codes for a small polypeptide that is toxic because of its effects on the cell membrane. In plasmid-containing cells, translation of *hok* mRNA is prevented by an antisense RNA molecule (*sok*, *s*uppression *o*f *k*illing) that is complementary to the leader region of the *hok* mRNA – it is transcribed from the same region of DNA but in the opposite direction. The *sok* RNA molecule decays rapidly, but the *hok* mRNA is very stable. So in the plasmid-free segregant, the *hok* mRNA persists and, in the absence of *sok* RNA, is translated to produce the toxin. A simplified model of this system is shown in Figure 5.18.

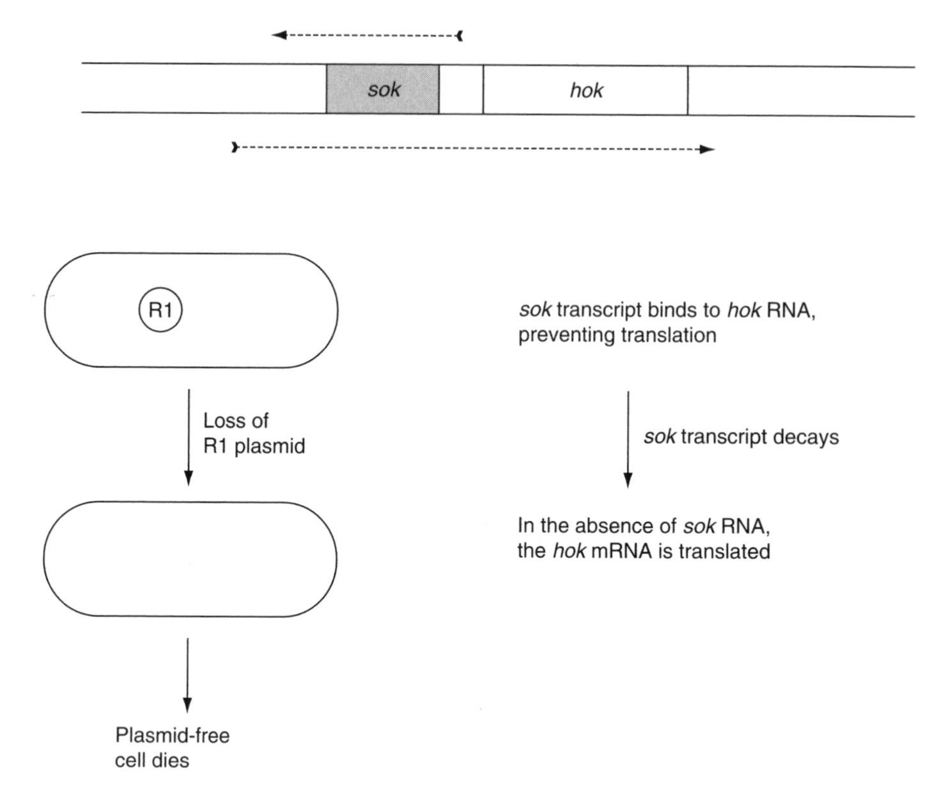

Figure 5.18 Role of *hok* and *sok* in maintenance of the R1 plasmid. The product of the R1 *hok* gene is toxic, but translation is prevented by the antisense *sok* RNA. If the plasmid is lost, the *sok* RNA decays rapidly, so the *hok* mRNA (which is more stable) can be translated, leading to the death of the cell

In summary, the stabilization of plasmids in partitioning rests on (1) active partitioning (for low copy plasmids) or random partitioning (for high copy-number plasmids), (2) resolution of dimers and multimers into monomeric plasmids and (3) post-segregational killing.

5.3.3 Differential growth rate

The third parameter is also important, i.e. is there any difference in the growth rate between cells that carry the plasmid and those that do not? If there is no difference, then a failure of partitioning at cell division will lead to only a slow increase in the proportion of plasmid-free cells. On the other hand, a substantial difference in growth rate will lead to a rapid elimination of the plasmid even if failure of partitioning occurs only rarely.

Differences in growth rate are expected to arise because of the metabolic load arising from replication of the plasmid and expression of its genes. With most wild-type plasmids, these demands are small, and hence the effect on growth rate is also small. On the other hand, genetically engineered plasmids are often maintained at very high copy numbers and express very high levels of a specific product. The burden on the cell is therefore very much greater and the problem of instability can become acute. Some ways of overcoming this problem are considered in Chapter 8.

5.4 Methods for studying plasmids

5.4.1 Associating a plasmid with a phenotype

The variability of a phenotype is often the first indication that a plasmid is involved. That is, some strains possess an unusual characteristic, which tends to be lost at a frequency higher than expected from mutation, and that loss is irreversible. With some plasmids it is possible to increase the rate at which the plasmid is lost by various treatments, such as acridine orange or growth at a higher temperature – a procedure known as 'curing' or plasmid elimination. It must be stressed that the terms 'elimination' and 'plasmid loss' do *not* mean the physical removal of the plasmid from a particular cell; the process works by interfering with the replication and/or partitioning of the plasmid so as to increase the rate at which plasmid-free segregants occur.

This evidence should be combined with detection of plasmid DNA by agarose gel electrophoresis (Figure 5.19; see also Box 2.2). The chromosomal DNA will show up as a rather diffuse band, since it is fragmented randomly by the extraction procedure, while any plasmid DNA will form separate, sharper bands at a position determined primarily by their size. The smaller the molecule, the faster it will run. Small plasmids will be found well ahead of the chromosomal DNA, while larger plasmids actually run slower than the chromosomal DNA. This may seem rather surprising (given that the chromosome is very much bigger than even the largest plasmid), but it should be remembered that the chromosome is broken into linear fragments while the plasmid will be present as intact circular molecules. Therefore they are not directly comparable.

The conformation of the plasmid will also affect its mobility. Usually the intact plasmid will be a covalently closed, supercoiled circle, which will migrate differently from a linear molecule of the same molecular weight (this must be taken into account when attempting to determine the size of a plasmid) and considerably faster than a nicked open circular form. In addition to these three major forms of the plasmid (supercoiled, nicked circles and linear), there may be dimers and higher multimers which will move more slowly. Since a single plasmid can give

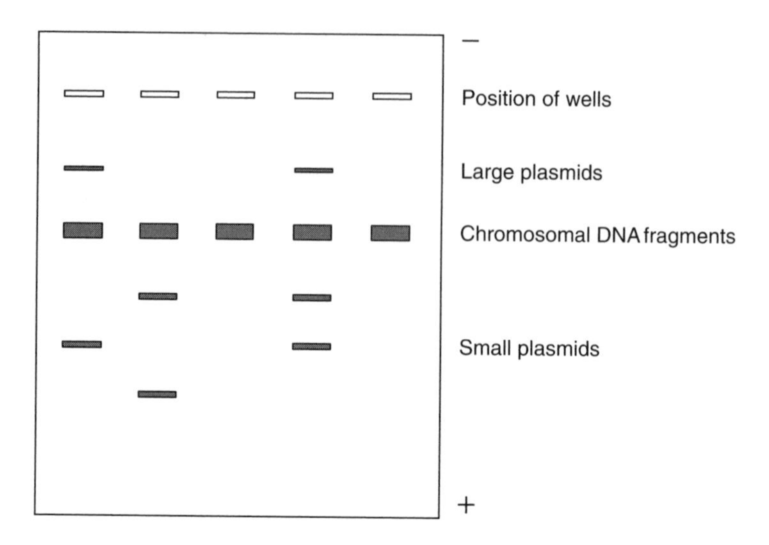

Figure 5.19 Demonstration of plasmids in cell extracts by agarose gel electrophoresis

rise to several bands, it is often difficult to determine with confidence how many plasmids are present in a given strain.

In order to study a plasmid in more depth it is necessary to purify it. Separating DNA from other cell components is relatively easy to accomplish. However, separating a plasmid from chromosomal DNA is less straightforward. Most commonly used methods rely on the different properties of the supercoiled circular plasmid and the linear fragments of chromosomal DNA. However, these procedures are not necessarily appropriate for larger plasmids, or for other forms of plasmid DNA such as the single-stranded and linear forms referred to earlier in this chapter.

The combination of the instability of the phenotype and the physical demonstration of plasmid DNA provides evidence that a plasmid is involved. The hypothesis can be strengthened by showing that colonies which have lost the characteristic in question have also lost the plasmid. But the evidence is still circumstantial. To obtain more conclusive evidence, the best procedure is to introduce the plasmid into a cell that does not have it (by conjugation or transformation; see Chapter 6), and to determine whether the acquisition of the plasmid leads to a corresponding change in the phenotype. Such experiments are easily carried out with selectable markers like antibiotic resistance. If the plasmid carries a resistance gene, the cells that have acquired the plasmid can be detected by simply plating them on a medium containing the antibiotic. In this way extremely rare events can be detected, e.g. one resistant transconjugant (or transformant) in 10^8 cells. Other characteristics are not so easy. To determine whether a plasmid carries genes that are necessary for virulence for example, it may be quite difficult to test it in such a simple manner.

5.4.2 Classification of plasmids

It is often useful to be able to compare plasmids identified in different bacterial isolates. This may for example provide information about the occurrence of a particular plasmid in strains causing an outbreak of infection in a hospital where antibiotic resistance is often a serious problem. The division into 'small' and 'large' plasmids is clearly much too crude to cope with the vast number of different plasmids that can be present in such a situation.

Analysis of the profile of antibiotic resistance genes carried by the plasmid has the virtue of being quick and easy to carry out, but it must be realized that plasmids are very fluid in their make-up; the presence of transposons (see Chapter 7) commonly leads to new resistance genes being acquired by a plasmid, even during the course of a specific outbreak. Conversely, plasmids can lose genes by deletions (as discussed earlier). Excessive reliance on the resistance pattern can therefore be misleading.

Incompatibility groups

Earlier in this chapter, it was shown that interactions between the replication control mechanisms of different plasmids may make them incompatible. This forms the basis of a widely used method of classifying plasmids. The test is carried out by transferring the unknown plasmid into each of a number of standard plasmid-carrying strains and testing for their incompatibility. A large number of incompatibility groups have been defined in this way; F for example belongs to the group IncFI, and is incompatible with plasmids such as ColV-K94. The R100 plasmid has a similar conjugation and transfer system but is compatible with F; it belongs to the group IncFII. There are a number of problems with this system, not the least of which is that it is extremely cumbersome. Also it is not possible to compare plasmids from different species, unless they can both be transferred to a common host. So there is one set of groups for *E. coli* plasmids, and another for plasmids from *Pseudomonas* species. However, the strength of the method is that it does classify plasmids on the basis of a fundamental characteristic – the nature of their replication control.

Host range

Many plasmids, including ColE1 and cloning vectors related to it, are only able to replicate in a limited range of bacterial hosts. However, this is not universally true and some plasmids have a remarkably broad host range. Notable amongst these are the P group of plasmids, such as RP4, which are able to replicate in some Gram-positive bacteria, as well as in most Gram-negatives. Plasmids such as RP4 are also promiscuous in the sense that not only can they replicate in a broad range of host bacteria, but they are also capable of promoting their own transmission by conjugation (see Chapter 6) between widely diverse bacterial species. Broad host-range plasmids can be useful as genetic tools (see Chapter 8), as well as their natural role in promoting gene flow between widely diverse bacterial species.

Molecular characterization

Plasmids can be compared more readily by the use of restriction endonucleases. Since these enzymes cut DNA at specific points, digestion of a plasmid will give rise to a characteristic set of fragments that can be separated on an agarose gel (Chapter 2). If two plasmids are the same, the restriction pattern will be identical. This procedure is however still subject to the same problem referred to above: plasmids can acquire additional genes by transposition or lose them by deletion. A different pattern of restriction fragments may therefore indicate the occurrence of events of this kind rather than the presence of unrelated plasmids.

Such an analysis can be extended by testing the ability of DNA fragments from one plasmid to hybridize to a second plasmid, which indicates the presence of related sequences. However, since two otherwise dissimilar plasmids may have acquired related antibiotic resistance genes (which will therefore cross-hybridize), these results also need to be interpreted with care.

Ultimately, of course, the best comparison of two plasmids at the molecular level is to determine the complete sequence of both plasmids.

6
Gene Transfer

The concept of the re-assortment of characteristics through sexual reproduction in animals and plants was a familiar one long before Mendel put it on a scientific footing. Not only do the features of individuals represent a combination of those of their parents (or grandparents), but the phenomenon has been used over the centuries to establish new strains of plants and animals that combine the best characteristics of different strains. How can we apply the same concept to organisms such as bacteria that do not exhibit sexual reproduction?

We now know that bacteria do exchange genetic information, not only in the laboratory but also in nature. There are three fundamentally distinct mechanisms by which such genetic transfer can occur.

(1) Transformation, in which a cell takes up isolated DNA molecules from the medium surrounding it.

(2) Conjugation, which involves the direct transfer of DNA from one cell to another.

(3) Transduction in which the transfer is mediated by bacterial viruses (bacteriophages).

Not all bacterial species exhibit all of these modes of genetic transfer. Conjugation is most readily demonstrated in Gram-negative bacteria but does occur in some Gram-positive genera such as *Streptomyces* and *Streptococcus*. Although some bacterial species are naturally transformable, in many other species transformation is only readily demonstrated after some form of artificial pre-treatment of the cells and therefore probably does not occur naturally in those organisms.

These mechanisms differ from true sexual reproduction in two main respects: there is no link with reproduction and the genetic contribution from the parents is

Molecular Genetics of Bacteria, 4th Edition by Jeremy Dale and Simon F. Park
 ISBN 0 470 85084 1 (cased) ISBN 0 470 85085 X (pbk)

unequal. The parents are thus referred to as donor and recipient cells; the recombinant progeny resemble the original recipient strain in most characteristics.

In this chapter the focus will be on the naturally-occurring transfer of genes between bacteria (horizontal gene transfer) and on the significance that this has for the evolution of bacteria. Mechanisms that are used in the laboratory, especially for genetic modification, and their application for both genetic manipulation and for genetic analysis, will be covered in Chapters 8 to 10.

6.1 Transformation

In a sense, it was the discovery of transformation, the uptake of DNA by a bacterial cell, that initiated the study of bacterial genetics and molecular biology as we know it today. It was in 1928 that Fred Griffith, working with pneumococcus (*Streptococcus pneumoniae*) discovered that avirulent strains could be restored to virulence by incubation with an extract from killed virulent cells. Sixteen years later, Avery, MacLeod and McCarty demonstrated that the 'transforming principle' was DNA, which established the role of DNA as the hereditary material of the bacterial cell. Transformation has been important in genetic analysis of some species and more recently (and to a much greater extent) because of its key role in gene cloning.

With the pneumococcus, cells spontaneously become competent to take up DNA. Such naturally-occurring transformation has been most studied in *Bacillus subtilis* and *Haemophilus influenzae* (as well as *Str. pneumoniae*) and was for some time thought to be limited to these and related species. It is now known to be much more widespread. In particular, transformation contributes extensively to the antigenic variation observed in the gonococcus (*Neisseria gonorrhoeae*) through the transfer of *pil* genes coding for the major protein subunit of the surface appendages (pili) by which the bacteria attach to epithelial cells. Although the number of species in which natural transformation has been demonstrated is still quite limited, it is likely that it occurs, albeit at a low level, in many other bacteria.

The details of the process vary between species, but some generalizations are possible. Competence generally occurs at a specific stage of growth, most commonly in late log phase, just as the cells are entering stationary phase. This may be a response to cell density rather than (or as well as) growth phase. For example, in *Bacillus subtilis*, some of the genes involved in the development of competence are also involved in the early stages of sporulation. The development of competence at this stage is associated not only with nutrient depletion but also with the accumulation of specific secreted products (competence factors) which act via a two-component regulatory system to stimulate the expression of other genes required for competence. Since the level of these competence factors is dependent on cell concentration, competence will only develop at high cell density. This is a form of quorum sensing, as described in Chapter 3, in which the response of an individual cell is governed by the concentration of bacteria in the surrounding medium.

Following the development of competence, double-stranded DNA fragments bind to receptors on the cell surface, but only one of the strands enters the cell. In some species, the process is selective for DNA from the same species, through a requirement for short species-specific sequences. For example, the uptake of DNA by the meningococcus (*Neisseria meningitidis*) is dependent on the presence of a specific 10-bp uptake sequence. The genome of *N. meningitidis* contains nearly 2000 copies of this sequence, which will only occur infrequently and by chance, in other genomes. Similarly, transformation of *Haemophilus influenzae* is facilitated by the presence of a 29-bp uptake sequence which occurs approximately 1500 times in the genome of *H. influenzae*. These organisms will therefore only be transformed efficiently with DNA from the same species.

On the other hand, *B. subtilis* and *Str. pneumoniae* can take up virtually any linear DNA molecule. But taking up the DNA is only the start. If the cell is to become transformed in a stable manner, the new DNA has to be replicated and inherited. As here fragments of chromosomal DNA (rather than plasmids) are being considered, replication of the DNA will only happen if the incoming DNA is recombined with the host chromosome. This requires homology between the transforming DNA and the recipient chromosome. This does not constitute an absolute barrier to transformation with DNA from other species. Provided there is enough similarity in some regions of the chromosome, those segments of DNA can still undergo recombination with the recipient chromosome. The closer the taxonomic relationship, the more likely it is that they will be sufficiently similar. One example of this, with considerable practical significance, is the development of resistance to penicillin in *Str. pneumoniae*. This appears to have occurred by the replacement of part of the genes coding for the penicillin target enzymes with corresponding DNA from naturally-resistant oral streptococci.

Natural transformation is of limited usefulness for artificial genetic modification of bacteria, mainly because it works best with linear DNA fragments rather than the circular plasmid DNA that is used in genetic modification. For introducing foreign genes into a bacterial host, various techniques are used to induce an artificial state of competence. Alternatively, a mixture of cells and DNA may be briefly subjected to a high voltage which enables the DNA to enter the cell (a process known as *electroporation*). Although the mechanisms involved are quite different, they all share the characteristic feature of the uptake of 'naked' DNA by the cells and are therefore also referred to as transformation. These methods are dealt with in Chapter 8.

6.2 Conjugation

Conjugation is the direct transmission of DNA from one bacterial cell to another. In most cases, this involves the transfer of plasmid DNA, although with some organisms chromosomal transfer can also occur. As with other modes of gene transfer in bacteria, there is a one-way transfer of DNA from one parent (donor) to the other (recipient).

In the simplest of cases, conjugation is achieved in the laboratory by mixing the two strains together and after a period of incubation to allow conjugation to occur, plating the mixture onto a medium that does not allow either parent to grow, but on which a transconjugant that contains genes from both parents will grow. For example, in the experiment illustrated in Figure 6.1, one strain (the donor) carries a plasmid that confers resistance to ampicillin, while the second strain does not have a plasmid but has a chromosomal mutation that makes it resistant to nalidixic acid. After incubating the mixed culture, a sample is plated onto a medium containing both antibiotics. Neither parent can grow on this medium, so the colonies that are observed are due to the transfer of a copy of the plasmid from the donor to a recipient cell. Although this event may be quite infrequent, the powerful selection provided by the two antibiotics means that plasmid transfer can be readily detected even if only, say, 1 in 10^6 recipients have received a copy of it.

Conjugation is most easily demonstrated amongst members of the Enterobacteriaceae and other Gram-negative bacteria (such as Vibrios and Pseudomonads). Several genera of Gram-positive bacteria possess reasonably well-characterized conjugation systems; these include *Streptomyces* species, which are commercially important as the major producers of antibiotics, the lactic streptococci, which are also commercially important because of their application to various aspects of the dairy products industry, and medically important bacteria such as *Enterococcus faecalis* (see later in this chapter).

The most obvious significance of conjugation is that it enables the transmission of plasmids from one strain to another. Since conjugation is not necessarily confined to members of the same species, this provides a route for genetic information to flow across wide taxonomic boundaries. One practical consequence is that plasmids that are present in the normal gut flora can be transmitted to infecting pathogens, which then become resistant to a range of different antibiotics.

6.2.1 Mechanism of conjugation

Formation of mating pairs

In the vast majority of cases, the occurrence of conjugation is dependent on the presence, in the donor strain, of a plasmid that carries the genes required for promoting DNA transfer. In *E. coli* and other Gram-negative bacteria, the donor cell carries appendages on the cell surface known as *pili*. These vary considerably in structure – for example the pilus specified by the F plasmid is long, thin and flexible, while the RP4 pilus is short, thicker and rigid. The pili make contact with

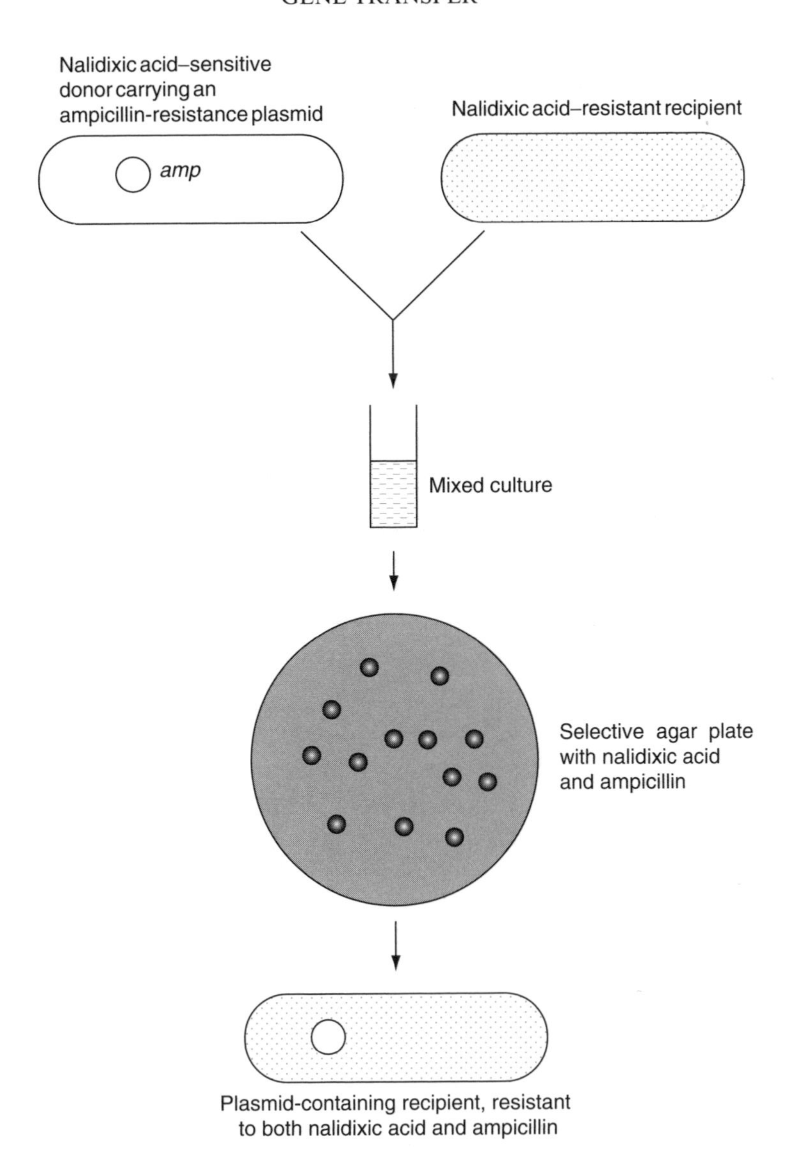

Figure 6.1 Conjugal transfer of a resistance plasmid. The donor strain is sensitive to nalidixic acid and carries a plasmid conferring ampicillin resistance (*amp*). The recipient is resistant to nalidixic acid, due to a chromosomal mutation, and sensitive to ampicillin. After growth of the mixed culture, plating on agar containing both ampicillin and nalidixic acid selects those recipients that have received the plasmid (transconjugants). The bacterial chromosome is omitted for clarity

receptors on the surface of the recipient cell, thus forming a mating pair (Figure 6.2). The pili then contract to bring the cells into intimate contact and a channel or pore is made through which the DNA passes from the donor to the recipient.

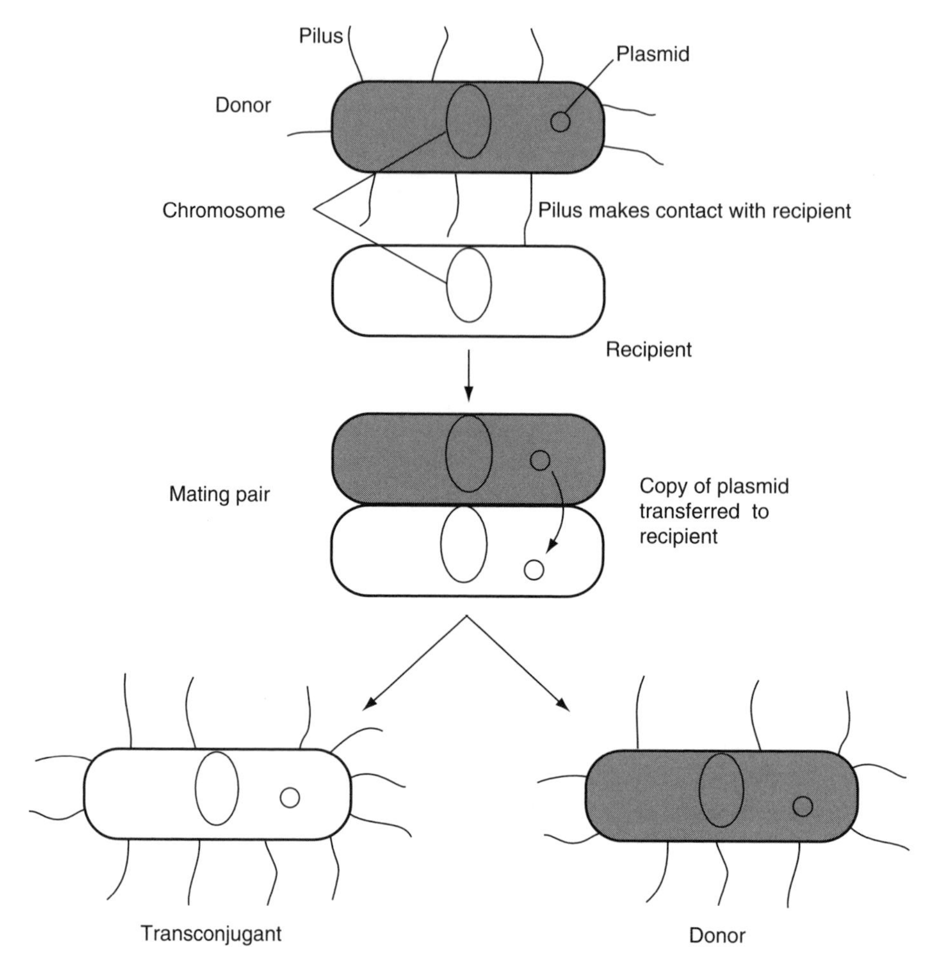

Figure 6.2 Transfer of DNA by conjugation

Interestingly, this mechanism has much in common with a protein secretion system (Type IV secretion, see Chapter 1) which is used by some bacteria to deliver protein toxins directly into host cells. Other mechanisms of conjugation that are important in Gram-positive bacteria will be discussed later in this chapter.

Transfer of DNA

The transfer of plasmid DNA from the donor to the recipient (Figure 6.3) is initiated by a protein which makes a single-strand break (nick) at a specific site in the DNA, known as the origin of transfer (*oriT*). A plasmid-encoded helicase

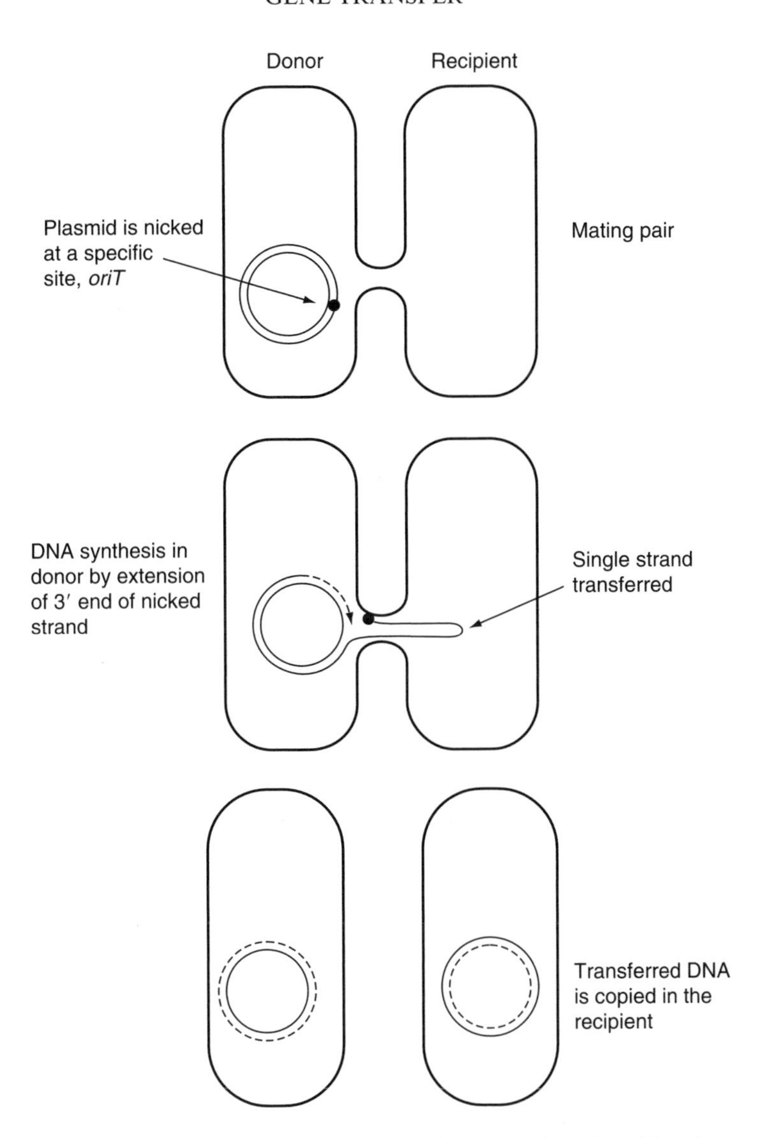

Figure 6.3 Mechanism of plasmid DNA transfer by conjugation. For clarity, only the plasmid is shown

unwinds the plasmid DNA and the single nicked strand is transferred to the recipient starting with the 5′ end generated by the nick. Concurrently, the free 3′ end of the nicked strand is extended to replace the DNA transferred, by a process known as rolling circle replication which is analogous to the replication of single-stranded plasmids and bacteriophages as described in Chapters 4 and 5. The nicking protein remains attached to the 5′ end of the transferred DNA. DNA synthesis in the recipient converts the transferred single strand into a double-stranded molecule.

Note that this is a replicative process. Thus although there is said to be a transfer of the plasmid from one cell to another, what is really meant is a transfer of *a copy of* the plasmid. The donor strain still has a copy of the plasmid and can indulge in further mating with another recipient. It is also worth noting that after conjugation the recipient cell has a copy of the plasmid and it can transfer a copy to another recipient cell. The consequence can be an epidemic spread of the plasmid through the mixed population.

Mobilization and chromosomal transfer

Not all plasmids are capable of achieving this transfer to another cell unaided; those that can are known as conjugative plasmids. In some cases a conjugative plasmid is able to promote the transfer of (*mobilize*) a second otherwise non-conjugative plasmid from the same donor cell. This does not happen by chance and not all non-conjugative plasmids can be mobilized.

In order to understand mobilization the plasmid ColEI can be taken as an example (see Figure 5.3). Mobilization involves the *mob* gene, which encodes a specific nuclease, and the *bom* site (=*oriT*, the origin of transfer), where the Mob nuclease makes a nick in the DNA. ColE1 has the genes needed for DNA transfer but it does not carry the genes required for mating-pair formation. The presence of another (conjugative) plasmid enables the donor to form mating pairs with the recipient cell and ColE1 can then use its own machinery to carry out the DNA transfer.

Some plasmids which can be mobilized do not carry a *mob* gene. Mobilization then depends on the ability of the Mob nuclease of the conjugative plasmid to recognize the *bom* site on the plasmid to be mobilized. This only works if the two plasmids are closely related. On the other hand, the *bom* site is essential for mobilization. This is an important factor in genetic modification as removal of the *bom* site from a plasmid vector ensures that the modified plasmids cannot be transferred to other bacterial strains (see Chapter 8).

In most cases, the DNA that is transferred from the donor to the recipient consists merely of a copy of the plasmid. However, some types of plasmids can also promote transfer of chromosomal DNA. The first of these to be discovered, and the best known, is the F (fertility) plasmid of *E. coli*, but similar systems exist in other species, notably *Pseudomonas aeruginosa*. In some cases, as described below, this involves integration of the conjugative plasmid into the donor chromosome (so that the chromosome is in effect transferred as part of the plasmid). However, in many cases chromosomal transfer occurs without any stable association with the plasmid, possibly by a mechanism analogous to mobilization of a non-conjugative plasmid.

When a plasmid is transferred from one cell to another by conjugation, the complete plasmid is transferred. In contrast, chromosomal transfer does not involve a complete intact copy of the chromosome. One reason for this is the time required for transfer. The process is less efficient than normal DNA replication and transfer of the whole chromosome would take about 100 min (in *E. coli*). The mating pair very rarely remains together this long. In contrast, a plasmid of say 40 kb is equivalent to 1 per cent of the length of the chromosome, – thus the transfer of the plasmid would be expected to be completed in 1 min. The incomplete transfer of the chromosomal DNA means that it does not constitute an intact replicon in the recipient cell. For this fragment to be replicated and inherited it must recombine with the host chromosome, usually replacing the corresponding recipient genes in the process.

6.2.2 The F plasmid

The F plasmid was originally discovered during attempts to demonstrate genetic exchange in *E. coli* by mixed culture of two auxotrophic strains, so that plating onto minimal medium would only permit recombinants to grow. It was shown quite early on that the recombinants were all derived from one of the parental strains and that a one-way transfer of information was therefore involved, from the donor ('male') to the recipient ('female'). The donor strains carry the F plasmid (F^+) while the recipients are F^-. One feature of this system which must have seemed curious at the time is that co-cultivation of an F^+ and an F^- strain resulted in the 'females' being converted into 'males'! This is of course due to the transmission of the F plasmid itself which occurs at a high frequency, in contrast to the transfer of chromosomal markers which is very inefficient with an F^+ donor.

Hfr strains

The usefulness of conjugation for genetic analysis was enormously enhanced by the discovery of donor strains in which chromosomal DNA transfer occurred much more commonly. These Hfr (High Frequency of Recombination) strains arise by integration of the F plasmid into the bacterial chromosome. An additional characteristic of an Hfr strain is that chromosomal transfer starts from a defined point and proceeds in a specific direction. The origin of transfer is determined by the site of insertion of the F plasmid and the direction is governed by the orientation of the inserted plasmid. This can be made clear by re-labelling the circular molecule in Figure 6.3 as chromosomal DNA containing an integrated F plasmid. Transfer is thus initiated from the *oriT* site on the integrated plasmid but now results in the transfer of a copy of the bacterial chromosome rather than just the plasmid. An F^+ donor, in contrast, transfers genes in a more or less random manner, since transfer does not start from a defined point on the chromosome. The combination of the partial transfer of chromosomal DNA with the ordered transfer of genes made conjugation an important tool in the mapping of bacterial chromosomes, which is covered more fully in Chapter 10.

Integration and excision of F: formation of F′ plasmids

Integration of the F plasmid occurs by recombination between a sequence on the plasmid and a chromosomal site. In Figure 6.4a this is shown occurring adjacent to the *lac* operon, but it can occur at a large number of sites. After integration, the F plasmid is found in a linear form at the integration site (Figure 6.4c). Integration is reversible since recombination between sites at the ends of the integrated plasmid will lead to its excision from the chromosome as an independent circular molecule (Figure 6.4, steps c–a). However, it is possible for this excision to occur inaccurately, i.e. recombination occurs at a different site. If this happens, the resulting plasmid will have incorporated a small amount of bacterial DNA. This forms what is known as an F′ (F-prime) plasmid. Figure 6.4c–e shows the formation of an F′*lac* plasmid, where the recombination event leading to excision has occurred at a site beyond the *lac* operon, rather than between the sites flanking the integrated F plasmid. The *lac* operon has therefore become incorporated into the plasmid (at the same time generating a deletion in the chromosome) and will be transferred with the plasmid to a recipient strain. This is one mechanism whereby a plasmid can acquire additional genes from a bacterial chromosome and transfer them to another strain or species.

Before the advent of gene cloning, F′ plasmids were useful in a number of ways, including the isolation of specific genes and their transfer to other host strains. This enabled the creation of *partial diploids*, i.e. strains with one copy of a specific gene on the plasmid in addition to the chromosomal copy. The use of partial diploids for the study of the regulation of the *lac* operon was described in Chapter 3, especially in distinguishing regulatory genes that operate *in trans* (i.e. they influence the expression of a gene on a different molecule) from those that only affect the genes to which they are attached (i.e. they operate *in cis*). Such experiments are still relevant, but would now use recombinant plasmids produced *in vitro*.

6.2.3 Conjugation in other bacteria

The above description of conjugation applies mainly to Gram-negative bacteria such as *E. coli* and *Pseudomonas*. Many Gram-positive species, ranging from *Streptomyces* to *Enterococcus*, also possess plasmids that are transmissible by conjugation and in many cases the mechanism of DNA transfer is quite similar to that described above. However there are substantial differences in other respects.

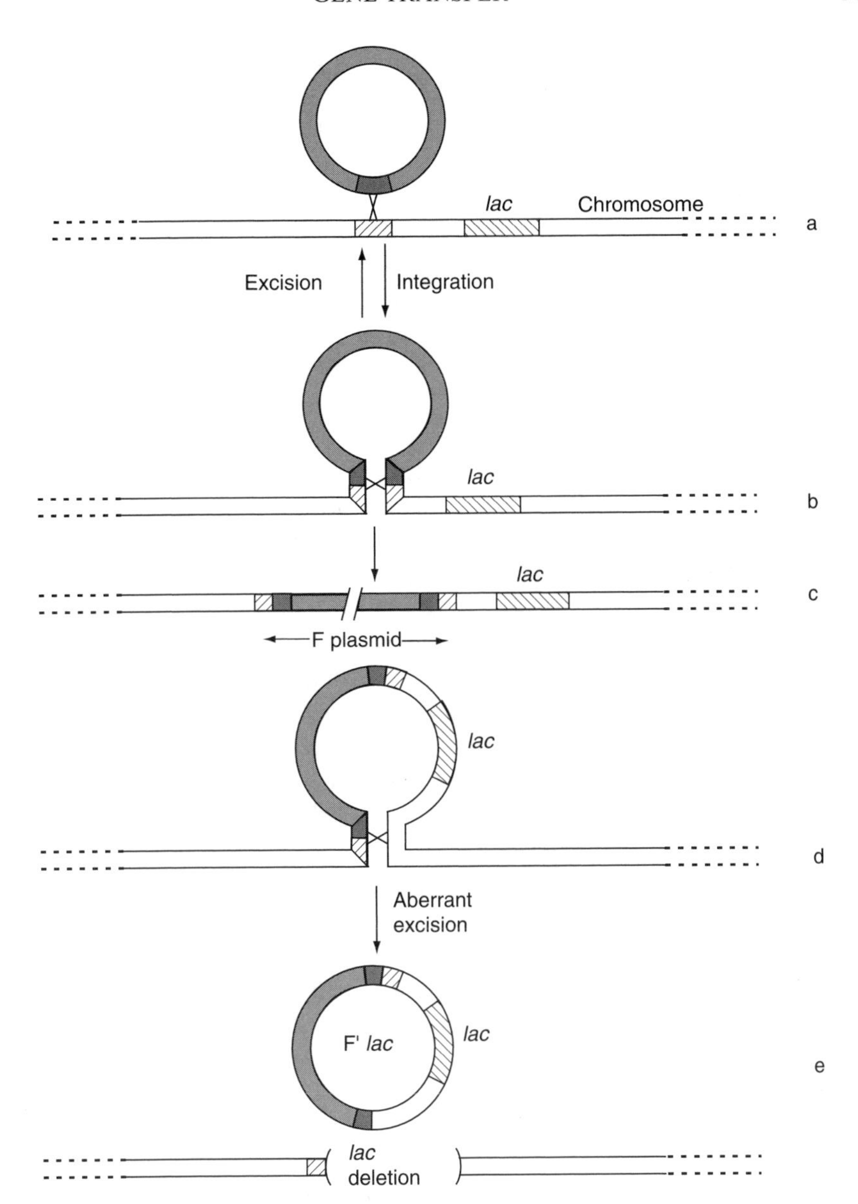

Figure 6.4 Integration and excision of F plasmid. Integration occurs by recombination with a chromosomal site (a–c). Accurate excision occurs by recombination at the same site, but recombination with a different chromosomal site results in incorporation of adjacent chromosomal DNA (in this case the *lac* gene) into the plasmid (d,e), forming an F′ plasmid and causing a chromosomal deletion

In general, the number of genes required for conjugative transfer, in some cases as few as five genes, is very much less than in Gram-negative bacteria where 20 or more genes are needed. Conjugative plasmids in Gram-positive bacteria can therefore be considerably smaller. One reason for a smaller number of genes being required is that there seems to be no need for production of a pilus. This is probably, at least in part, a reflection of the different cell-wall architecture in Gram-positive bacteria which lack the outer membrane characteristic of the Gram-negatives.

One group of Gram-positive bacteria where conjugation systems have been studied in detail are the enterococci, principally *Enterococcus faecalis*. Some strains of *E. faecalis* secrete diffusible peptides that have a pheromone-like action that can stimulate the expression of the transfer (*tra*) genes of a specific plasmid in a neighbouring cell. Note that, rather surprisingly, it is the *recipient* cell that produces the pheromones. The donor cell, carrying the plasmid, has a plasmid-encoded receptor on the cell surface to which the pheromone binds. Different types of plasmid code for different receptors and are therefore stimulated by different pheromones. However the recipient produces a range of pheromones and is therefore capable of mating with cells carrying different plasmids.

After the pheromone has bound to the cell-surface receptor it is transported into the cytoplasm, by a specific transport protein, where it interacts with a protein called TraA. This protein is a repressor of the *tra* genes on the plasmid and the binding of the peptide to it relieves that repression, thus stimulating expression of the *tra* genes. One result is the formation of aggregation products which cause the formation of a mating aggregate containing donor and recipient cells bound together. A further consequence of expression of the *tra* genes is stimulation of the events needed for transfer of the plasmid which occurs by a mechanism similar to that described previously.

One advantage of this system is that the cells containing the plasmid do not express the genes needed for plasmid transfer unless there is a suitable recipient in the vicinity. Not only does this reduce the metabolic load on the cell but it also means that they are not expressing surface antigens (such as conjugative pili) that could be recognized by the host immune system.

Conjugative transposons

E. faecalis also provides an example of an exception to the general rule that conjugation is plasmid-mediated. Some strains of *E. faecalis* contain a *transposon* known as Tn*916*. Transposons will be covered more generally in Chapter 7, but for the moment it is sufficient to know that they are mobile genetic elements which are able to move from one DNA site to another. What sets Tn*916* apart from other transposons is its ability to transfer from one cell to another by conjugation.

Conjugative transposons such as Tn*916* differ from plasmids in that they are replicated and inherited as part of the chromosome. There is no stable independently replicating form as there is with a plasmid. However, closer inspection of the method of transfer (Figure 6.5) shows that there is a significant similarity to

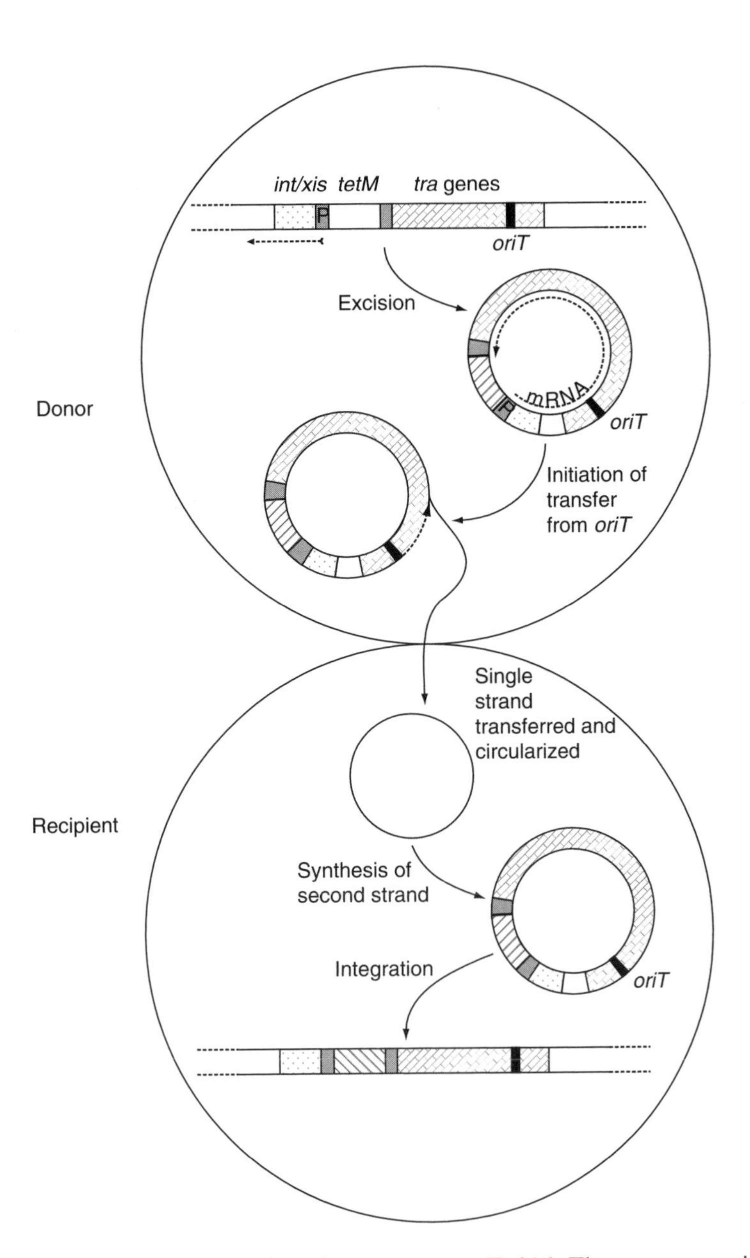

Figure 6.5 Transfer of the conjugative transposon Tn916. The transposon is excised from the chromosome, using the Int and Xis enzymes and circularized. This enables the *tra* genes to be expressed from the promoter shown and conjugative transfer is initiated from *oriT*. In the recipient, the transferred DNA is circularized, the second strand is made and the transposon is integrated into the chromosome

plasmid transfer. In particular, Tn*916* contains a origin of transfer (*oriT*) which is quite similar to that found in many plasmids. The first step in transfer is the excision of the transposon from the chromosome, using transposon-encoded enzymes (Int and Xis) which are related to those responsible for the integration and excision of bacteriophage lambda (see Chapter 4). This produces a circular molecule that resembles a plasmid in all but one vital feature – it does not have an origin of replication so is unable to be copied in the normal way. However since it does have an *oriT* site and carries the *tra* genes needed for conjugal transfer, it can be transferred to a recipient cell.

As with the plasmid transfer systems described previously, transfer of Tn*916* involves single-stranded DNA synthesis initiated at *oriT* and transfer of the displaced strand to the recipient. The transferred single strand is then circularized and converted to a double-stranded circular form which is inserted randomly into the recipient chromosome by the action of the integrase.

One additional feature of Tn*916* is worth considering. If transfer occurred *without* excision from the chromosome then mobilization of the chromosome would be expected to occur with incomplete transfer of the transposon. Transfer would start from *oriT* and would have to work right round the chromosome before reaching the rest of the transposon. This does not seem to happen. The reason is that the promoter for expression of the *tra* genes is found towards the left-hand end of the transposon (in Figure 6.5) and faces away from the *tra* genes. In the integrated linear form the *tra* genes will not be expressed. However, when the transposon is excised from the chromosome and circularized, this brings the promoter into the correct position and orientation for transcription of the *tra* genes. They will therefore be expressed from the circular intermediate, but not from the integrated form. This ensures that the transfer system will only be activated after excision has occurred.

Tn*916* is the prototype of a family of related conjugative transposons that are especially widespread in Gram-positive cocci, although related elements also occur in Gram-negative bacteria (e.g. *Bacteroides*). For many of these elements, including Tn*916*, conjugative transmission is promiscuous in that they can transfer to other species or genera. It can be assumed therefore that conjugative transposons have played a significant role in the dissemination of genetic material, including antibiotic resistance genes, throughout the bacterial kingdom. In particular. many of these transposons, including Tn*916*, carry a tetracycline-resistance gene (*tetM*) which is found in a wide range of bacterial species, suggesting that they have played a role in the dispersal of this particular gene.

6.3 Transduction

Transduction is the phage-mediated transfer of genetic material. The key step in transduction is the packaging of DNA into the phage heads during lytic growth of the phage (see Chapter 4). This process is normally highly specific for phage DNA. However, with some phages, errors can be made and fragments of bacterial DNA (produced by phage-mediated degradation of the host chromosome) are occasionally packaged by mistake leading to phage-like particles that contain a segment of bacterial genome (see Figure 6.6). These *transducing particles* are

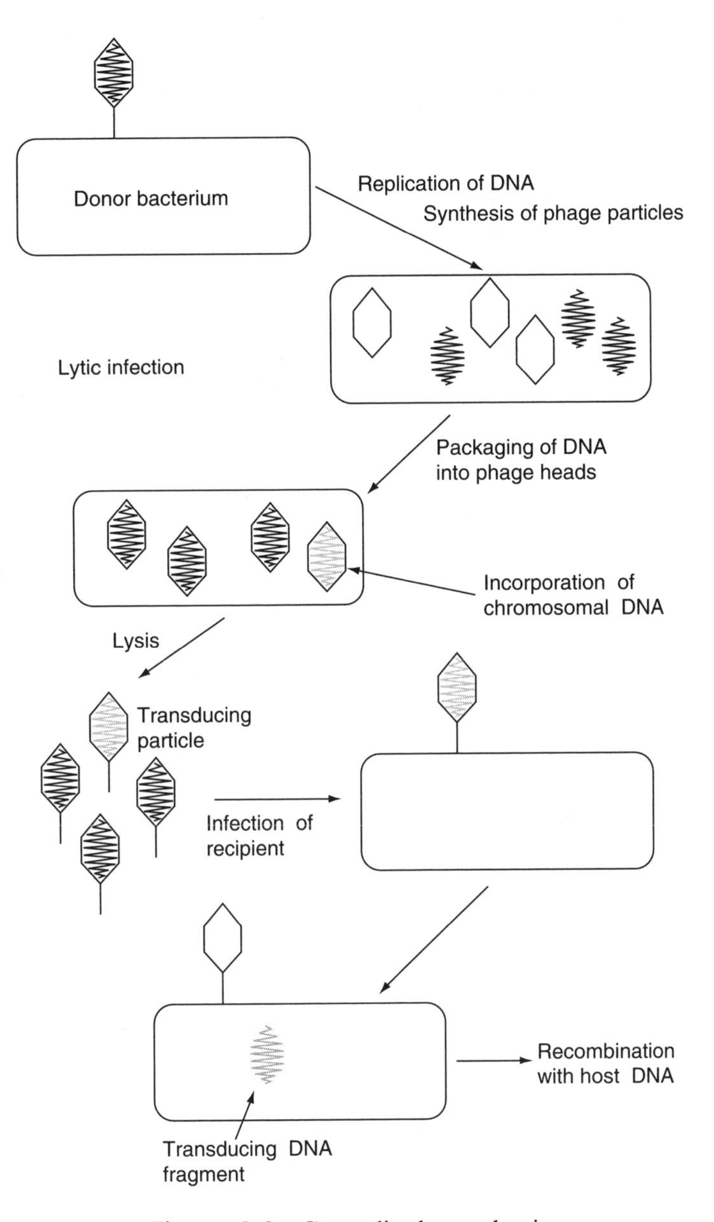

Figure 6.6 Generalized transduction

capable of infecting a recipient cell, since the information necessary for attachment and injection of DNA is carried by the proteins of the phage particle, irrespective of the nucleic acid it contains. The transduced segment of DNA will therefore be injected into the new host cell.

Not all bacteriophages are capable of carrying out transduction. The basic requirements of an effective transducing phage are that infection should result in an appropriate level of degradation of the chromosomal DNA to form suitably sized fragments at the right time for packaging and that the specificity of the packaging process should be comparatively low.

In some cases, the transduced DNA is a bacterial plasmid, in which case the injected DNA molecule is capable of being replicated and inherited. More commonly the DNA incorporated into the transducing particle is a fragment of chromosomal DNA which will be unable to replicate in the recipient cell. For it to be replicated and inherited, it must be incorporated into the recipient chromosome (by homologous recombination), as is the case with other mechanisms of gene transfer.

This process is known as *generalized* transduction (as opposed to specialized transduction – see below) since essentially any gene has an equal chance of being transduced.

6.3.1 Specialized transduction

As described in Chapter 4, some phages (temperate phages) are able to establish a state known as lysogeny, in which expression of phage genes and replication of the phage is repressed. In many cases the prophage is inserted into the bacterial DNA and replicates as part of the chromosome. When lysogeny breaks down and the phage enters the lytic cycle, it is excised from the chromosome by recombination between sequences at each end of the integrated prophage. If this recombination event happens in the wrong place, an adjacent region of bacterial DNA is incorporated into the phage DNA. All the progeny of this phage will then contain this bacterial gene which will therefore be transduced at a very high frequency (effectively 100 per cent per phage particle) once the transducing phage has been isolated. Since the DNA transferred is limited to a very small region of the chromosome, the phenomenon is known as *specialized* (or restricted) transduction. This is very similar to the formation of F′ plasmids referred to earlier (see Figure 6.4). As with the F′ plasmids, it is now much easier to add genes to λ DNA by creating recombinants *in vitro* (Chapter 8).

Another phage that has been employed in a similar way is the phage Mu (see Chapter 7) which has the advantage of inserting at multiple sites in the chromosome by a transposon-like mechanism. It is therefore much easier to create a wide range of specialized transducing phages with Mu which can be used both in genetic mapping and in mutagenesis.

6.4 Recombination

6.4.1 General (homologous) recombination

A common feature of all the forms of gene transfer between bacteria, except for the transfer of plasmids (which can replicate independently), is the requirement for the transferred piece of DNA to be inserted into the recipient chromosome by breaking both DNA molecules, crossing them over and rejoining them. This process, known as *recombination*, was introduced briefly in Chapter 2. There are several different forms of recombination, but the mechanisms that require the presence of homologous regions of DNA which must be highly similar but do not have to be identical are of specific interest in this context. It is therefore known as *homologous recombination*. Of the alternative forms of recombination, site-specific recombination is particularly important, for example in the integration and excision of bacteriophage λ (Chapter 4) and conjugative transposons as described above.

It should be noted that recombination mechanisms have other roles within the cell apart from the incorporation of foreign DNA. In particular, recombination mechanisms are involved with some types of DNA repair (see Chapter 2). These may actually be of more fundamental importance to the cell and may be the real reason why bacteria have evolved to contain several mechanisms for recombining DNA molecules.

A model of the recombination process

One model of the process of homologous recombination envisages firstly a pairing of the two DNA molecules in the homologous region (Figure 6.7). This is followed (ii) by a nick in one of the strands, which leads to that strand displacing part of the corresponding strand from the second molecule. The displaced strand is in turn nicked (iii) to produce an intermediate form with partially exchanged strands and the nicks are sealed to produce a structure with interlinked strands.

In Figure 6.8, structure iii is redrawn in alternative forms, first by bending the arms to produce the X-shaped structure iiib and then by rotating the lower half by 180° yielding the structure iiic, which is known as a Holliday junction, after Robin Holliday who first suggested the model from which this scheme is derived. This structure can be resolved by cutting the DNA strands in structure iiic at the positions marked with arrows. Ligation of the ends will then produce the recombinant structures shown in iv. (One of the simplifications used in this representation is the omission of other pathways that lead to alternative products). These structures contain some genetic markers from each parent and are therefore recombinant in both the genetic and molecular senses. Note that in this diagram a short region, containing the marker Q/q, is a heteroduplex, i.e. one strand is from one parent and the second strand is from the other parent. This

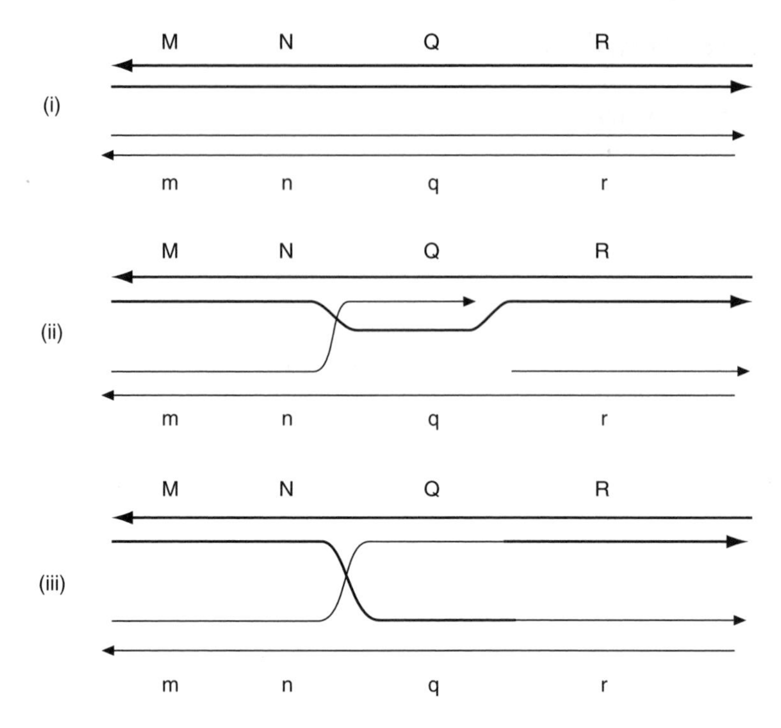

Figure 6.7 Initial stages of homologous recombination. (i) Pairing of the homologous regions. (ii) Nicked strand invades the opposite DNA molecule, displacing the corresponding strand. (iii) The displaced strand is nicked and the exchanged strands are re-joined

heteroduplex will either be repaired (i.e. q will be converted to Q, or vice versa), or if replication occurs first, the progeny will be mixed for this character.

Enzymes involved in recombination

One of the key enzymes in this process is the RecA protein which was described in Chapter 2 as playing a key role in the induction of the SOS response. In the context of recombination however, its role is to stimulate the interaction between the recombining DNA molecules. RecA protein can polymerize on DNA strands forming regular helical filaments in which the DNA helix is in a stretched conformation, thus facilitating an interaction with another DNA molecule. A second protein which is involved is an endonuclease with three subunits coded for by the *recB, recC* and *recD* genes (and hence known as the RecBCD endonuclease). This is a multifunctional enzyme with both endonuclease and exonuclease activity and is also able to unwind DNA molecules to provide the necessary single-stranded regions. As it unwinds the DNA, one strand is degraded, until the enzyme reaches a specific sequence known as a *chi* (χ) site. Further nuclease

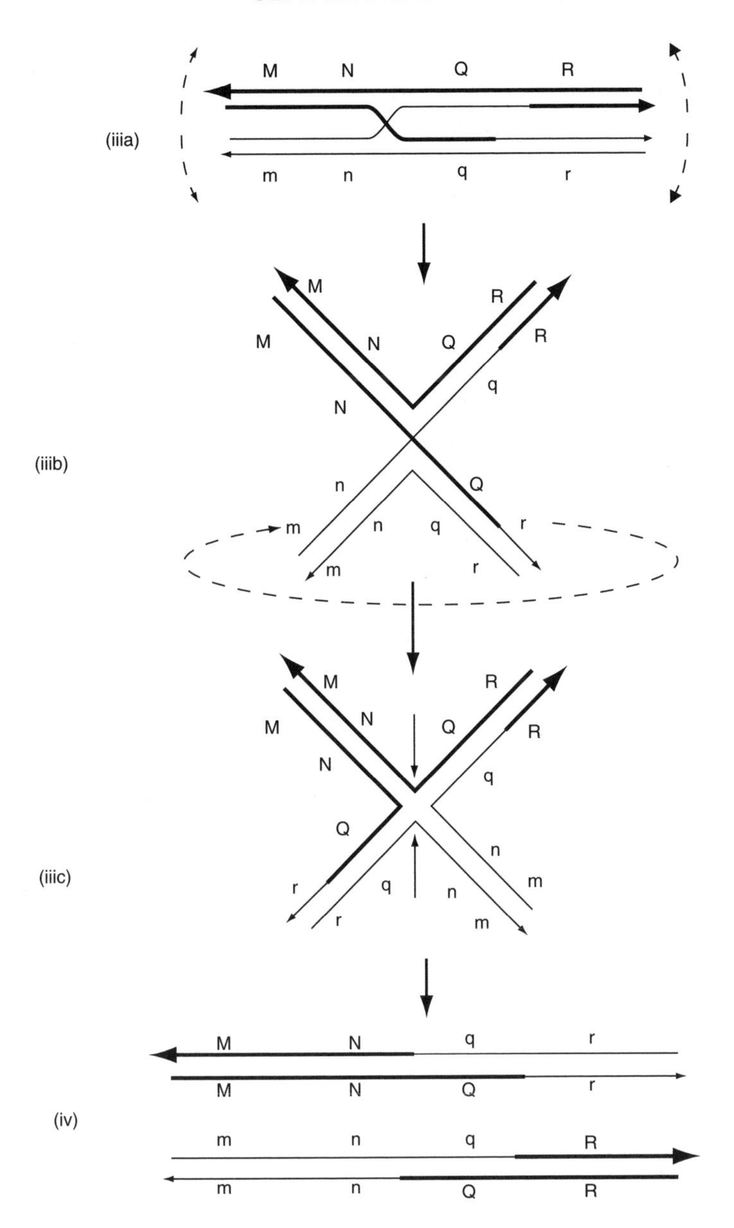

Figure 6.8 Homologous recombination: the Holliday junction. Structure iii from Figure 6.7 (iiia) is bent to an X shape (iiib) and the lower half is rotated to produce the Holliday junction (iiic). Resolution occurs by cutting the DNA at the arrowed points, producing the recombinant molecules shown (iv)

degradation is then inhibited, leaving a single-stranded tail that is able to participate in strand invasion (with the assistance of RecA). These χ sites are therefore hot-spots for recombination. In *E. coli*, this sequence is

5′GCTGGTGG3′ (but may be different in other bacteria). An eight-base sequence would be expected to occur within 65 kb on average if randomly distributed and most of the fragments generated during chromosome transfer by conjugation will be large enough to be likely to contain a χ site. However, when smaller fragments are involved the absence of a χ site in the DNA may limit the amount of recombination observed. This may be the case during transduction for example, and even more so during genetic manipulation experiments such as gene replacement (see Chapter 10).

Three proteins, RuvA, RuvB and RuvC, are responsible for events at the Holliday junction. RuvA binds to the Holliday junction and stabilizes the structure needed for the subsequent events, while RuvB is a helicase that unwinds the adjacent DNA, enabling the junction to migrate along the DNA (thus increasing the extent of the heteroduplex). RuvC is the nuclease responsible for cutting the DNA strands as required for resolution of the Holliday junction.

A different pathway (although still RecA-dependent) is required for repair of single-stranded gaps in the DNA, as described in Chapter 2. RecBCD is not able to participate in this system which uses instead RecF and several other proteins to prepare the single-stranded DNA for the loading of RecA which is needed for invasion of the sister strand.

Consequences of recombination

Finally, it is necessary to relate these mechanisms back to the events described earlier in this chapter. For recombination between a linear DNA fragment (introduced by transformation, transduction or conjugation) and the recipient chromosome, it is necessary for two such events to occur (see Figure 6.9) so that a portion of the linear fragment will become integrated into the circular chromosome replacing the corresponding region of the chromosome.

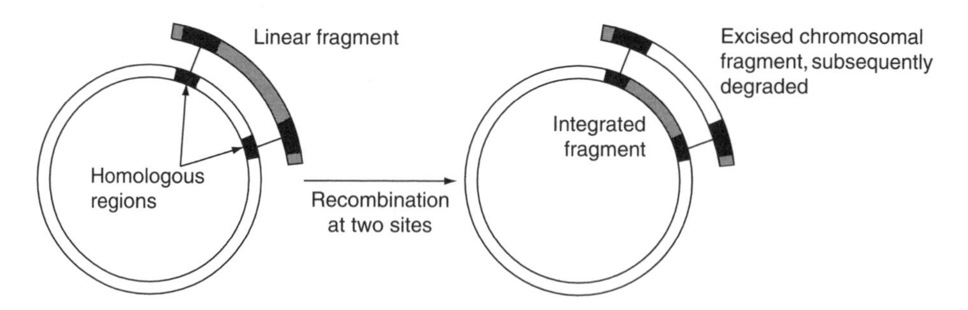

Figure 6.9 Recombination between a linear DNA molecule and a circular molecule. Recombination at two sites leads to replacement of a portion of the circular molecule. The reaction therefore requires at least two regions of homology

On the other hand, if both participating molecules are circular (e.g. two plasmids or a plasmid and the chromosome), then a single recombination event will suffice (see Figure 6.10) producing a fusion of the two original circles. This may be a reversible event. Recombination between the two ends of the inserted plasmid will lead to excision of the plasmid, as discussed earlier in relation to the integration of the F plasmid and also the integration and excision of phage λ (Chapter 4).

In this discussion it has been assumed that two different molecules were involved, i.e. *intermolecular* recombination. What happens if the two homologous regions are on the same molecule, i.e. recombination is *intramolecular*? This will happen if there are two copies of a repetitive element such as an insertion sequence (see Chapter 7). The consequences will depend on the relative orientation of the two homologous sequences. As can be seen from Figure 6.11a, if the homologous regions are in the same orientation (direct repeats), recombination between them will result in separation into two separate circular molecules. (This can be followed by tracing the course of the DNA in the intermediate form where the two regions are paired, starting from point A and switching strands when the paired region is reached again. The course of the DNA thus misses B and C and continues straight to D and back to A). In general, one of these circular molecules will not contain a replication origin and so will be lost. The consequence is therefore a deletion of that portion of the original DNA.

On the other hand, if the homologous regions are in opposite orientations (inverted repeats), as in Figure 6.11b, recombination leads not to separation of two molecules but to inversion of the region between the inverted repeats. The circular molecule remains intact.

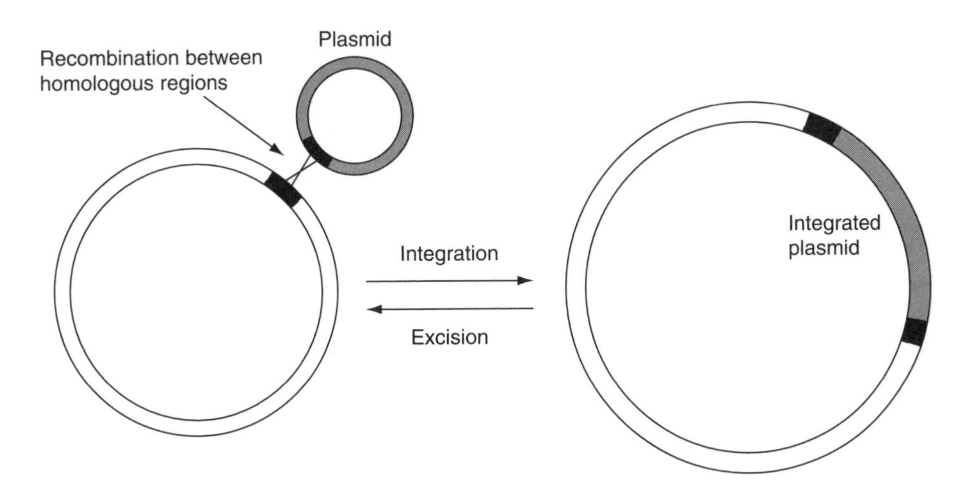

Figure 6.10 Recombination between two circular DNA molecules leads to integration. Recombination at a single site (a single crossover) is sufficient for integration

(a) Recombination between direct repeats leads to deletion

A B C D

C B A D

C B

Deletion of B-C

A D

(b) Recombination between inverted repeats leads to inversion

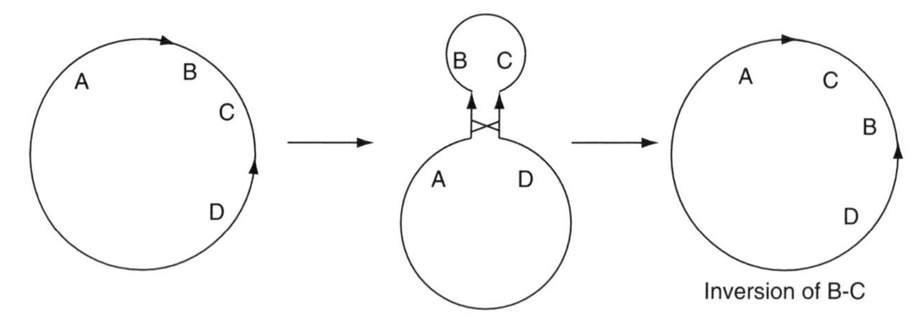

Figure 6.11 Intramolecular recombination between repeat sequences. (a) Recombination between two direct repeats (arrowheads) produces two separate molecules, i.e. deletion of B–C (if this structure is unable to replicate), or resolution, if B–C forms a viable replicon. (b) Recombination between two inverted repeats leads to inversion of the region (B–C) between the repeated sequences

The presence of repetitive elements can cause deletions or inversions of chromosomal regions in this way, as well as being a cause of plasmid rearrangements as described in Chapter 5. For example, if the chromosome contains two copies of an insertion sequence in the same orientation and quite close together, then recombination between the two IS elements will lead to deletion of the region of the chromosome between them. This can be a significant cause of variation between bacterial strains, as in *Mycobacterium tuberculosis* where much of the variation between strains arises from deletions due to recombination between insertion sequences.

6.4.2 Site-specific and non-homologous (illegitimate) recombination

Recombination between DNA molecules can occur in a variety of other ways which are not dependent on the presence of extensive regions of homology nor on the action of RecA. Examples of these, dealt with elsewhere in this book, include the integration of bacteriophage λ DNA into the chromosome, which involves a site-specific recombination between a defined sequence on the phage DNA and a specific chromosomal site (Chapter 4) and the transposition of mobile elements (insertion sequences and transposons) as described in Chapter 7.

6.5 Mosaic genes and chromosome plasticity

Gene transfer and recombination are not confined to whole genes, but can also involve parts of genes. If the transferred gene segment comes from a different species, this will result in a *mosaic* gene in which one segment is radically different from that normally seen (see Figure 6.12). One example, encountered earlier in this chapter, is the development of penicillin resistance in *Str. pneumoniae*. Examination of the *pbp* genes (encoding the penicillin-binding proteins which are the target for penicillin action) in penicillin-resistant strains shows that some regions are substantially different from the corresponding sequences in penicillin-sensitive strains. The differences are too great to be explained by simple mutation. However these regions are highly similar to sequences in the naturally resistant streptococci that are found in the mouth, such as *Streptococcus mitis*. From

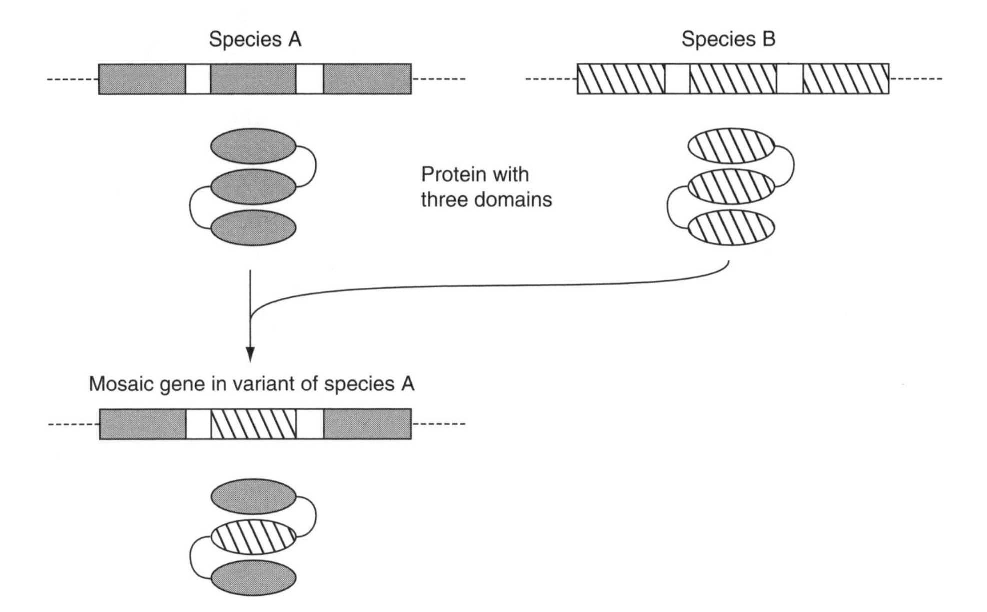

Figure 6.12 Mosaic genes and domain shuffling. The gene shown codes for a protein which folds into three domains. The mosaic gene contains two domains from species A and one domain from species B

this, we infer that parts of the *pbp* genes have been replaced by DNA from these other species.

The ease with which a part of a gene can be replaced by another piece of DNA relates to the structure of the corresponding protein. Many proteins fold into several relatively independent structures known as *domains* which are connected by flexible loops (see Figure 6.12). Each domain has its own function in the activity of the enzyme. So, for example, a phosphorylating enzyme (kinase) may have one domain that binds ATP and a second domain that binds the specific substrate. It is therefore possible to mix and match the genes concerned without disrupting the overall structure of the resulting enzyme. This is sometimes referred to as *domain shuffling*.

The overall concept of variation in bacteria and the structure of the bacterial genome can now be refined. The simple view – that bacteria (as asexually-reproducing organisms) undergo variation merely through the gradual acquisition of mutations – is clearly inadequate. Horizontal gene transfer is rife (in most species) and not only between strains of the same species but between species and genera – sometimes across quite wide taxonomic boundaries. Technically, a species that varies only by mutation without horizontal gene transfer is referred to as *clonal.* All members of a clone are descended from a single individual, so although there will be some gradual variation, examination of a single characteristic (such as the serotype) will give a reasonable prediction of other characteristics of members of that clone. Horizontal gene transfer breaks down this relationship so that two strains can be identical in many respects but radically different in others. This concept will be referred to later on when considering the use of molecular techniques for typing bacteria (Chapter 9).

Later in the book (Chapter 10) the analysis of bacterial genome structure will also be discussed. Comparative genomics shows that in some species, in addition to variation through mutation and through the acquisition of DNA from other species, there has been considerable rearrangement of the genome. Regions of DNA carrying a number of genes are found in entirely different locations in different strains or in closely related species. These rearrangements of the genome through recombination (possibly mediated by the action of transposable elements, as will be shown in the next chapter) provides evidence of *genome plasticity* which contributes greatly not only to the variation that exists amongst bacteria but also to the excitement of investigating it.

7

Genomic Plasticity: Movable Genes and Phase Variation

The traditional view of the DNA of a cell has been that of a fixed sequence of bases that is only subject to occasional changes by means of mutation or by recombination when cells exchange genetic material. There has therefore been a tendency to think of all cells in a pure bacterial culture as having an identical genetic make-up. It is now known that this is far from the truth. The genetic material is much more fluid than that and is subject to a range of larger-scale alterations in its structure, including insertions, transpositions, inversions and deletions. Some of these variations are readily reversible and generate high levels of genetic diversity which allows bacteria to survive in hostile and ever-changing environments.

7.1 Insertion sequences

Insertion of a DNA fragment into a gene will usually result in the inactivation of that gene, and it is by the loss of that function that such events were initially recognized. A number of genetic elements, including some phages and plasmids (see earlier chapters), can be inserted into the bacterial chromosome. However, this chapter is concerned with elements that do not usually have any independent existence but are only found as a part of some other DNA molecule. The simplest of these genetic elements are known as Insertion Sequences (IS). As will be shown later, many of the other elements that participate in genetic rearrangements share key features with IS elements.

7.1.1 Structure of insertion sequences

There are many IS elements known. They differ in size and other details, but the overall structure of most such elements is similar. One example (IS*1*) is shown in

Molecular Genetics of Bacteria, 4th Edition by Jeremy Dale and Simon F. Park
 ISBN 0 470 85084 1 (cased) ISBN 0 470 85085 X (pbk)

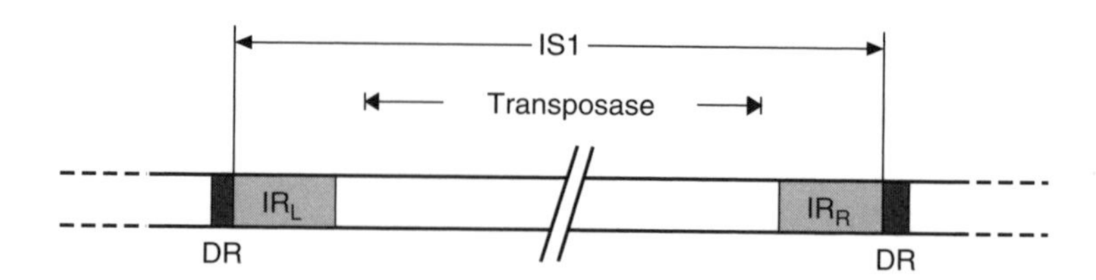

Figure 7.1 Structure of the insertion sequence IS*1*. DR, direct repeat (duplicated target sequence); IR, inverted repeats

Figure 7.1; IS*1* is 768 bases long but many other IS elements are longer (usually 1300–1500 bases). The central region of an IS element codes for a protein (known as a *transposase*) which is necessary for the movement of the element from one site to another. At the ends of the insertion sequence are almost perfect inverted repeat (IR) sequences, which in IS*1* consist of 23 nucleotides. A minority of elements, such as IS*900* from *Mycobacterium paratuberculosis* do not have inverted repeat ends.

It must be stressed that reference to an inverted repeat of a DNA sequence does NOT mean that the sequence on an individual strand is repeated backwards, but that the sequence from left to right on the 'top' strand is repeated from right to left on the 'bottom' strand so that reading either copy of the IR in the 5′ to 3′ direction will result in the same sequence of bases. Since DNA sequences are often presented as just one of the two strands, an inverted repeat of the sequence CAT will appear as ATG.

In addition to the inverted repeats, inspection of a DNA region containing an insertion sequence usually shows a further short sequence that is duplicated – but this sequence is repeated in the same orientation and is therefore referred to as a *direct* repeat (DR). This is not part of the IS, but arises from duplication of the DNA at the insertion site (Figure 7.2) and therefore different copies of IS*1* will have different target sequence repeats depending on the point of insertion. Transposition of IS*1* generates rather long direct repeats (nine base pairs). With other insertion sequences, the direct repeats are commonly as short as two to three base pairs. The presence of these direct repeats is linked to the mechanism of transposition which is considered later in this chapter.

7.1.2 Occurrence of insertion sequences

Insertion sequences have been identified in most bacterial genera, although the presence and the number of copies of any one element often varies from strain to strain. A typical laboratory strain of *E. coli* for example might contain six copies of IS*1* as well as a number of copies of other insertion sequences.

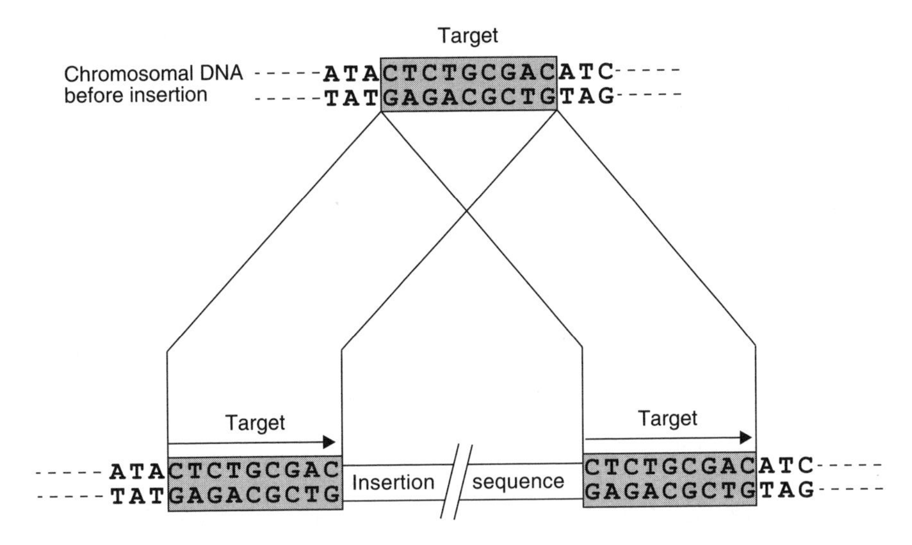

Figure 7.2 Target duplication following insertion of IS*1*. Flanking IS*1*, there are direct repeats of a 9-bp sequence, derived from duplication of the target region. The top part of the diagram shows the target structure before insertion of IS*l*, while the lower part shows the insertion of IS*l* and the duplication of the target sequence

Hybridization of a Southern blot (see Chapter 2, Box 2.4) with a probe for a specific IS will produce a banding pattern which depends on the number of copies of the element and their position on the chromosome. Since both the copy number and the chromosomal distribution of an IS may vary from one strain to another, a different pattern may be seen for different isolates of the same species. This is exploited in a technique known as *restriction fragment length polymorphism* (or RFLP) for typing bacterial strains which is described more fully in Chapter 9.

IS elements are also commonly found on bacterial plasmids. For example, in Chapter 5 it was shown that the antibiotic resistance plasmid R-100 (Figure 5.5) carries two copies of IS*1* and two copies of a different insertion sequence, IS*10*. The presence of IS elements can be a major cause of plasmid instability since recombination between two insertion sequences on the same plasmid will lead to inversion or deletion of the intervening region. Insertion of the plasmid into the chromosome or recombination between two plasmids can also arise by homologous recombination between IS elements present on each DNA molecule. Similarly, the presence of two copies of an IS element in the chromosome can result in inversion or deletion of the region between them due to homologous recombination. IS elements can therefore play a significant role in the variation of genomic structure between one strain and another.

Although IS elements can affect the phenotype by inactivation or deletion of genes, they do not carry any genetic information other than the transposase needed for transposition. They are therefore of little or no direct benefit to the bacterium. Why then does the cell tolerate their presence? The Darwinian notion of the 'survival of the fittest' would suggest that evolutionary pressure will tend to eliminate such elements that are not beneficial since their presence will constitute a metabolic drain on the cell (even if small). The simplest answer is that these sequences are essentially parasitic and have devised strategies that prevent their elimination.

For some IS elements, insertion into another DNA site is a replicative event: the original copy remains and a duplicate is inserted at the new site. The number of copies carried by the cell will therefore tend to increase. When there are a certain number of copies present in the cell, this process is repressed, which prevents the cell from being overwhelmed and dying (thus eliminating the 'parasite'). However, if the cell does lose one or two copies, the number is now lower than that needed for repression, leading to a renewed round of replication and insertion. Thus, like any well-adapted parasite, an insertion sequence 'colonizes' its host while refraining from doing sufficient damage to seriously weaken it.

7.2 Transposons

When resistance plasmids were first discovered, there was much speculation as to how a single element could have evolved to carry a number of different antibiotic resistance genes and in particular how apparently related plasmids could have different combinations of such genes (or, conversely, how otherwise dissimilar plasmids could carry related resistance genes). It was assumed that a basic plasmid, having the ability to replicate independently but not carrying any other information, had somehow picked up a resistance gene from the chromosome of a resistant host strain. Transfer of this plasmid to an otherwise sensitive strain then produces a selective advantage for that strain, and therefore indirectly a selective advantage for this 'new' plasmid. As the plasmid moves from one organism to another it has the opportunity to acquire additional resistance genes, thus giving rise to a family of plasmids containing different combinations of resistance genes.

Since this model implies that unrelated plasmids could pick up the same gene independently, this would explain the widespread distribution of certain resistance genes, notably a type of β-lactamase (the enzyme that destroys penicillin and hence confers resistance to penicillins). This particular enzyme, the TEM β-lactamase, is the commonest type amongst plasmids in the Enterobacteriaceae and is also present in many members of the genus *Pseudomonas*. The same gene has also been found in connection with plasmid-mediated penicillin resistance in species as diverse as *Haemophilus influenzae* and *Neisseria gonorrhoeae*.

The reason behind the ubiquity of the TEM β-lactamase became apparent from the discovery that this gene could move (transpose) from one plasmid to another. This is exemplified by the conjugation experiment shown diagrammatically in

Figure 7.3. A strain of *E. coli* containing two different plasmids, one with an ampicillin resistance gene and one conferring resistance to kanamycin, is used as a donor. The recipient strain is sensitive to both drugs (but resistant to nalidixic acid). Plating the mixed culture on a medium containing nalidixic acid, ampicillin and kanamycin will therefore select for recipient cells that have received both resistance genes from the donor. It was found that resistance to both antibiotics was transferred at a rate much higher than would be predicted from the rate of independent transfer of the two plasmids. It was also found that the recipients which were resistant to both drugs contained a single plasmid carrying both resistance genes.

This effect was not due to ordinary recombination between the two plasmids since it occurred equally well in recombination deficient (*recA*) strains. From additional evidence, it was deduced that the ampicillin resistance gene had moved (transposed) from one plasmid to the other. The term *transposon* was coined to signify an element that was capable of such behaviour, i.e. a mobile genetic element containing additional genes unrelated to transposition functions.

This movement of resistance genes can occur not only between two plasmids but also from plasmid to chromosome and vice versa. It therefore provides part of the explanation for the observed rapid evolution of resistance plasmids and also of plasmids that carry genes other than antibiotic resistance. Although resistance

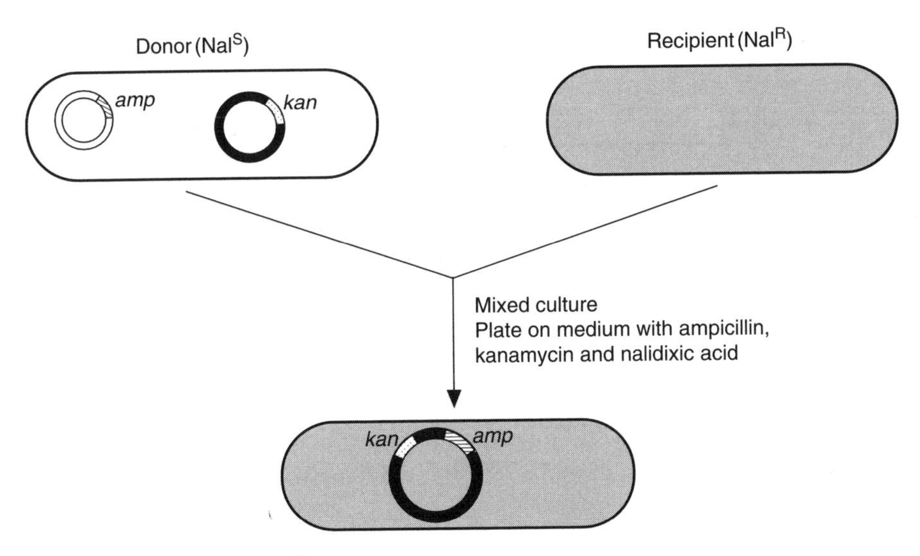

Figure 7.3 Transposition of a resistance gene between plasmids. The donor strain has two plasmids, carrying ampicillin- and kanamycin-resistance genes (*amp* and *kan*) respectively. Conjugation with a nalidixic acid-resistant recipient, using a selective medium containing all three antibiotics, leads to colonies with a single plasmid carrying both *amp* and *kan*, due to transposition of *amp* to the second plasmid

transposons have been the most studied, other plasmid-borne genes are also known to be transposable on occasions.

7.2.1 Structure of transposons

The structure of a simple transposon, Tn*3*, is shown in Figure 7.4; it consists of about 5000 base pairs and has a short (38 bp) inverted repeat sequence at each end. It is therefore analogous to an insertion sequence, the distinction being that a transposon carries an identifiable genetic marker – in this case the ampicillin resistance gene (*bla*, β-lactamase). Tn*3* codes for two other proteins as well: a transposase (TnpA), and TnpR, a bifunctional protein that acts as a repressor and is also responsible for one stage of transposition known as resolution (this is explained more fully later on). As with the insertion sequences, there is a short direct repeat at either end of the transposon (five base pairs in the case of Tn*3*).

Some transposable elements have a more complex structure than Tn*3*. These *composite transposons* consist of two copies of an insertion sequence on either side of a set of resistance genes. For example the tetracycline resistance transposon Tn*10*, which is about 9300 bp in length, consists of a central region carrying the resistance determinants flanked by two copies of the IS*10* insertion sequence in opposite orientations (Figure 7.5; see also Figure 5.5). IS*10* itself is about 1300 bp long with 23-bp inverted repeat ends and contains a transposase gene.

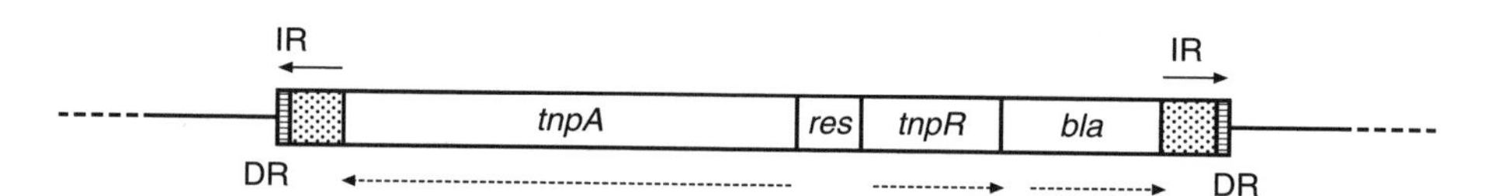

Figure 7.4 Structure of the transposon Tn*3*. DR, five-base pair direct repeat (target duplication); IR, 38-base pair inverted repeats; *res*, resolution site; *tnpA*, transposase; *tnpR*, resolvase; *bla*, β-lactamase (ampicillin resistance)

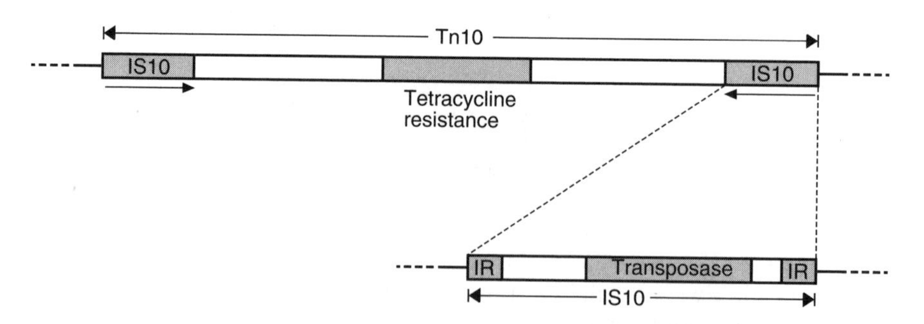

Figure 7.5 Structure of a composite transposon Tn10

Composite transposons may have their flanking IS regions in inverted orientation or as direct repeats. For example Tn*10* and Tn*5* (see Figure 7.6) both have inverted repeats of an IS (IS*10* and IS*50* respectively) at their ends, while Tn*9* has direct repeats of IS*1*. The transposition behaviour of such composite elements can be quite complex; the insertion sequences themselves may transpose independently or transposition of the entire region may occur. Furthermore, recombination between the IS elements can occur, leading to deletion or inversion of the region separating them (see Chapter 6, Figure 6.11).

Even more complex arrangements can occur. For example, Tn*4* appears to be related to Tn*21* but contains a complete copy of Tn*3* within it. The ampicillin resistance gene of Tn*4* can thus be transposed as part of the complete Tn*4* transposon or by transposition of the Tn*3* element. Thus several layers of transposons can occur, nested within one another.

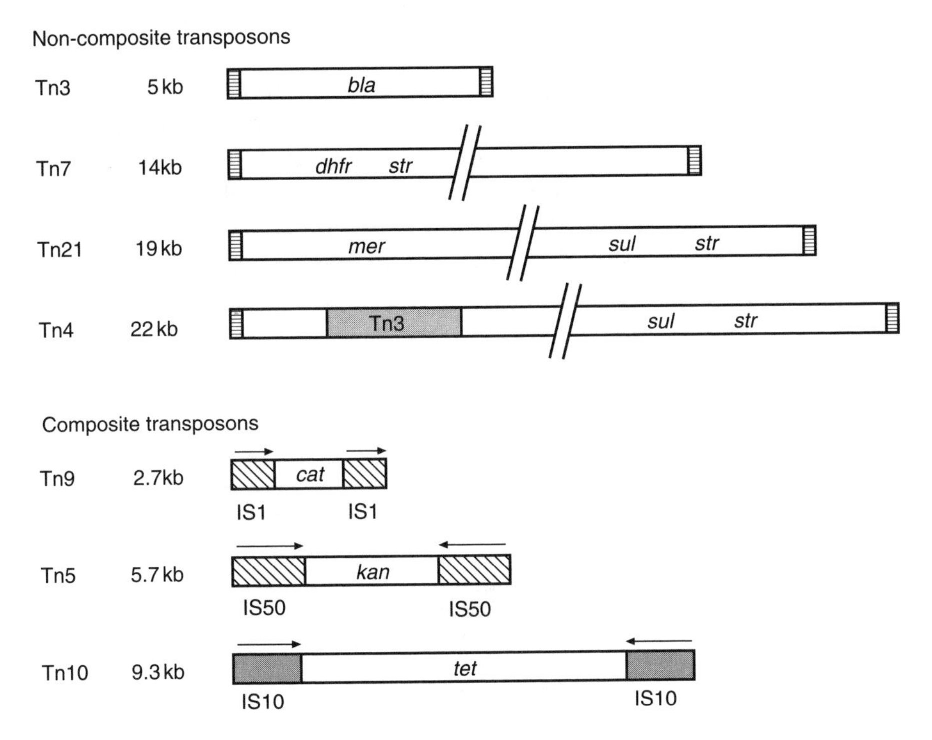

Figure 7.6 Structure of selected transposons. *bla*, β-lactamase (ampicillin resistance); *dhfr*, dihydrofolate reductase (trimethoprim resistance); *str*, streptomycin resistance; *mer*, mercuric ion resistance; *sul*, sulphonamide resistance; *cat*, chloramphenicol resistance (chloramphenicol acetyltransferase); *kan*, kanamycin resistance; *tet*, tetracycline resistance

The composite transposons such as Tn*10* are also known as class I transposons, while transposons such as Tn*3*, flanked by inverted repeats rather than IS elements, are referred to as class II transposons.

7.2.2 Integrons

Extremely complex and large transposons can also be built up by insertion of additional genes within an existing transposon. Many large transposons have been identified which are related to Tn*21* which has a structure analogous to class II transposons such as Tn3: it has inverted repeats (38 bp) at each end and carries genes for transposition functions (Figure 7.7). Tn*21* may have developed from a smaller transposon (such as Tn*2613*) by acquisition of additional genes. Tn*2603* and Tn*1696* (and a family of other transposons) are also very similar to Tn*21* but contain additional resistance genes.

It is now known that the transposons in the Tn*21* family have acquired resistance genes by a specific mechanism. Each individual gene has been inserted separately, as a *gene cassette* which contains a single gene and a recombination site. Tn*21* contains a site known as an *integron* into which such gene cassettes can be inserted by site-specific recombination. The integron region in Tn*21* also contains a gene coding for an integrase which is responsible for the site-specific recombination (and is related to the bacteriophage λ integrase; see Chapter 4). After insertion of a gene cassette into the integron, the recombination site remains available for insertion of a further gene cassette, enabling the build-up of an array of several cassettes within the integron. A further twist to the story is that the gene cassettes do not normally contain a promoter. However there is a promoter region within the integron itself,

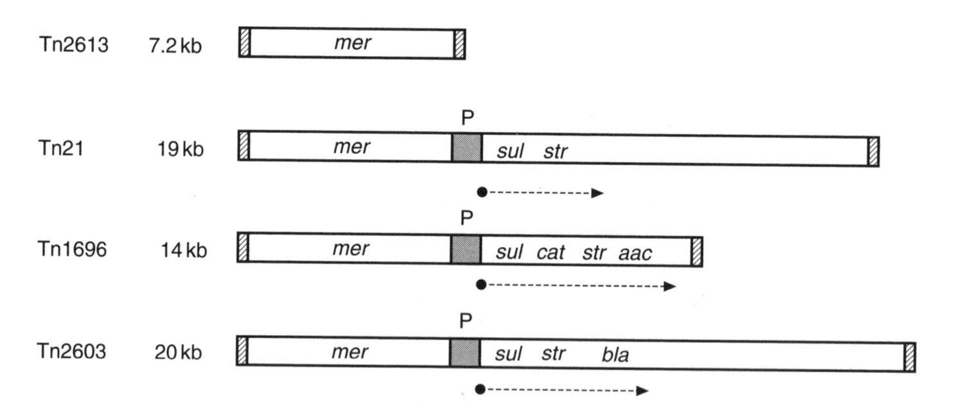

Figure 7.7 Tn21 family of transposons: integrons. Gene cassettes are inserted to the right of the promoter (P), by means of a transposon-encoded integrase (position not shown). *aac*, aminoglycoside acetyltransferase (gentamicin resistance; see Figure 7.6 for the identity of the other genes)

upstream from the insertion site, so each of the gene cassettes is transcribed from the integron promoter. Integrons are thus a naturally-occurring analogy to the expression vectors that will be discussed in Chapter 8, for obtaining expression of foreign genes by inserting them into a vector adjacent to a promoter.

Integrons are not only found in the Tn*21* family of transposons, several classes of recognizably distinct integrons have been found in other transposons as well as in plasmids which do not carry functional transposons. They therefore represent a significant additional mechanism for the evolution of bacterial plasmids and the spread of antibiotic resistance.

7.3 Mechanisms of transposition

7.3.1 Replicative transposition

In considering mechanisms of transposition, it is not necessary to distinguish between insertion sequences and transposons as the same mechanisms apply to both types of transposable elements. Transposons such as Tn*3* transpose by a replicative mechanism: a copy of the element is inserted at a different site (on the chromosome or on a plasmid) while the original copy is retained. The stages of replicative transposition are outlined in Figure 7.8 which depicts transposition from one plasmid (A) to a second plasmid (B). The transposase mediates a form of recombination between plasmid A carrying the transposon and the target plasmid B. With some transposons, this target site appears to be more or less random, i.e. there is no requirement for a specific sequence, while other transposons do have a degree of specificity in their target sequences. An extreme example is Tn*7* which has only one insertion site in the *E. coli* chromosome.

The outcome of this stage is the formation of a larger plasmid known as a *cointegrate*, which consists of the complete sequence of both plasmids fused together, but now with two copies of the transposon, in the same orientation. With some naturally-occurring plasmids, cointegrate molecules such as this can be readily isolated; in other cases, the intermediate is rapidly resolved into two separate plasmids, each of which contains a copy of the transposon, with the target sequence on the recipient plasmid being duplicated on either side of the inserted transposon.

Since the two copies of the transposon are in the same orientation (i.e. it is a *direct* repeat), resolution of the cointegrate can occur by recombination between the two copies as described in Chapter 6 (see Figure 6.11). This can be achieved by general recombination using host recombination systems. However, some transposons (including Tn*3*) encode their own resolution system. The *tnpR* gene of Tn*3* (see Figure 7.4) codes for a *resolvase* which mediates a site-specific recombination at the resolution site within the transposon, thus ensuring an efficient resolution of the cointegrate, independent of host recombination activity.

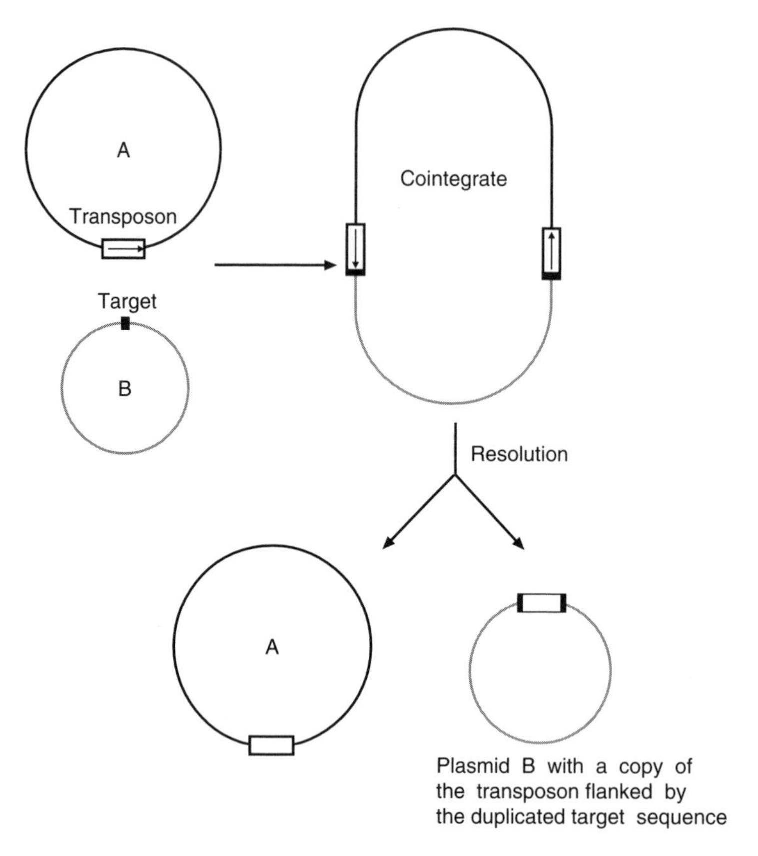

Figure 7.8 Schematic diagram of replicative transposition. The first step shows formation of a cointegrate structure, carrying two copies of the transposon, with duplication of the target sequence. Recombination between the two transposon copies leads to separation into two plasmids, each having a copy of the transposon. The direct repeats flanking the original copy of the transposon are not shown

Molecular basis of replicative transposition

The molecular basis of replicative transposition can be considered in two stages: formation of the cointegrate, followed by resolution. A simplified model of cointegrate formation is shown in Figure 7.9 in which the transposon is on one plasmid (A) and the target sequence on a second plasmid (B) as in Figure 7.8. It is important to remember that the linear structures shown are parts of circular molecules and of course the strands are twisted around each other rather than lying side by side. The steps shown are as follows.

(a) Single strand breaks (nicks) in the DNA are produced at each 3′ end of the transposon (in opposite strands). The recipient plasmid is also nicked on either

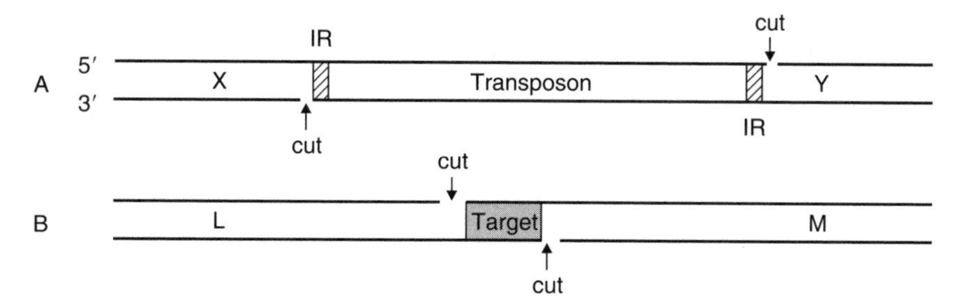

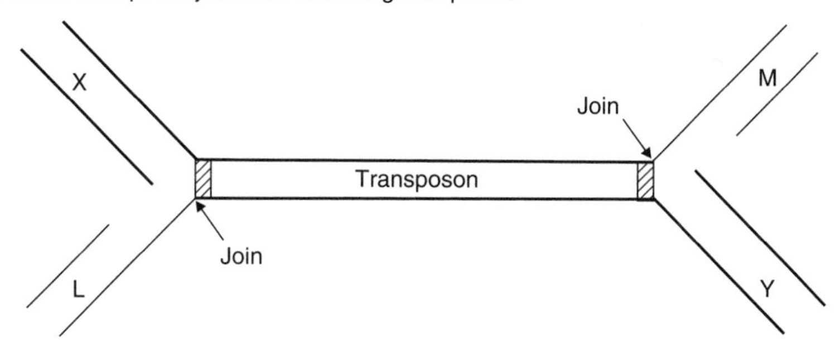

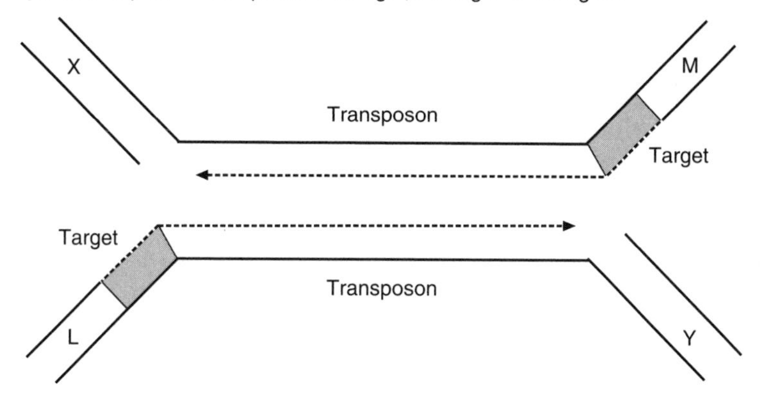

Figure 7.9 Model of the mechanism of replicative transposition: formation of a cointegrate (see the text for details)

side of a short target sequence. The staggered nature of the nicks is the ultimate cause of the duplication of the target sequence.

(b) The free ends of the transposon are joined to the free ends generated by the nicks in the recipient plasmid. At no stage in this process is the transposon itself liberated as an independent molecule.

(c) The free 3′ ends of the recipient plasmid sequence act as primers for the synthesis of DNA strands using host DNA polymerase. This synthesis will proceed through the transposon, separating the two strands until it reaches the existing complementary strand to which the new strand will be joined by the action of DNA ligase. This replicative step is responsible for the duplication of the transposon itself and of the target sequence. This produces the cointegrate structure (see Figure 7.8), i.e. the structure in (c) is now one large circle).

The cointegrate plasmid (Figure 7.10) contains two copies of the transposon in the same orientation (direct repeat). Recombination between the two copies gives rise to the end products of transposition: two plasmids each containing a copy of the transposon. Plasmid A is as it started while plasmid B has acquired a copy of the transposon flanked by a direct repeat of the target sequence.

7.3.2 Non-replicative (conservative) transposition

Not all transposons show exactly the same behaviour. In particular, some transposons and some insertion sequences, do not replicate when they transpose, exhibiting a mode of transposition known as conservative (or non-replicative) transposition. This occurs with the insertion sequence IS*10* and the related transposon Tn*10*.

It is possible to use a modified version of the model in Figure 7.9 for non-replicative transposition. If, following the joining of the free ends of the transposon to the nicked recipient plasmid (step b), the previously unbroken donor strands are cut, this will release the recipient replicon which now contains the transposon flanked by a single-stranded target sequence. These single-stranded regions are filled in by repair synthesis while the donor replicon, now linearized by excision of the transposon, is degraded.

An alternative model of non-replicative transposition is the 'cut and paste' process presented in Figure 7.11. In this case, the transposon is completely excised from the donor molecule before being attached to the target site. With some transposons, there is evidence for the existence of small amounts of free transposon DNA, often in a circular form.

The difference between these models is not as great as appears at first, since it consists mainly in the relative timing of the events concerned. Many transposons appear to use both replicative and non-replicative transposition; indeed, this may be the norm rather than the exception.

Although, with conservative transposition the plasmid molecule that acted as the transposon donor is degraded in the process, this does not necessarily mean the complete loss of that plasmid from the cell. Since there may be a number of copies of the plasmid, the loss of one copy is easily rectified by replication. This can make it very difficult to determine with certainty which type of transposition has occurred.

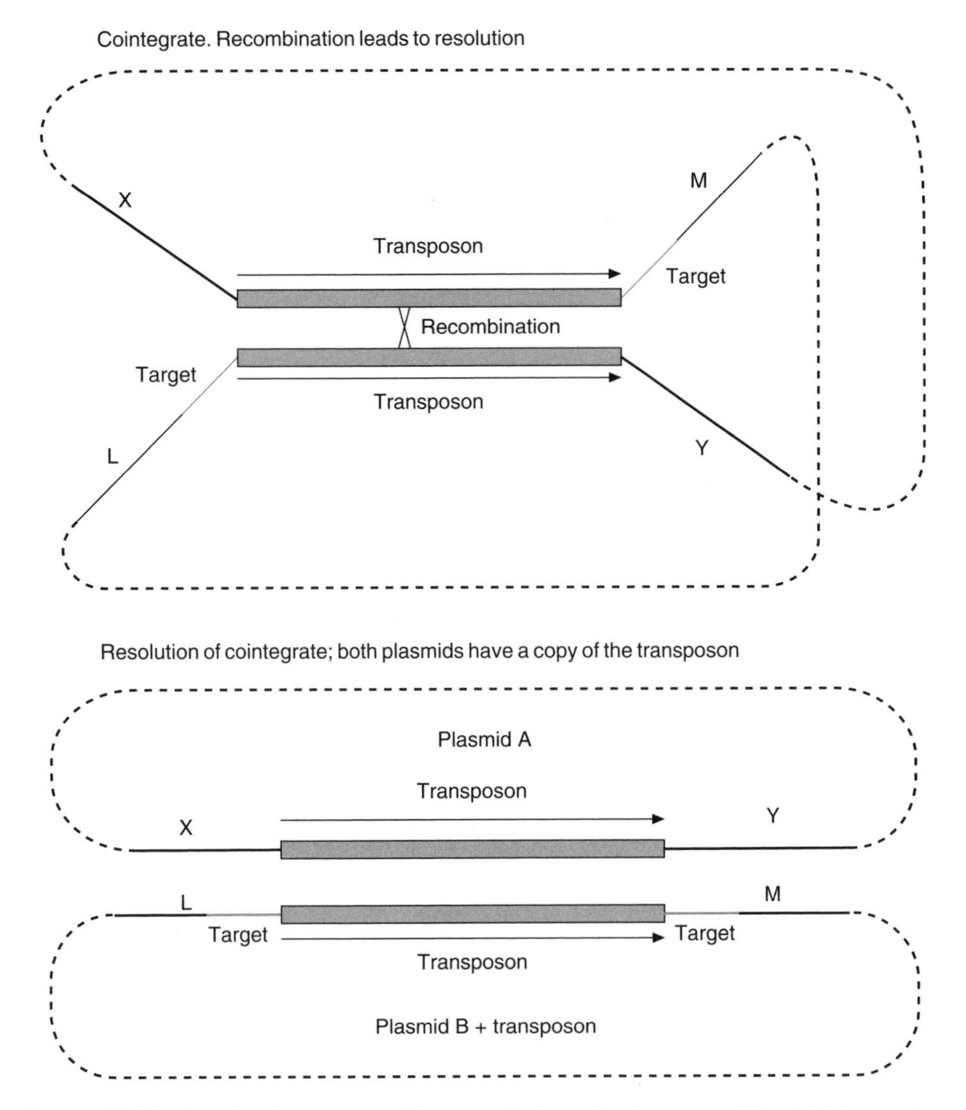

Figure 7.10 Replicative transposition: resolution of cointegrate. The initial structure (cointegrate) is another representation of the end-product of Figure 7.9 containing two copies of the transposon in the same orientation. Resolution occurs by recombination between the two copies of the transposon

7.3.3 Regulation of transposition

An excessive level of transposition is likely to be extremely damaging to the host cell, due to the frequent occurrence of insertional mutations, plus (in the case of replicative transposition) the accumulation of a large number of transposons or

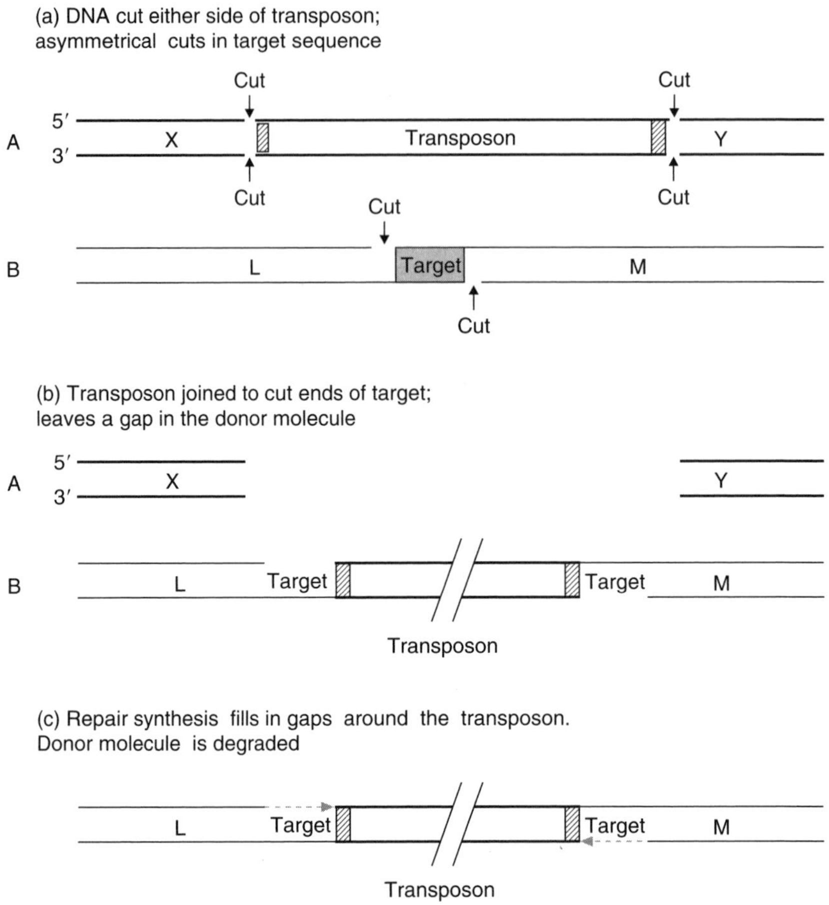

Figure 7.11 Non-replicative (conservative) transposition. The mechanism shown is known as 'cut and paste'; the transposon is moved from one molecule to another and the donor molecule (which now has a double-stranded gap) is degraded. However, if the donor plasmid is present in more than one copy, replication of the remaining copy will replace the lost plasmid and the overall effect is still replicative.

insertion sequences. As with conventional parasites, it is not in the best interests of the element to damage its host too much.

Transposable elements use a variety of mechanisms to control the level of their transposition. One has already been mentioned: the Tn*3* TnpR protein not only acts as a resolvase but also functions as a repressor of transcription of the transposase gene, *tnpA*. Other transposable elements, although lacking an identified repressor of transcription, are usually transcribed at a low level due to the lack of a strong promoter.

A less conventional mechanism has been demonstrated with IS*1* and with elements related to IS*3*. In these cases, a key protein required for transposition

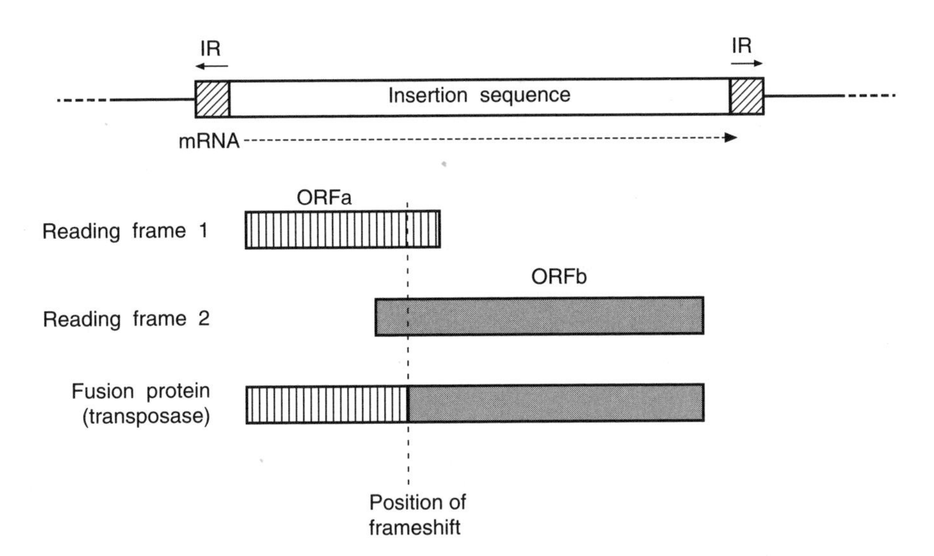

Figure 7.12 Regulation of transposition by ribosomal frameshifting. With some insertion sequences, production of functional transposase requires the ribosomes to change reading frame at an intermediate point. Since this is an infrequent event, very low levels of the transposase are made

is translated from two different reading frames on the mRNA: the ribosomes start reading the mRNA in one frame and then at a defined point are required to shift back one base and continue reading in a different frame (see Figure 7.12, and also Figure 3.28). Since this ribosomal frameshifting will occur infrequently, it ensures that very little functional enzyme will be made. Yet another regulatory mechanism is exhibited by IS*10*, which produces an antisense RNA that is complementary to part of the transposase mRNA and thus inhibits translation of the message.

Whichever method is used, the consequence is the same. Transposition of insertion sequences and transposons is usually a rare event. The outcome can be detected quite readily in an experiment such as that shown in Figure 7.3, because the rare events that lead to the production of a new plasmid carrying both antibiotic resistance markers can be selected. However, attempts to identify the movement of an insertion sequence without phenotypic markers and relying only on changes in the banding pattern on a Southern blot, may lead to the examination of thousands of colonies before finding one with an altered pattern.

7.3.4 Activation of genes by transposable elements

Up to this point it has been assumed that apart from the movement of any genes carried by the transposon, the only consequence of transposition will be the

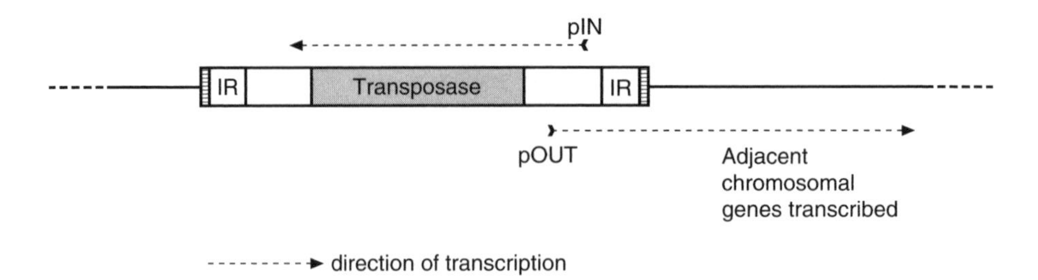

Figure 7.13 Activation of chromosomal genes from an insertion sequence

inactivation of the gene into which the element is inserted. With some transposable elements it is known that the converse effect can occur, i.e. insertion of the element actually promotes the expression of genes adjacent to the site of insertion. The reason for this is that some insertion sequences (such as IS*10*) contain a promoter (identified as pOUT in Figure 7.13) which is directed outwards, i.e. away from the transposase gene and towards any genes that may be found in the flanking chromosomal DNA. These genes, if in the correct orientation, will therefore be turned on by the presence of IS*10*.

7.3.5 Mu: a transposable bacteriophage

The name Mu is short for mutator, which is derived from the fact that *E. coli* cells which carry this phage show an abnormally high rate of mutation.

The Mu phage particle contains about 38 kb of DNA as a linear structure which carries variable ends that are derived from bacterial DNA. After injection into an *E. coli* cell, the Mu DNA becomes inserted into the bacterial DNA at a random position. In common with transposons and insertion sequences, this insertion involves the duplication of a short region (5 bp in this case) of host sequence at the target site. Thereafter replication occurs by repeated replicative transposition events with the transposed copies being inserted at different sites around the chromosome. The insertion of copies of Mu DNA at various positions on the *E. coli* chromosome causes a loss of function of those genes, hence the high rate of mutation. Eventually, the productive phase of phage infection involves the excision of Mu DNA copies from the chromosome starting from a fixed point to the left of the Mu DNA insert (thus including some bacterial DNA). Packaging proceeds until the phage head is full (cf. bacteriophage T4; Chapter 6); since this requires more DNA than the length of Mu itself, some bacterial DNA is also included at the right-hand end.

7.3.6 Conjugative transposons and other transposable elements

In Chapter 6, *conjugative transposons* were described, which are not only able to transpose from one site to another within a cell, but can also transpose to other cells by a conjugative mechanism. Transposition of these elements differs radically from the models described above since it involves excision of the transposon to generate a covalently closed circular intermediate. This structure is unable to replicate but is a substrate for subsequent integration into a different site, as well as being able to be transferred by conjugation to a recipient cell.

Some other mobile elements also transpose via circular intermediate structures. These elements are noticeably different from typical transposons and insertion sequences in lacking inverted repeat ends and in not generating direct repeats of the target sequence at the site of insertion.

7.4 Phase variation

In addition to the transcriptional and translational regulatory mechanisms described in Chapter 3, many bacteria have evolved additional adaptive mechanisms to enable them to respond to changing environments. In these bacteria a genetically diverse population is generated by reversible genetic changes known as *phase variation*. The principle is illustrated in Figure 7.14 which envisages a reversible (but inherited) switch in the state of a specific gene. In phase 1, the gene is not expressed, but when a cell switches to phase 2, the gene is expressed. Initially, the population is mainly in phase 1 (expression off) with a small minority of cells in the alternative state (expression on). When this population is subjected to changed conditions (such as the immune response of the host) to which phase 1 cells are susceptible, all the cells in phase 1 will be killed, but the phase 2 cells will survive and multiply. This gives rise to a population that is expressing this gene (phase 2). Since the change is reversible, when conditions change again, the population will become diverse again or even switch back to being predominantly phase 1. Some examples of phase variation, and the mechanisms involved, are considered below.

Reversible variation offers considerable advantages over conventional (irreversible) mutation. For example, a pathogen may be killed either by the immune response of the host or by antibiotic treatment. With conventional mutations, a tiny minority of the original population might escape from the immune response, but would still be sensitive to the antibiotic and a different minority would become resistant to the antibiotic but would be killed by the immune response. However, if the antigenic variation is due to a reversible phase variation, then this is a characteristic of *all* the cells in the population (even though it is only expressed

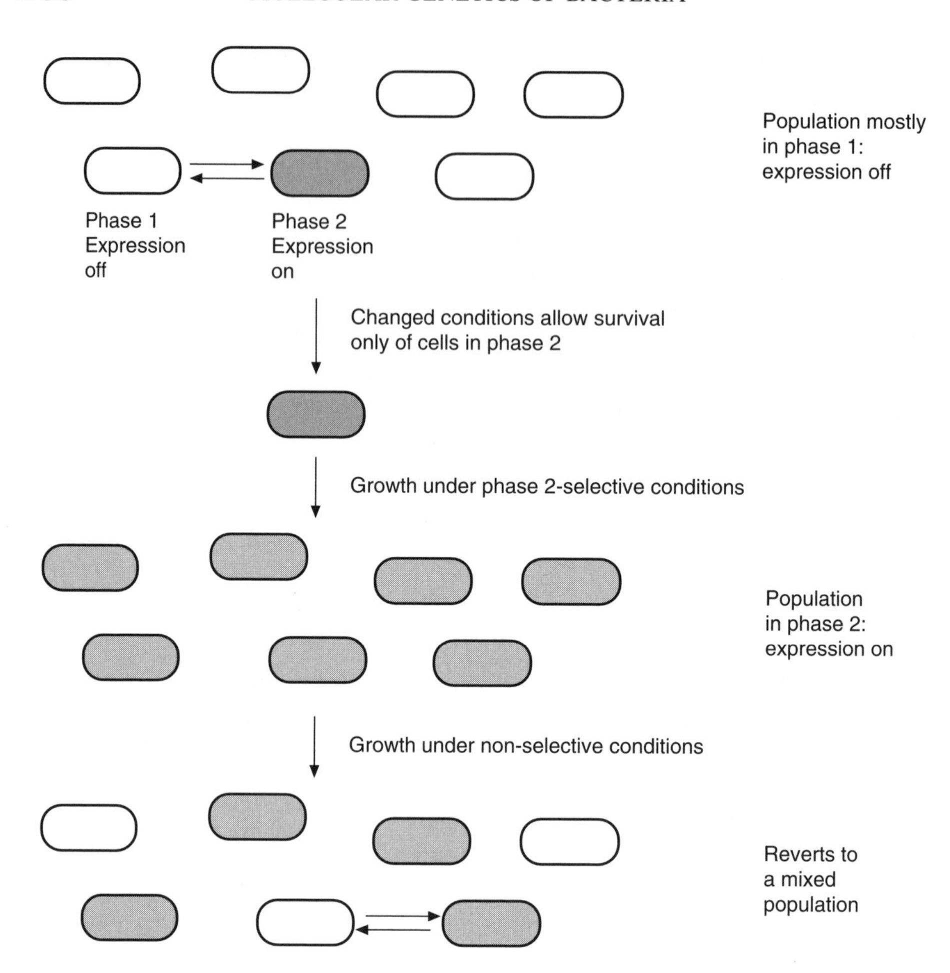

Figure 7.14 Phase variation. In the original state, the population is mainly in phase 1 (not expressing a specific gene), but a few cells are in phase 2 (expression on) due to a reversible genetic change. Exposure to conditions under which only phase 2 cells will survive will produce an homogenous phase 2 population. When changed to non-selective conditions, the population will gradually revert to a mixture

by a minority at any one time), so the cells that survive the antibiotic will still be capable of expressing different antigens. Furthermore, reversible phase variation enables the population to retain potential genetic characteristics that are actually disadvantageous under certain circumstances – such as expression of an antigen that is necessary for adhesion and thus for colonization, but also provides a target for the immune response. In these respects, evolutionary pressures will operate at the population level rather than on individual cells.

7.4.1 Variation mediated by simple DNA inversion

One of the simplest phase variation systems is that which controls the expression of type 1 fimbriae in *E. coli*. These are external filamentous appendages which enable adherence to host cells. The bacteria switch between expression and non-expression of this structure at a significant frequency such that any culture of the strain, although predominantly in one phase, will contain some cells in the opposite phase. This ensures that, if the host mounts an immune response to the fimbriae, at least some cells survive – or conversely that if the structure is required to initiate infection then at least some cells are able to do this. The promoter for the *fimA* gene, which encodes the fimbrial structural subunit, is located upstream of *fimA* in a 314-bp region of DNA which is bordered by two 9-bp inverted repeats (Figure 7.15). In the 'ON' position the promoter is directed towards *fimA* and expression of this gene occurs. Two other genes in this system encode integrases, FimE and FimB, which can act upon the inverted repeats to flip or invert this DNA region. If this occurs then the promoter is no longer directed towards *fimA* and expression of this gene no longer occurs. The inversion of this sequence is carefully controlled and at 37°C, a temperature appropriate for growth in humans, the ON state is favoured. Nevertheless, because of the inversion mechanism, there will always be a small population of cells that do not produce fimbriae even at this temperature.

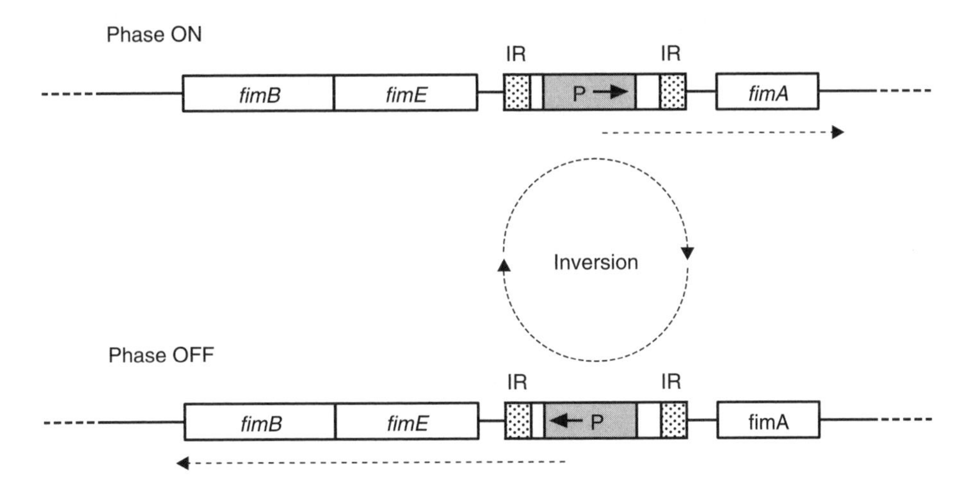

Figure 7.15 Phase variation of type I fimbriae expression in *E. coli*. *fimA* codes for the structural subunit of the fimbriae and is expressed from a promoter located on an invertible DNA region. In the first phase expression is on, but when the control region is inverted the promoter faces in the wrong direction and expression is off. The products of *fimB* and *fimE* carry out the inversion

Whilst this system results in the simple 'ON/OFF' expression of a gene, other more complicated systems of phase variation are present in bacteria. For example, most species of *Salmonella* are capable of producing two different, and antigenically distinct, types of flagellum (H-antigens) in a phase-variable manner. Whilst this mechanism is controlled by the inversion of a DNA region, as is the *E. coli fim* system, the result of the inversion is more complicated. Expression of the H2 flagella is mediated by a promoter that is located in a 996-bp region of DNA bounded by two 14-bp inverted repeats (Figure 7.16). In the H2 phase, this promoter drives the expression of the *H2* flagellin gene and also another gene called *rh1*, which encodes a repressor that inhibits transcription of the alternative flagellin gene *H1*. So, in this orientation, H2 flagellin expression is promoted whilst that of H1 is repressed. Inversion of the region containing the promoter is mediated by a site-specific recombinase (Hin, standing for H inversion) which is encoded within the invertible region and which acts upon the inverted repeats. When inversion occurs, the promoter faces in the opposite direction and consequently the *H2* flagellin gene is no longer expressed. Since the *rh1* repressor is also no longer produced, the H1 flagellin is expressed instead.

7.4.2 Variation mediated by nested DNA inversion

Campylobacter fetus is an economically important pathogen of farm animals. One of its defences against the host response is a surface layer (the S-layer) of a protein called SapA. Although this renders cells resistant to serum killing, it is also antigenic and so provides a target for antibody attack. *C. fetus* is able to cope with this problem because it possesses up to nine different *sapA* genes that encode antigenically-distinct proteins. Each cell expresses only one of these genes, as the others lack a promoter and are therefore silent. However, rearrangement of the DNA can move the promoter to a site adjacent to a different *sapA* gene, thus allowing a switch of antigen expression. It can be seen from Figure 7.17 that the promoter is found between two *sapA* genes which are in different orientations. Looping of the chromosome allows recombination between conserved regions at the 5′ end of each gene, thus delivering the promoter to a different *sapA* gene by inversion of the promoter-containing region. As can be seen from the figure, if the inverted region contains one or more *sapA* genes in addition to the promoter, then the promoter can be moved to any of the previously silent genes. This switch of gene expression results in antigenic variation that allows the bacterium to escape from the immune response.

7.4.3 Antigenic variation in the gonococcus

The Gram-negative bacterium *Neisseria gonorrhoeae*, familiarly known as the gonococcus, which is the causative agent of gonorrhoea, also evades the host

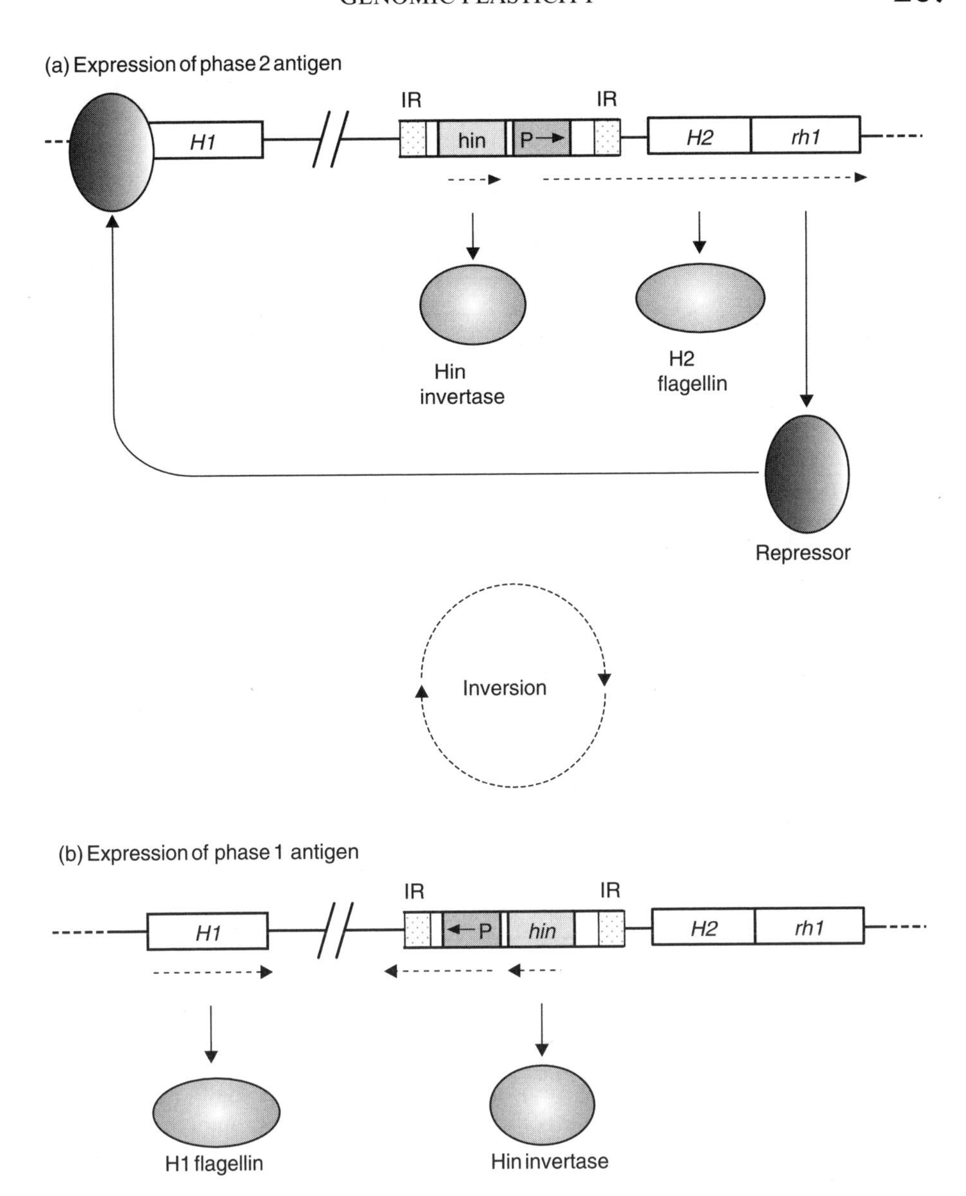

Figure 7.16 *Salmonella* phase change by inversion of a control element. *Salmonella* can express two types of flagella (H antigens). (a) In phase 2 the H2 gene is expressed and the H1 gene is repressed by the product of the gene *rh1*. (b) Inversion of the control region switches off both *H2* and *rh1* allowing production of phase 1 flagella

immune response by antigenic variation. This bacterium produces pili (or fimbriae) which are thin, proteinaceous structures that protrude from the cell surface. Pili are important virulence determinants since they enable the bacterium to attach to the mucosal surface during the initial stages of infection. Gonococcal pili are polymers of a polypeptide (pilin) of M_r about 20 kDa. The 50 or so amino

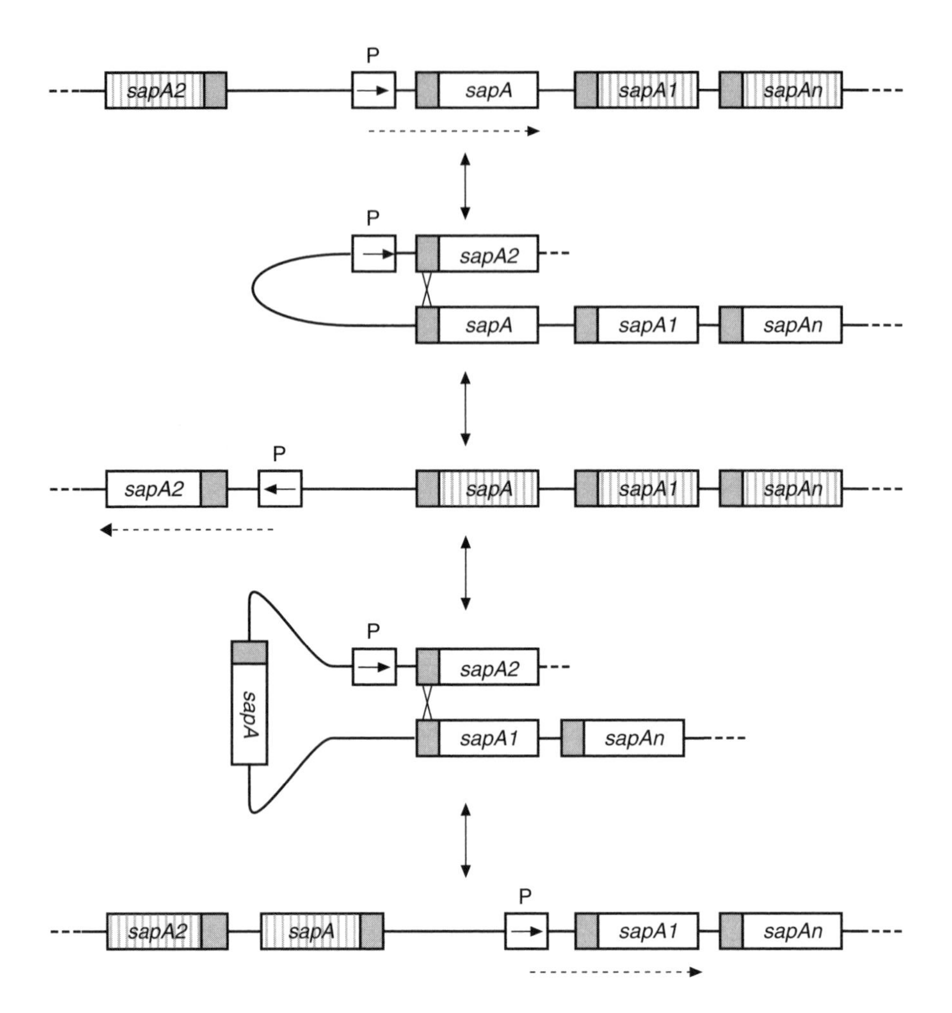

Figure 7.17 Antigenic variation in *Campylobacter fetus*. *C. fetus* has a number of different versions of the *sapA* gene, only one of which is expressed at a time, as the others lack a promoter. Recombination between the conserved regions at the 5′ end of inverted copies of *sapA* causes inversion of the control region, putting the promoter adjacent to a different copy of *sapA*. ▒, conserved region

acids at the N-terminus of pilin are the same in all pili, but the remainder of the molecule (about 100 amino acids) varies considerably.

As with the *C. fetus sapA* system described above, the gonococcus contains a number of *pil* genes, but only one is expressed at a time. However the mechanism is different from that of *sapA*. A simplified model is presented in Figure 7.18. Silent copies of the pilin genes are maintained at a locus known as *pilS*, while one *pil* gene is found at an expression locus, *pilE*. The switch in gene expression that causes antigenic variation arises by replacement of the *pil* gene at the

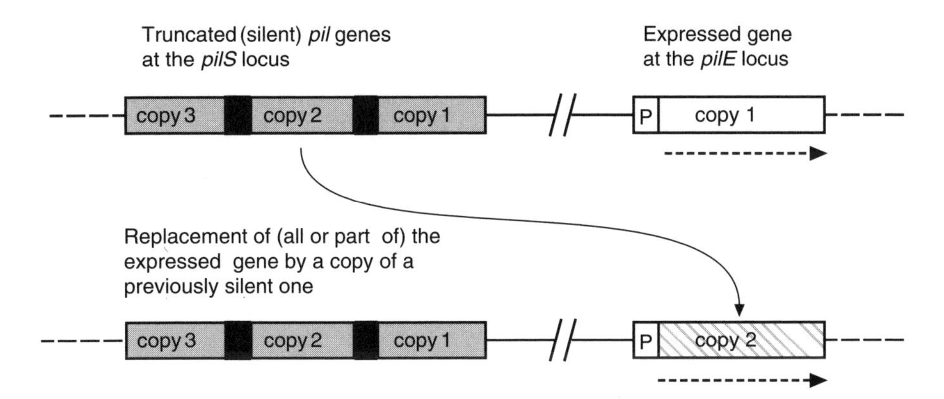

Figure 7.18 Antigenic variation of gonococcal pili. The *pilS* locus contains several different copies of the gene, all lacking a promoter; hence they are not expressed. Antigenic change occurs by replacement of part of the expressed gene (at *pilE*) with a copy of the equivalent part of a previously silent gene. This is a highly simplified representation of part of a complex system

expression locus with a copy of a different *pil* gene, an event known as *gene conversion*.

The variation arising from this system is actually much more extensive than simply switching expression between alternative genes. Since the gene conversion may involve not a whole gene but only part of one, the newly expressed *pil* gene will include sequences from two different copies. It has been calculated that this can give rise to 10^7 antigenically-distinct varieties of the pilus. In addition, since *N. gonorrhoeae* is naturally transformable (see Chapter 6), transfer between cells of DNA containing *pil* genes adds to the versatility of this organism in evading the immune response.

7.4.4 Phase variation by slipped strand mispairing

Runs of a single nucleotide, e.g. AAAAAAAAAAA (homopolymeric tracts), or repeated units of more than one nucleotide, called multimeric repeats (e.g. **AT**ATATATATATAT or **GCC**GCCGCCGCCGCC), provide another mechanism for phase variation through errors during DNA replication which result in the loss or gain of one or more of the repeat units. This process is called *slipped strand-mispairing* (Figure 7.19). When the chromosome replicates, whilst in many cases the sequence remains unchanged, a small population of cells arises which have a different number of repeat units. This can affect gene expression at either the translational or transcriptional level. If the repeat is within a coding sequence, the reading frame of the gene will be altered, leading to premature termination of translation. This mechanism of phase variation controls the expression of many

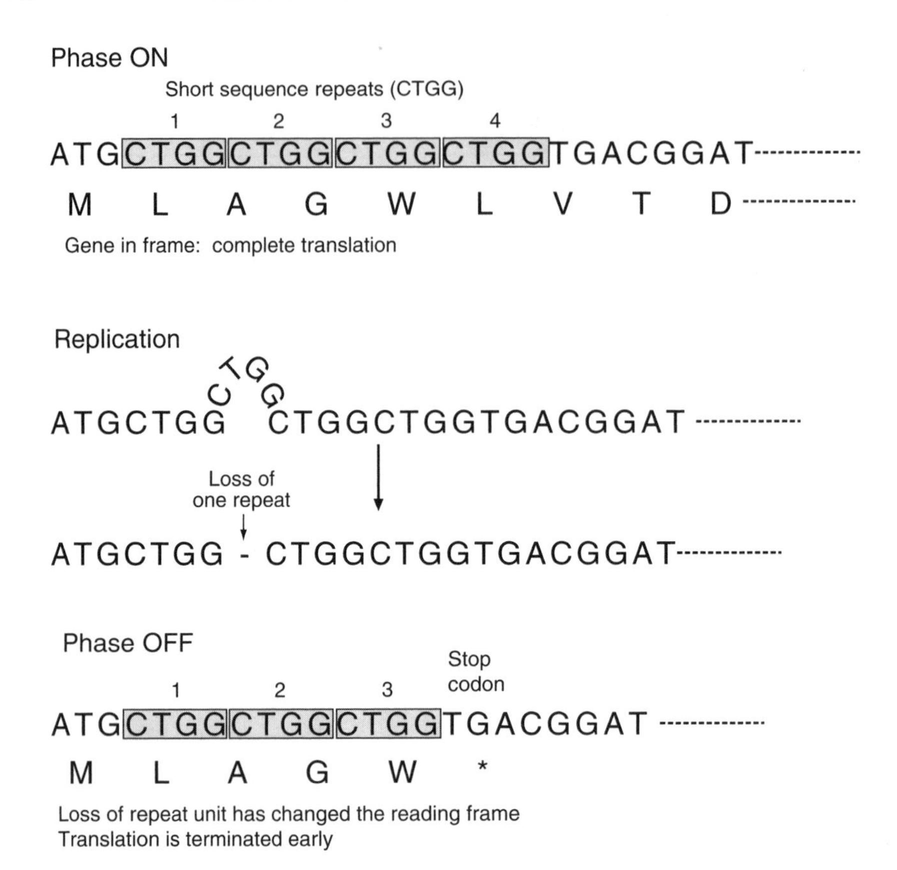

Figure 7.19 Phase variation by slipped strand mispairing. The gene shown contains four copies of a four-base repeat (CTGG). If one copy of this repeat is lost (or gained) during replication, the reading frame will be altered, usually leading to premature termination of protein synthesis

genes involved in virulence including capsular polysaccharide production in *Neisseria meningitidis* and lipopolysaccharide antigenicity in *Haemophilus influenzae*, *Campylobacter jejuni* and *Helicobacter pylori*.

Alternatively the repeat sequence may be in a regulatory region and changes in its length may alter the interaction of the DNA with regulatory proteins or RNA polymerase. The Opc outer membrane protein of *Neisseria meningitidis* undergoes this form of regulation. In this situation phase changes are brought about in a repeat of cytosine bases next to the promoter and this changes the efficiency of transcriptional initiation by RNA polymerase. When the number of bases in this repeat is 10, no transcription of the gene occurs. In cells containing 14 cytosines in this region, the gene is weakly expressed but when cells have 12 cytosines, the

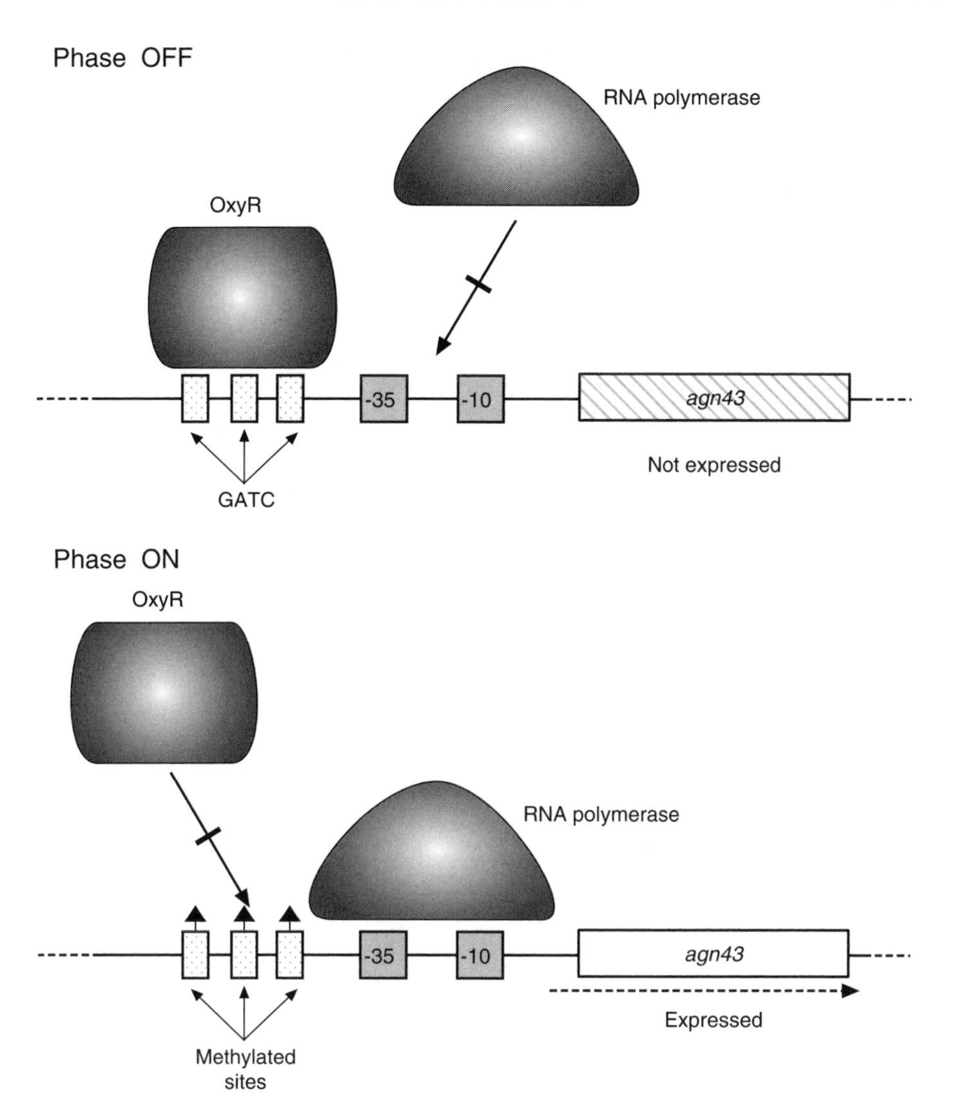

Figure 7.20 Antigen 43 phase variation by differential DNA methylation. OxyR can block transcription of the *agn43* gene. However if the three GATC sequences within the OxyR binding site are methylated, OxyR cannot bind and *agn43* will be expressed

spacing is optimal for RNA polymerase recognition and expression is fully on. It should be noted that this form of phase variation is very unusual in that it not only gives ON or OFF regulation but also exhibits control over the *amount* of protein produced.

7.4.5 Phase variation mediated by differential DNA methylation

The mechanisms of phase variation described above all result from changes in the genome of a bacterium. One type of phase variation differs from this, in that the variation generates altered phenotypes whilst the genome sequence remains unaltered. In other words it is *epigenetic*. An example is the mechanism regulating expression of the *agn43* gene which codes for the major phase-variable outer membrane protein of *E. coli* (antigen 43). Although the exact function of this protein is not known, by its self-association characteristics it mediates autoaggregation and flocculation of cells of *E. coli*.

A key role in the phase variation of antigen 43 is played by the enzyme deoxyadenosine methylase (DAM), which adds a methyl residue to the N-6 position of adenine in the DNA sequence GATC. DNA methylation has been historically associated with restriction modification systems (see Chapter 4) but it can also alter the interactions of regulatory proteins with DNA and consequently is used by the cell to regulate the expression of certain genes in a phase-variable manner. The regulator OxyR, which is normally an activator of genes involved in the response to oxidative stress, is able to bind to a DNA region upstream of *agn43* and acts, unusually in this case, to *prevent* transcription (Figure 7.20). This binding site for OxyR contains three GATC sites which are targets for DAM methylation. If all three of these sites are methylated, OxyR is no longer able to bind to the DNA and *agn43* is expressed. Conversely, in the absence of methylation, OxyR is able to bind and prevent transcription. After DNA replication, these sites are randomly methylated and this results in phase variation, as some cells will express antigen 43 and others will not.

8
Genetic Modification: Exploiting the Potential of Bacteria

The term 'genetic modification' has become associated specifically with *in vitro* genetic manipulation technology ('gene cloning'), but it should be considered alongside older, *in vivo*, methods of altering the genetic composition of bacteria. In this chapter, we will review the methods available for producing useful bacterial strains. These include enhanced formation of natural products such as antibiotics, making non-bacterial products such as human growth hormone and the development of vaccines against infectious diseases. In Chapter 9 the role of genetics in developing our understanding of the biochemical and physiological processes that are at work within the cell will be discussed, while Chapter 10 continues the story of genome sequencing and bioinformatics. There is a considerable degree of interaction between the concepts and techniques in these three chapters and the distribution of material between them is to some extent arbitrary.

8.1 Strain development

Natural evolution is based on three processes: the generation of variants, the selection of variants with desirable properties and the re-assortment of characteristics between strains by genetic exchange. The application of strain improvement programmes to commercially useful microorganisms, mostly for increased product formation, has mainly involved the first two processes.

8.1.1 Generation of variation

In earlier chapters, it has been shown that variation occurs naturally but at a low frequency. For practical purposes the frequency needs to be increased by using

Molecular Genetics of Bacteria, 4th Edition by Jeremy Dale and Simon F. Park
 ISBN 0 470 85084 1 (cased) ISBN 0 470 85085 X (pbk)

mutagenic agents. Early studies used powerful mutagens such as X-rays or nitrogen mustard, but these tend to cause additional undesirable effects, as well as being hazardous to use. Milder, more easily controlled agents such as ultraviolet irradiation are generally to be preferred (see Chapter 2).

Since mutation is most likely to be deleterious to the gene affected, it might be expected that it would be difficult to isolate derivatives with increased product formation. This is only partly true. As will be shown later on, enhanced product formation can be obtained by abolishing the regulation of the pathway or by eliminating other metabolic activities which reduce the accumulation of the desired product.

Such mutations can be produced empirically, in the absence of information about the pathways or their control by screening for the required phenotype. Alternatively, gene cloning can be used, as described later in the chapter. However, this requires more knowledge of the genes concerned.

8.1.2 Selection of desired variants

In evolutionary terms, improved strains are selected because of their increased fitness. Similarly in microbial genetics we usually speak of selection as meaning the application of conditions under which only the desired strain is able to grow. Selection of antibiotic-resistant mutants is the clearest example.

However, when applying strain improvement programmes to commercially useful microorganisms, it is rarely possible to do this. Producing higher levels of an antibiotic does not confer a selectable advantage on the producing strain. Instead it is usually necessary to pick individual colonies and test the level of production for each isolate. This process is best referred to as screening, but, rather confusingly, it is also often referred to as 'selection'.

Characteristics other than the level of product are also important for commercial strains. These characteristics include, for example, growth rate, substrate utilization, response to different fermentation conditions and the absence of undesirable by-products. Such by-products not only detract from formation of the required material but also may contaminate the final product, thereby increasing the cost of downstream processing (i.e. the extraction and purification of the product).

8.2 Overproduction of primary metabolites

In microbial biotechnology, it is usual to distinguish between primary and secondary metabolites. A primary metabolite can be considered as a substance (an amino acid for example) that is formed as part of normal growth and occupies a readily identifiable role in cell metabolism. A secondary metabolite (such as an antibiotic) on the other hand is usually produced during stationary phase and does not seem to be central to the growth of the cell.

8.2.1 Simple pathways

A primary metabolite can be considered as the end-product of a series of reactions (Figure 8.1) which converts a precursor substrate (S) into the final product (P). The first of the major factors that are likely to limit the production of P is the supply of the initial substrate (S); increased intracellular levels of S may be expected to increase the throughput of the pathway. However, since S is also used for other metabolic pathways within the cell, it may not be easy to influence its availability. One example (glutamate production) is considered later in this chapter.

Secondly, we come to the rates of the individual enzymic reactions. In each case, this will be influenced by the number of enzyme molecules, the catalytic activity of the enzyme and its affinity for the substrate. Theoretically, it is possible to alter the enzyme structure to increase its maximum activity and/or its substrate affinity, but in practice such mutations will be extremely rare. A more likely target is an increase in the rate of production of the enzyme, most commonly by an alteration in the promoter site so as to increase the transcription of the gene.

Feedback regulation

The third (and often the most important) factor is the regulation of the pathway. The pathway may be capable of high throughput, but as the level of product increases, feedback effects reduce the rate of product formation. This may be due to repression where the production of the enzyme is diminished or to inhibition where the activity of the enzyme is reduced. Either or both of these effects will mean that attempts to increase the production of P will simply lead to shutting down the pathway.

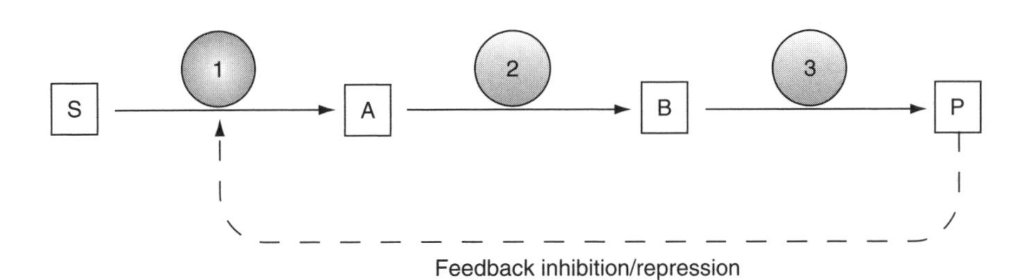

Figure 8.1 Regulation of primary metabolite production: a simple unbranched pathway

Therefore, if colonies are screened for overproduction of the final product, it is most likely that mutants which are deficient in the feedback regulation of the pathway will be obtained. If the production of tryptophan is taken as an example, the whole operon of five genes (in *E. coli*) is repressed by the presence of tryptophan (Chapter 3). Mutations that abolish these controls will prevent the bacterium from responding to the presence of tryptophan, leading to increased synthesis of the product.

Feedback inhibition on the other hand is (usually) caused by the end-product binding to a site on the enzyme which alters the shape of the protein so that it can no longer carry out its enzymic function. While it is possible to obtain mutants that are resistant to feedback inhibition, the required alterations to the enzyme are much more subtle than those that inactivate a repressor protein and are therefore more difficult to isolate by random screening.

Antimetabolites

Mutants resistant to feedback regulation can sometimes be isolated with the aid of *antimetabolites*. These are analogues of the end-product of the pathway that are lethal because they substitute for the genuine product in the feedback effect but are unable to substitute for it in its metabolic role within the cell. For example, the tryptophan analogue 5-methyl tryptophan represses the tryptophan pathway (so no tryptophan is made) but it does not substitute for tryptophan in protein synthesis. Mutants that are deficient in feedback regulation are able to grow in the presence of 5-methyl tryptophan, since the *trp* operon will be expressed and tryptophan will thus be produced for protein synthesis. Regulatory mutants can therefore be selected by plating the mutagenized culture on a medium containing the antimetabolite.

8.2.2 Branched pathways

Many amino acids are products not of a simple pathway but of a branched pathway. In Figure 8.2 the production of R is diverting resources from the product P. A mutant defective in enzyme 4, and thus unable to produce R, will be expected to make higher levels of product. An additional advantage of the absence of R is that a branched pathway often exhibits concerted (or multivalent) repression. In this example, enzyme 1 is only repressed if both P and R are present in sufficient quantity. For example, the attenuation of the *ilv* operon requires the simultaneous presence of leucine and valine (which are the end-points of a branched pathway), as well as isoleucine which requires the same enzymes as valine. Another example is the production of lysine, which is considered in detail below.

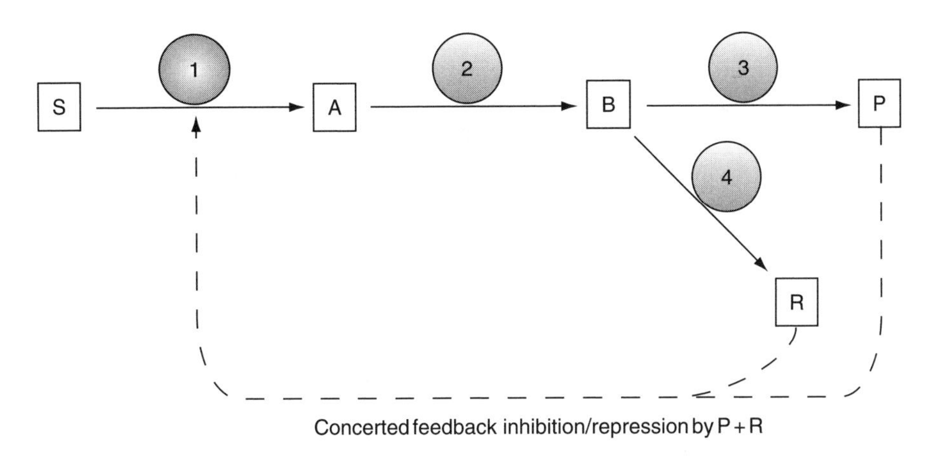

Figure 8.2 Regulation of primary metabolite production: a branched pathway

In the general outline in Figure 8.2, if R, the unwanted by-product, is essential for growth of the cell, then a mutant that does not make enzyme 4 will be auxotrophic for R and can be isolated by replica plating (Chapter 2). Growth of these mutants of course requires the addition of R to the growth medium. It might be expected that this would result in feedback inhibition/repression of enzyme 1. However, it is usually possible to add such substrates in amounts that are sufficient for growth but do not result in feedback effects.

Lysine production

A specific example is the commercial production of lysine which is widely used for supplementation of cereal-based animal feeds. A simplified representation of the lysine production pathway of *Corynebacterium glutamicum* (Figure 8.3) shows that, as with other bacteria, the initial steps of lysine synthesis are shared with those of the pathways for the synthesis of threonine, isoleucine and methionine.

Mutants of *C. glutamicum* that are defective in the enzyme homoserine dehydrogenase are auxotrophic but can grow if homoserine (or a mixture of threonine and methionine) is provided. Such mutants produce large amounts of lysine (over $50\,g\,l^{-1}$) due partly to the diversion of metabolites away from the other amino acids, but also to relief from feedback control. The first enzyme in the common pathway (aspartokinase) is subject to concerted feedback regulation by lysine and threonine. However, in the auxotroph, fed with limiting amounts of homoserine, the level of threonine will be too low to cause inhibition of the aspartokinase.

Amino acid analogues can be used to obtain feedback-resistant mutants of this system. For example, the lysine analogue S-(2-aminoethyl)-L-cysteine mimics the feedback inhibition by lysine of aspartokinase. Mutants that are resistant to this

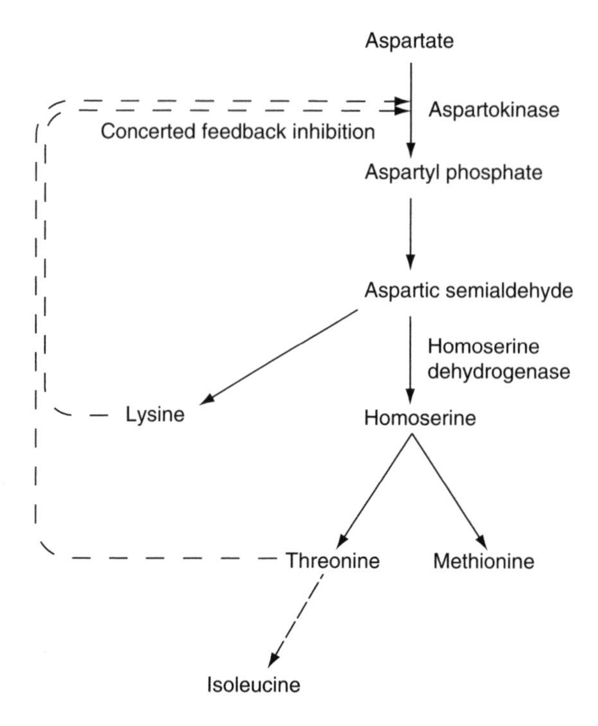

Figure 8.3 Lysine synthesis in *Corynebacterium glutamicum*

analogue possess an altered aspartokinase which is much less sensitive to feedback inhibition by lysine and therefore are able to accumulate high levels of lysine.

Glutamic acid production

Another commercially important amino acid is glutamic acid, which is employed as a flavour enhancer in the form of its monosodium salt (monosodium glutamate). This is also produced by *C. glutamicum* from the TCA cycle intermediate α-ketoglutarate by the action of glutamate dehydrogenase (Figure 8.4). This reaction competes for its substrate with the next enzyme in the TCA cycle, α-ketoglutarate dehydrogenase. Mutants defective in α-ketoglutarate dehydrogenase activity tend to accumulate glutamate instead.

8.3 Overproduction of secondary metabolites

Antibiotics are the most important secondary metabolites produced by microorganisms. The first true antibiotic, penicillin, was a product of a fungus (*Penicillium*), but the major sources of current naturally produced antibiotics are the

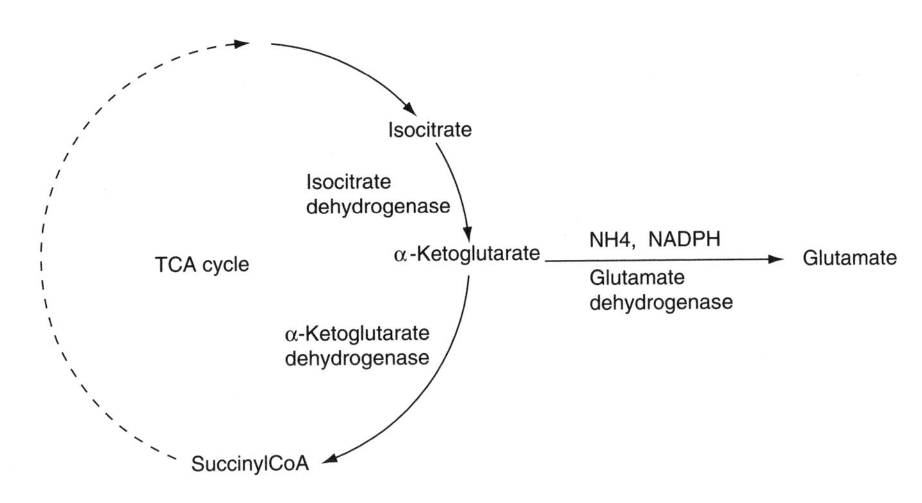

Figure 8.4 Synthesis of glutamic acid

filamentous bacteria known as actinomycetes (especially *Streptomyces*). Generally the level of antibiotic production by natural strains is much too low for a commercially viable process and so a strain improvement programme is needed. However the metabolic pathways for synthesis of these secondary metabolites are more complex, and more diverse, than those for primary metabolite production, therefore, the strain improvement programme has to be carried out empirically in the absence of any clear knowledge of either the pathways or the regulatory processes.

This has not proved to be a difficult obstacle to overcome. Screening large numbers of colonies is generally successful in identifying variant colonies with enhanced production. These are then subjected to a further round of mutagenesis and screening. Repeated cycles of this process can lead to a strain producing a level of antibiotic several orders of magnitude higher than the original isolate.

It is worth emphasizing that most antibiotic production employs strains that have been largely or entirely derived from such empirical processes. Although many of the pathways have now been elucidated and genome sequencing is enhancing our understanding of the processes involved, random mutation and screening may still be the most effective way of achieving higher production levels. On the other hand, a detailed knowledge of the pathways and their control does open up new prospects for manipulating the pathway for the production of different derivatives of the antibiotic in question.

8.4 Gene cloning

Classical (*in vivo*) genetic techniques are essentially limited by two factors. Firstly, they can only be applied to the existing genetic complement of an organism, i.e.

they are restricted to naturally-occurring genes or relatively minor modifications of these genes. It is not possible using these techniques to make a product totally foreign to that organism.

Secondly, with classical techniques, work can only be carried out on the basis of the phenotype, i.e. mutants are selected by their effect on the observable characteristics of the organism. This severely limits the changes that can be selected.

The advent of gene cloning (also referred to as *in vitro* genetic manipulation or genetic engineering) has dramatically changed the picture in both respects. The basis of these techniques is the use of enzymes to cut and rejoin fragments of DNA. In this way, foreign DNA fragments can be inserted into a vector (a plasmid or a bacteriophage) which enables the DNA to be replicated within a bacterial cell. This section provides an introduction to the concepts and applications of gene cloning.

8.4.1 Cutting and joining DNA

The ability of restriction endonucleases to cut DNA at specific sites as described in Chapter 4, is a key factor in gene cloning since it enables the attainment of specific fragments of DNA. Such fragments can be joined together by DNA ligase. Any two ends generated with the same restriction enzyme can be joined in this way, but an *Eco*RI fragment cannot be ligated to say, a *Bam*HI fragment. (A BglII fragment can however, be joined to one generated by *Bam*HI or *Sau*3A; although the enzymes recognize different sequences, the sticky ends generated are identical: see Box 4.1). DNA ligase can also join together blunt-ended fragments. Although less efficient, blunt-end ligation can be useful because it does not require the fragments to have been generated with the same enzyme.

Replication of the DNA fragment is achieved by transforming a suitable host strain after ligation with a vector that is itself able to replicate, i.e. a plasmid or bacteriophage. Using a plasmid vector, it would be possible to recover a single colony from an agar plate and to use this to produce a bacterial culture in which each cell carries a copy of the original DNA fragment. Strictly speaking *cloning* is this process of purifying a single colony from the mixture of transformed cells (or a phage preparation from a single plaque). However, the term 'gene cloning' is often used to describe the whole process or even just the step of recombining the DNA fragment with the vector. The power of the technique arises firstly from the fact that the source of the DNA fragment is immaterial; DNA is DNA, and can be cloned and replicated in *E. coli*, irrespective of its source or its sequence. Secondly, it is not necessary to purify the fragment to be cloned. Provided that there is a method available for identifying which bacterial colonies carry the gene of interest, isolating those colonies in pure form (cloning them) achieves that purification in a very simple manner.

8.4.2 Plasmid vectors

Many of the features of a plasmid vector for gene cloning are illustrated by the plasmid pUC18 (Figure 8.5) which is widely used for gene cloning.

Origin of replication

The first requirement is that the plasmid must be able to replicate in the chosen host (in this case, *E. coli*). The basic replicon from which pUC18 was constructed is related to the plasmid ColE1; this is the region containing the origin of replication (*oriV*). This plasmid therefore follows the mode of replication described for ColE1 (see Chapter 5).

Selectable marker

Secondly, it is necessary to be able to select those cells which have received the plasmid (transformants). Plasmid cloning vectors are therefore constructed so as to carry one or more antibiotic resistance genes; with pUC18 this is a β-lactamase gene which confers resistance to ampicillin.

There is a second marker carried by pUC18, namely a β-galactosidase gene. (To be strictly accurate, pUC18 codes for a fragment of β-galactosidase with the rest being provided by a host gene so that a functional enzyme can be made; i.e. pUC18 complements the otherwise inactive host gene). Transformation of a suitable host strain with pUC18 will therefore yield cells that produce functional β-galactosidase. These can be detected using the chromogenic substrate X-gal which gives a blue product when hydrolysed by β-galactosidase. Thus, colonies containing pUC18 will be blue, while those without the plasmid will be white. Note that this is not a selectable marker – both types of cell will grow – but it does provide an additional and useful form of discrimination.

Cloning site

The third essential requirement of a cloning vector is that it must contain suitable recognition sites for cleavage by one or more restriction endonucleases. This is so that the circular plasmid can be opened up at that point and the ends ligated with the ends of the DNA fragment to be cloned (see Figure 8.5).

For example, if the DNA fragment to be cloned has been generated by *Bam*H1 digestion and mixed with pUC18 DNA that has been linearized with the same enzyme, DNA ligase will join the *Bam*H1 ends of the fragments, with the result being the recombinant plasmid shown in Figure 8.5.

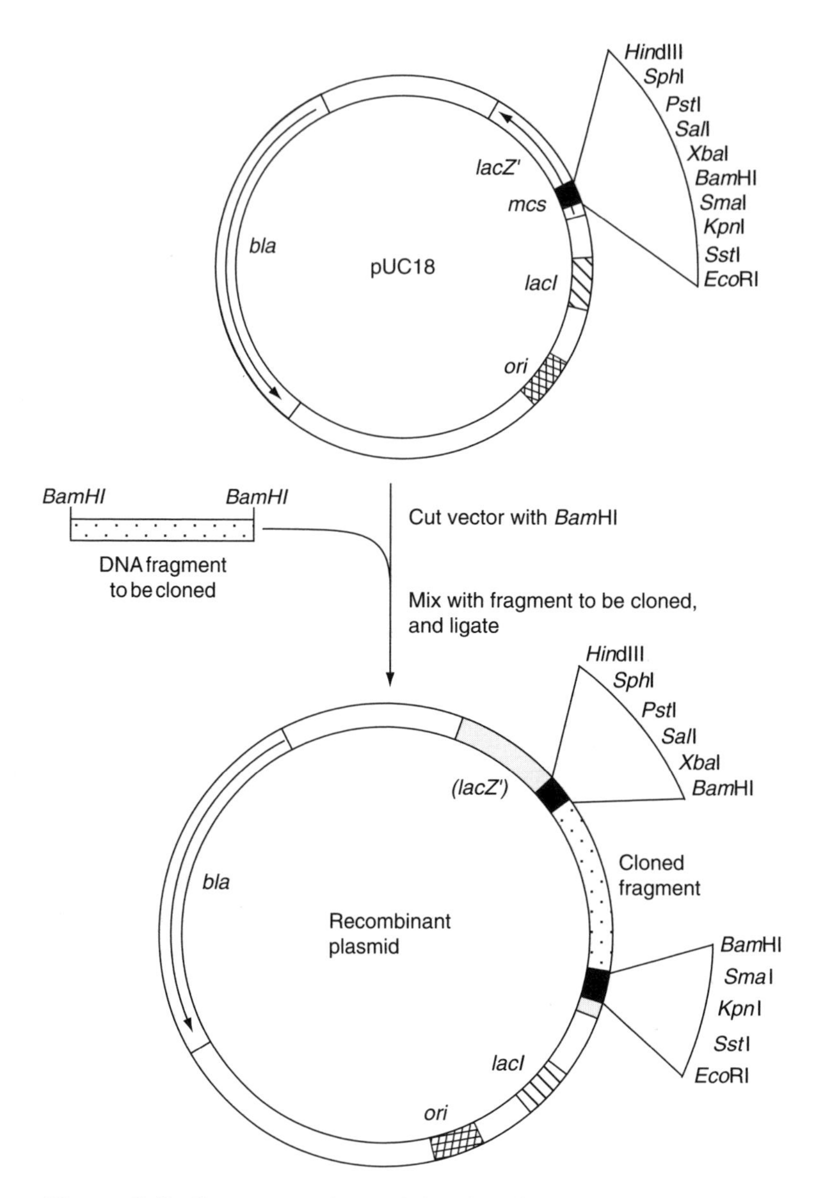

Figure 8.5 Structure and use of the plasmid cloning vector pUC18

In pUC18, a piece of synthetic DNA has been inserted which contains recognition sites for a number of restriction enzymes. This *multiple cloning site* allows a considerable degree of flexibility in the choice of restriction enzyme. The plasmid has been engineered so that each of these sites is unique, i.e. it contains no other sites recognized by any of these enzymes.

Insertional inactivation

The position of the cloning site is important. Insertion of a DNA fragment will (usually) result in inactivation of the gene in which that site is found. The cloning site must therefore not be within an essential region of the plasmid. It can be seen in Figure 8.5 that in pUC18 the multiple cloning site is within the initial sequence of the β-galactosidase gene (but does not inactivate the β-galactosidase gene). If a DNA fragment is inserted into the multiple cloning site, it will (usually) prevent the production of β-galactosidase, either by interrupting transcription or by altering the reading frame; this is known as *insertional inactivation.* Genuine recombinants will therefore be white on a medium containing X-gal and can be distinguished from blue colonies containing the original pUC18.

8.4.3 Transformation

Having produced a recombinant plasmid, it is necessary to put it into a host cell. Naturally-occurring transformation systems, as described in Chapter 6, are not suitable for this purpose, partly because they often do not work well with circular plasmid DNA and partly because in many bacteria (including *E. coli*) competent cells, able to take up DNA, do not occur naturally. For these bacteria, transformation requires the artificial induction of competence. There are numerous ways of doing this, the simplest of which (Figure 8.6) involves washing the cells repeatedly with cold calcium chloride solution. The competent cells are mixed with the DNA solution and subjected to a heat shock treatment, such as heating them at 42° for 1–2 min and then transferring them back onto ice. They are then diluted into broth and incubated at 37° to allow expression of the newly acquired DNA before plating onto an appropriate medium.

Much effort has been put into optimizing this system for transforming *E. coli*, including the development of special strains that show high transformation efficiency, up to 10^9 transformants per μg of DNA. Even with such a system, only a proportion of the host cells will have taken up the DNA, so it is necessary to have a good selectable marker on the transforming plasmid, as described above.

Other forms of transformation, notably electroporation, are especially useful for host bacteria other than *E. coli* and are considered later in this chapter.

8.4.4 Bacteriophage lambda vectors

Insertion vectors

Vectors derived from bacteriophage λ have been widely used for gene cloning in *E. coli*. In the simplest of these, insertion vectors, the DNA of the vector has a

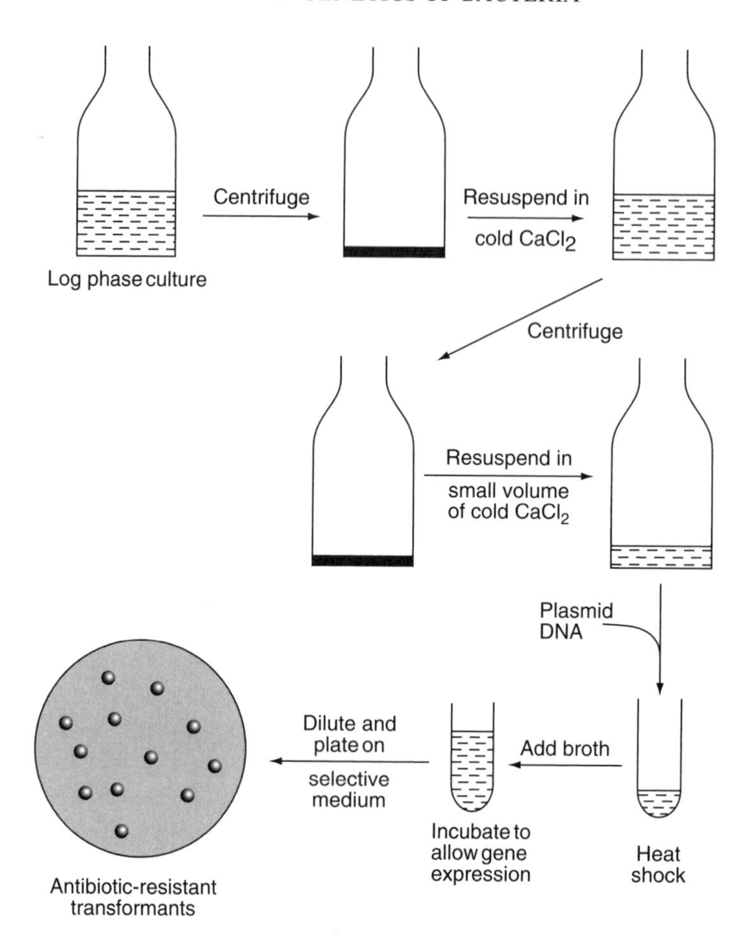

Figure 8.6 Basic procedure for plasmid transformation of *E. coli*

single recognition site for the chosen restriction enzyme and the fragment to be cloned is inserted at that position (Figure 8.7).

The most obvious difference from a plasmid vector is that after introducing the DNA into a bacterial cell, the success of the procedure is determined by the appearance of phage plaques in a bacterial lawn (see Chapter 4) rather than by the selection of antibiotic-resistant colonies. Another important difference concerns the way in which the cells take up this DNA. In the natural lytic cycle of λ, lengths of DNA cut from a multiple length DNA molecule are packaged into the empty phage heads, as described in Chapter 4. It is possible to do this *in vitro*, using cell extracts which contain phage heads and tails and the required enzymes. This process, known as *in vitro* packaging, produces virus particles that are able to inject the DNA into a host cell.

The natural packaging limits of λ impose constraints on the size of fragments that can be cloned using insertion vectors. Regions of the λ genome that are not

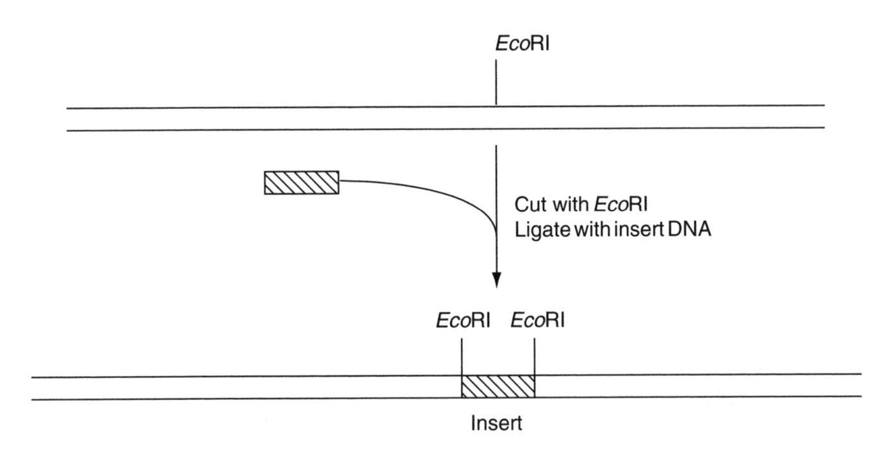

Figure 8.7 Structure and use of λ insertion vectors

essential for lytic growth can be deleted from the vector to make room for the insert. However, if more than about 10 kb is removed, the size of the vector DNA falls below the minimum packaging limit, so that the vector itself would not be viable. This is unfortunate since one of the main uses of λ vectors, the construction of gene libraries (see below), often requires the cloning of fragments that are as large as possible.

Replacement vectors

The cloning capacity of λ vectors can be increased by using a different approach. In the example shown in Figure 8.8, the vector contains two restriction sites rather than one. Cutting the DNA of this vector will therefore result in three fragments: the left and right arms and a central piece of DNA known as the *stuffer* fragment. The stuffer contains no genes needed for growth of the phage, so it can be separated and discarded. The left and right arms together are too small to allow viable phage particles to form; such particles require an additional inserted piece of DNA which must be, in this case, at least 7 kb and not more than 22 kb. Therefore, not only is the cloning capacity increased, but such vectors are selective for reasonably large inserts. Vectors of this type are known as *replacement* vectors.

8.4.5 Cloning larger fragments

Packaging of λ DNA requires only the DNA region around the *cos* sites (see Chapter 4). It is therefore possible to use *in vitro* packaging with a class of vectors known as *cosmids* (Figure 8.9). These are simply plasmids into which a *cos* site has

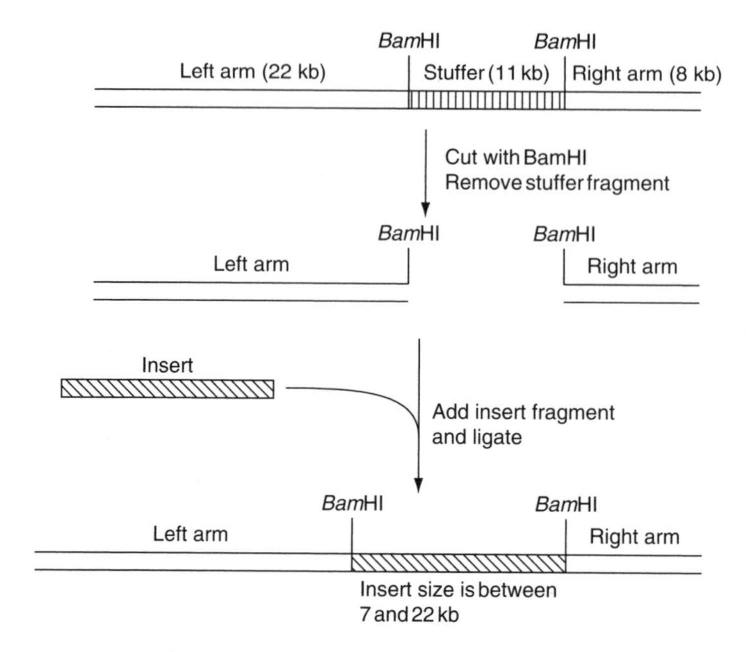

Figure 8.8 Structure and use of λ replacement vectors

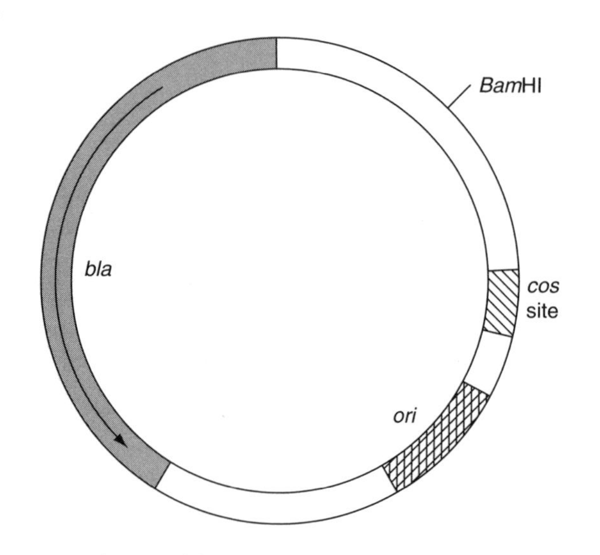

Figure 8.9 Structure of a cosmid. *ori*, origin of replication; *bla*, β-lactamase (ampicillin resistance as a selectable marker); *Bam*H1, restriction enzyme recognition site where DNA can be inserted. The presence of the *cos* site enables the plasmid to be packaged into λ phage particles provided that a sufficiently large piece of DNA has been inserted

been introduced. The cosmid itself is much too small to be packaged successfully. However, if the insert is large (say 40 kb), then the recombinant cosmid will be large enough for viable particles to be produced. These particles can then infect a sensitive host, but in this case antibiotic-resistant colonies rather than phage plaques will indicate the success of the procedure.

Larger fragments can be cloned using vectors based on the bacteriophage P1, which have a cloning capacity of 95 kb, or the so-called bacterial artificial chromosome (BAC) system which is based on the F plasmid and is able to maintain inserts of greater than 300 kb. These vectors have been particularly useful in genome sequencing projects (Chapter 10).

8.4.6 Bacteriophage M13 vectors

The filamentous bacteriophage M13 (see Chapter 4) provides the basis for another group of vectors. With M13, the phage particles contain single-stranded DNA while the replicative form is double stranded. Fragments of DNA can therefore be inserted into the replicative form of the M13 DNA. Transformation of a suitable host will produce plaques from which the recombinant phage can be obtained. Since the phage particles contain single-stranded DNA, this is a convenient way of obtaining cloned DNA in a single-stranded form, which is useful for some applications such as site-directed mutagenesis. Gene sequencing originally depended heavily on obtaining single-stranded templates using M13 vectors. Although modern sequencing methods can use double-stranded templates, M13 clones are still occasionally used for this purpose.

8.5 Gene libraries

8.5.1 Construction of genomic libraries

A gene library (or more precisely a genomic library, to contrast it with a cDNA library – see below) is a collection of recombinant clones each of which carries a different piece of DNA from the organism of interest, so that between them they represent the complete genome of that organism.

To construct a genomic library, the DNA from the source organism is fragmented randomly and the mixture of pieces joined to the vector DNA (see Figure 8.10). This will produce a number of recombinant molecules, each having a different piece of source DNA. No attempt is made to separate these molecules; instead the whole mixture is used for transformation of *E. coli*. Each individual clone carries a different piece of DNA and the whole collection of clones constitutes the gene library.

The number of clones needed for a complete genomic library depends on the size of the genome and the average size of fragment cloned. For example, for a bacterium with a genome of 4×10^6 base pairs and an average insert size of 18 kb (using a λ vector), a library of 1000 clones gives a 99 per cent probability of

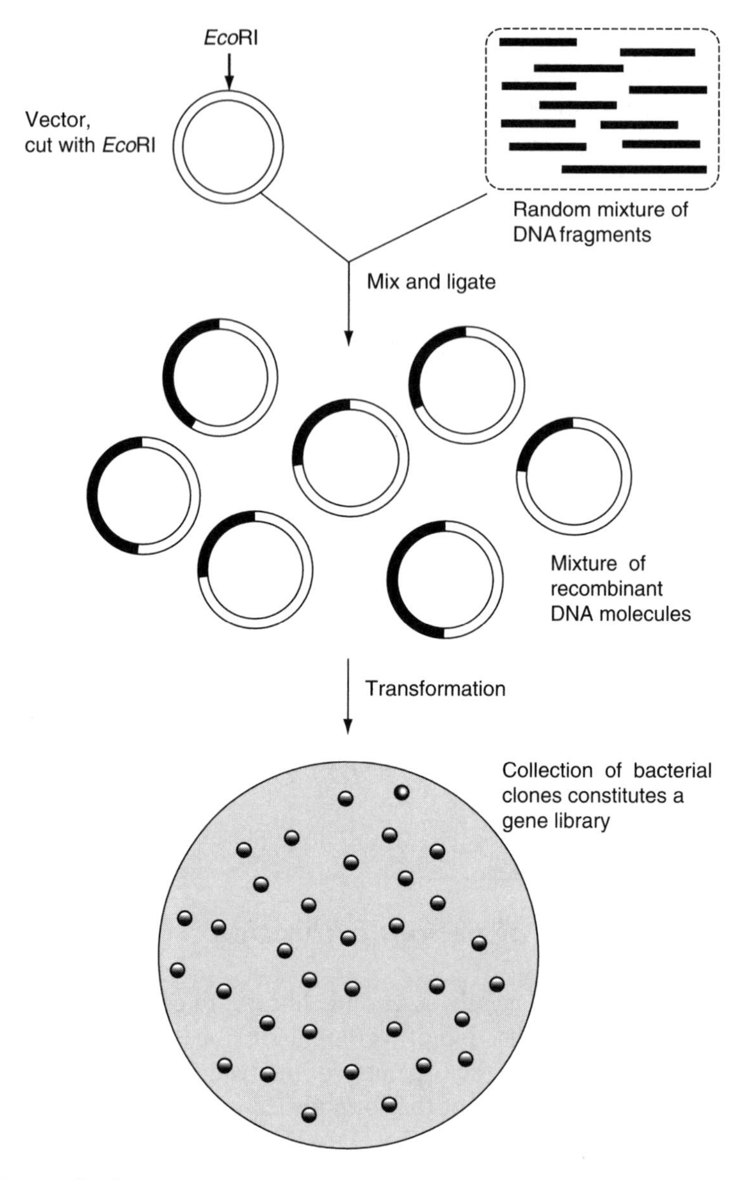

Figure 8.10 Construction of a genomic library using a plasmid vector

recovering any specific gene. With an average insert size of say 1 kb, about 18 000 clones would be required to achieve 99 per cent representation of the same genome. (The calculation has to take account of the overlap and duplication that will occur in a random library). Using the techniques described in the next section, it is not difficult to screen several thousand clones, so even a small insert library is quite usable for a bacterial genome. But with larger genomes, the need for larger inserts becomes important, unless machines are available which can handle hundreds of thousands of clones.

8.5.2 Screening a gene library

Making a genomic library of bacterial DNA is comparatively straightforward. However, it is necessary to have some way of finding what is required. Rapid screening methods have been developed to enable large numbers of colonies to be tested simultaneously, commonly using either gene probes or antibodies.

Screening with gene probes

The use of gene probes relies on the hybridization of complementary single-stranded nucleic acid sequences (see Chapter 1). Screening a gene library involves transferring a print of the colonies (or phage plaques) to a filter, which is subsequently hybridized with a labelled gene probe. A variety of non-isotopic labels are now commonly used, rather than radioactive labels. The position of a positive reaction on the filter can be correlated with the position of a specific clone which can be picked off and purified.

The main factor is how to obtain the required probe. If it is known, or suspected, that the gene required is related to one that has already been characterized from another source, then that DNA sequence can be used as a heterologous probe, using a low stringency of hybridization (to allow for a degree of mismatching between the DNA sequences).

Alternatively, if a small amount of purified protein can be isolated, then it is possible to determine the amino acid sequence of a small part of that protein (usually the N-terminal sequence). From that information, the likely sequence of the DNA can be deduced and a short probe corresponding to that sequence can be synthesized.

Polymerase chain reaction

An alternative way to generate the required probe is by using the polymerase chain reaction (PCR; see Chapter 2). For this purpose, the PCR primers are derived from the sequence of related genes in the databanks, using portions of the genes that are very similar in all known sequences of this family of genes (conserved regions). If the target gene has sufficiently similar sequences, the primers will anneal and the PCR will therefore amplify a DNA fragment from that gene (Figure 8.11). This product can then be labelled and used to screen a genomic library.

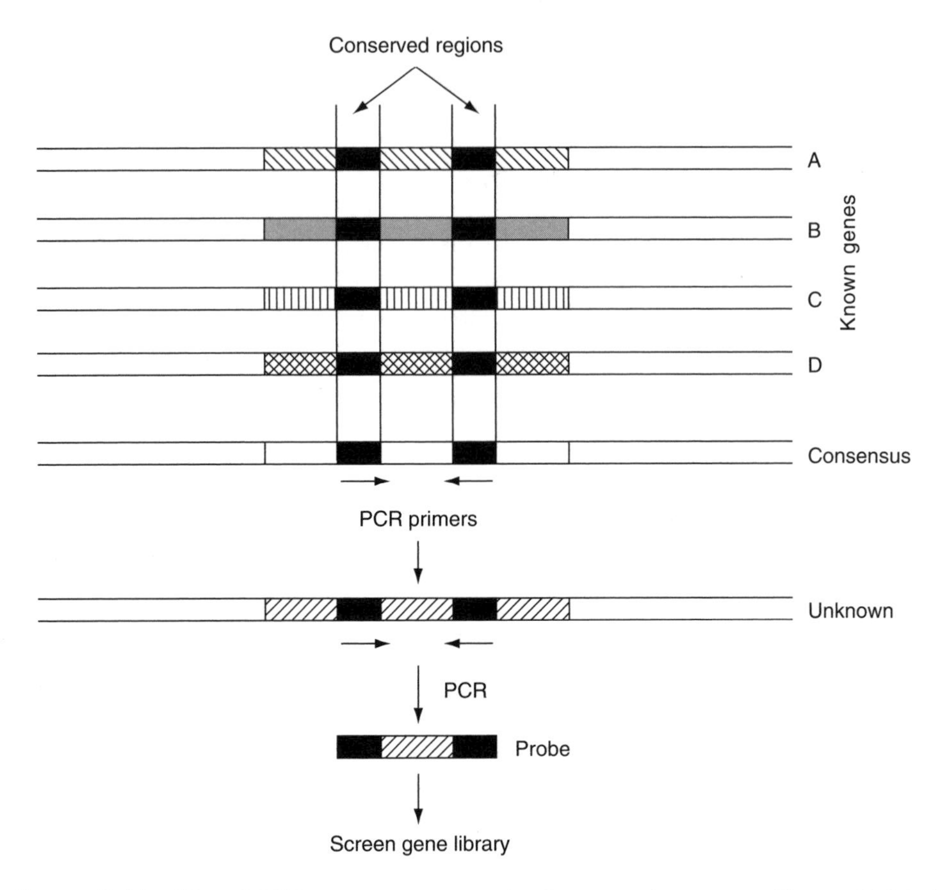

Figure 8.11 Use of PCR to generate a probe for screening a gene library. Related genes A–D, from different sources, have two conserved regions (i.e. they are similar in all four genes). PCR primers designed to recognize these conserved sequences can be used to amplify part of the corresponding gene from another source. The amplified product can be used as a probe for screening a gene library so as to isolate the complete gene

The use of PCR can bypass the need for making a gene library at all. If we are studying a specific gene in closely related species and the sequence of the gene from one species is known, then the primer sequences from the known gene can be simply derived and used to amplify the corresponding gene from the other organisms. This application will be considered further in Chapter 10.

Screening gene libraries with antibodies

If antibodies to the protein of interest can be produced, these can be used to screen a gene library in much the same way as described for a gene probe. There are certain limitations to this approach: the gene concerned must be expressed (although not necessarily in an active form), the antibody should recognize the protein in a denatured state (and hence should recognize a linear epitope rather than a conformational one) and it must recognize the primary product of translation rather than some modified form (e.g. a glycosylated epitope), since this post-translational modification may not occur in the recombinant bacterium. Despite these limitations, antibody screening has been widely used, especially for genes coding for key antigens of specific pathogens (see the description of λ gt11 below).

Screening gene libraries by complementation

It is also possible to screen a gene library by testing for the ability to complement specific mutant strains of *E. coli*. For example, a leucine auxotroph (which is unable to grow without added leucine) may be converted to prototrophy by a recombinant plasmid carrying a *leu* gene from another bacterium. The required clone can therefore be identified by plating the library onto a medium lacking leucine.

Of course, this procedure only works if the cloned gene is expressed and only if the product is functional (and readily detected) in *E. coli*. It is therefore not the most powerful way of screening gene libraries, but it is particularly useful for confirming the identity of cloned genes when a possible function has been identified by sequence comparisons.

8.5.3 Construction of a cDNA library

In some circumstances, it is advantageous to make a gene library from the mRNA rather than from the DNA. This requires the synthesis of double-stranded DNA using the mRNA as a template. There are many variations in the details of this procedure, which will not be considered here, but the central feature is the use of an enzyme known as *reverse transcriptase* (since it is an RNA-directed DNA polymerase, as opposed to the enzyme responsible for transcription proper which is a DNA-directed RNA polymerase). The DNA produced (complementary DNA, or cDNA) can be ligated to a vector to produce a cDNA library.

Libraries of cDNA are commonly used for animals and other eukaryotes. Since a cDNA library contains only those genes that are expressed in the starting material, it will be much smaller than a genomic library. Furthermore, eukaryotic genes usually contain introns (see Chapter 1); these are removed during processing of the mRNA, so the size of a cDNA clone needed to contain the complete gene is much smaller than the corresponding length of genomic DNA.

Bacterial cDNA libraries are less commonly used. Since bacterial genomes are smaller, and bacterial genes do not often contain introns, there is less of an advantage in using a cDNA library (and making a genomic library is much easier). Nevertheless, bacterial cDNA libraries are used occasionally, for example to identify and isolate genes that are selectively expressed under specific growth conditions. Other applications of reverse transcriptase will be discussed in Chapter 10.

8.6 Products from cloned genes

8.6.1 Expression vectors

If a gene from one organism is put into another, it may not be expressed very well if at all. The most common reason for this is that the promoter may not be recognized by the RNA polymerase of the new host. So, if expression of the product is required, an expression vector (see Figure 8.12) would be used which contains an *E. coli* promoter adjacent to the cloning site. Insertion of a DNA fragment at this site, in the correct orientation, places it under the control of the promoter provided by the vector, thereby ensuring expression in the recombinant *E. coli* clone.

A further development of this concept is to use the vector to supply translational signals as well as a promoter. The result (if the construct is designed to maintain the correct reading frame) will be a *fusion protein* which has an N-terminal sequence provided by the vector and a C-terminal region which is the product that is to be detected. This can improve the expression and stability of the product.

Often once a protein has been expressed in *E. coli*, the next step would be to purify the protein and separate it from the other proteins present in the cell. Protein purification can be simplified by using specialized expression vectors that incorporate an affinity tag at the N-terminal sequence of the cloned protein. Peptides with short stretches of histidine residues bind specifically to nickel, for example. If the gene is manipulated so that the protein has a string of histidines (usually six to 10) at its N-terminus, the His-tagged protein can be purified on a column containing a solid phase with bound nickel ions.

λ gt11

One example of an expression vector, although used in a rather different way, is the λ vector gt11 (Figure 8.13). This is an insertion vector with a single *Eco*RI site within a β-galactosidase gene, so that an inserted DNA fragment will be under the control of the *lac* promoter. Furthermore, if the insert is in the correct reading frame, the peptide coded for by that fragment is actually linked to the β-galactosidase. This vector is useful for gene libraries that are to be screened with antibodies, for example in the identification and isolation of genes coding for protein antigens of pathogenic bacteria.

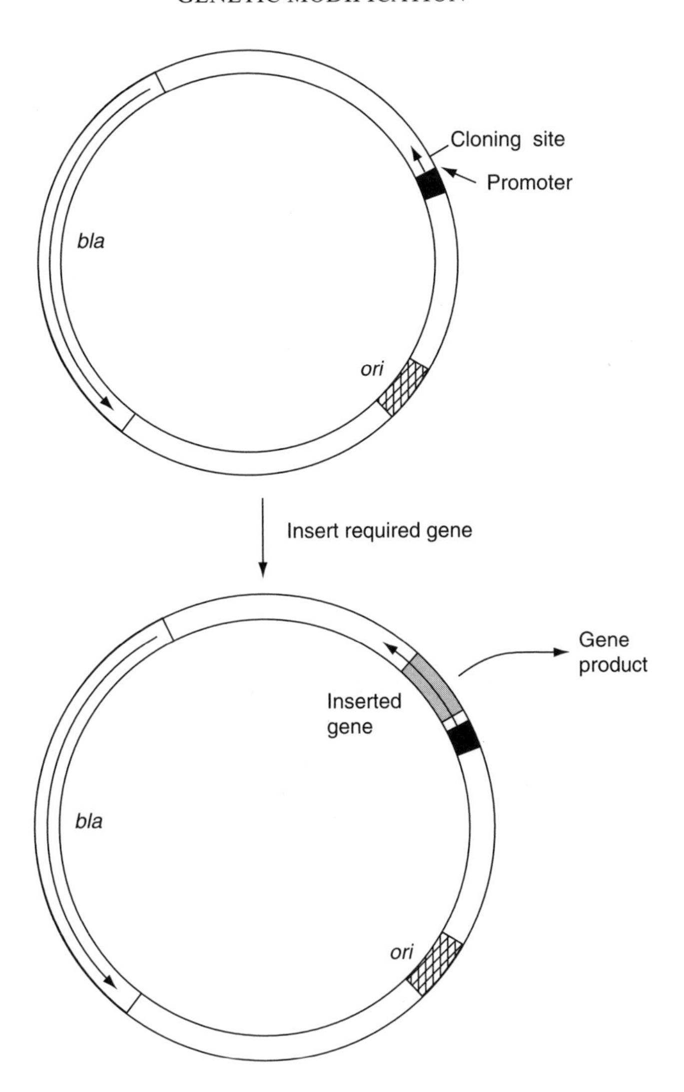

Figure 8.12 Basic features and use of an expression vector. The cloning site is adjacent to a promoter on the vector so that inserted DNA will be transcribed from that promoter. With some expression vectors, translational signals are also included

Maximizing product formation

The application of gene cloning which received most publicity, before the advent of gene therapy and transgenic animals, was the production of foreign proteins by microbial hosts. For example, human growth hormone (HGH or somatotropin) is a polypeptide hormone which is used to treat the condition known as pituitary dwarfism. Prior to gene cloning, the only source of this material was human pituitary glands removed at autopsy. The limitations on supply coupled with

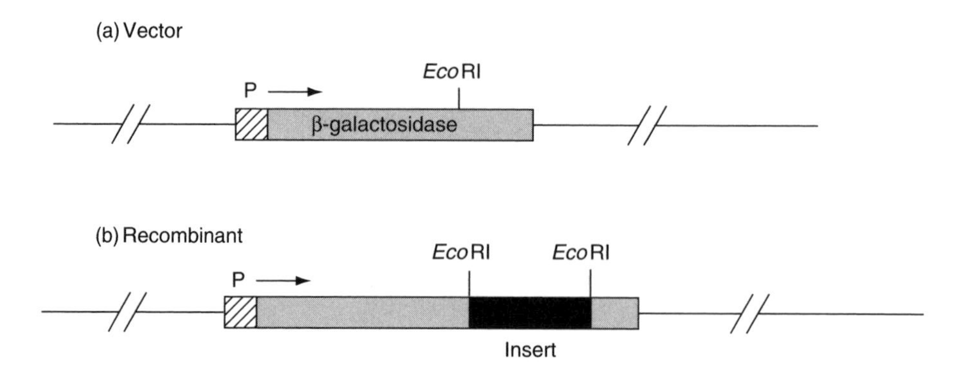

Figure 8.13 Use of λ gt11 for generation of fusion proteins. (a) The vector carries a β-galactosidase gene with an *Eco*RI site within it. (b) Insertion of DNA at the *Eco*RI site disrupts the *lacZ* gene. Functional β-galactosidase is not made but (if the insert is in the correct reading frame) a fusion protein is produced

the risks of transmission of latent infections, were a powerful stimulus for efforts to develop an alternative supply. Cloning and expressing the gene in *E. coli* provided a safe and plentiful supply of the hormone.

In order to take full advantage of bacterial expression of foreign proteins, sophisticated expression vectors are used to maximize yields of the recombinant product. Extremely high levels of specific protein, representing perhaps 50 per cent of total cell protein, can be produced using such a vector. However, diverting so much of the cell's resources into a product that is useless as far as the cell is concerned will slow down the growth rate (which will also exacerbate plasmid instability as discussed in Chapter 5). The usual strategy for overcoming this problem is to use a promoter that is subject to repression so that the bacteria can be grown to high cell density before product formation is induced by altering the growth conditions. For further details and specific examples, see the reading list in Appendix A.

8.6.2 Making new genes

DNA synthesis

Gene cloning need not be restricted to DNA sequences which occur in nature. It is possible to synthesize stretches of DNA (oligonucleotides) of any sequence using a DNA synthesizer. The operator merely types in the required sequence and the machine does the rest. Since this is a sequential process, the accuracy tends to decline with longer oligomers, but sequences of up to 100 bases can be constructed with a reasonable degree of accuracy. Longer sequences can be made by enzymatic joining of these shorter oligomers.

Site-directed mutagenesis

It is possible, using synthetic DNA, to construct a totally new gene of any sequence. In practice, it is more common to introduce defined alterations in a naturally-occurring gene, a procedure known as *site-directed mutagenesis*. There are many ways of doing this; the basic concept of one method is shown in Figure 8.14.

A short synthetic oligonucleotide is made that contains the altered sequence and this is hybridized to a plasmid containing the complementary strand of the old gene. Although the sequences are not identical, there is sufficient similarity for binding to occur. The synthetic oligonucleotide (the mismatch primer) can then prime synthesis of the complete DNA strand which now contains the altered sequence. After transformation, the mismatched sequences will both be replicated, so that some cells will contain the new gene and some the original. These can be distinguished using the mismatch primer as a probe. The example in Figure 8.14 represents part of the active site of a β-lactamase in which the serine residue plays an essential role in the hydrolysis of the β-lactam bond of penicillins. Changing this residue to a cysteine results in a novel enzyme with some unusual properties.

Required mutation changes a serine to a cysteine:

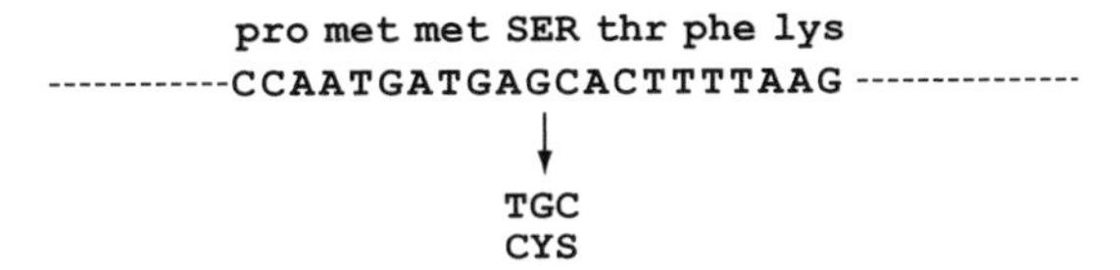

Synthetic oligonucleotide containing the altered base:

CCAATGATGTGCACTTTTAAG

Hybridize oligonucleotide to circular (complementary) single-strand DNA: acts as a primer for synthesis of new DNA strand containing the mutated gene

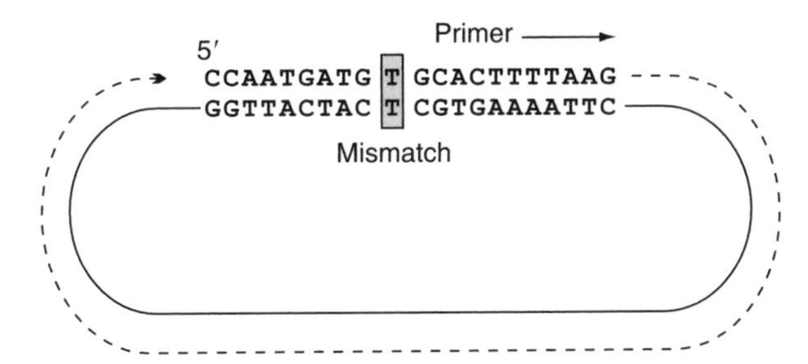

Figure 8.14 Site-directed mutagenesis. The diagram illustrates the basic principle of the procedure; there are a number of different methods for achieving site-directed mutagenesis

Protein engineering

One application of these techniques is in the production of proteins with novel properties. For example, some commercially used enzymes are limited in their application because of temperature instability. If the enzyme were more stable, it would be possible to carry out that process at a higher temperature, thus increasing reaction rates and the overall efficiency of the process. Alternatively minor changes in the amino acid sequence around the substrate binding site could alter the substrates recognized by the enzyme. Changes of this sort are not easy to obtain by conventional genetic means.

This technique, and many refinements thereof, forms the basis of an exciting area of research known as protein engineering. It should be said however, that replacing even a single amino acid in a protein chain can have many untoward effects on the structure of that protein. A completely rational use of protein engineering demands not only a detailed knowledge of the crystal structure of that protein but also an intimate understanding of what determines that conformation and the relationship between the structure and function of the enzyme. This understanding is still essentially fragmentary.

Directed in vitro *evolution of protein function*

Ways of introducing specific mutations into genes so as to produce enzymes with new functions or improved characteristics, have been discussed in the sections above. However, the specific methods of site-directed mutagenesis suffer not only from limited understanding of the structure and function of the protein, but also become very laborious if a number of changes to the gene are required. An alternative technique, DNA shuffling, allows the accelerated and directed evolution of proteins *in vitro* through recombination which can be carried out in a test tube using PCR. In this procedure, related genes from different species, or genes with related function, are fragmented. The fragments are then mixed and PCR is used to reassemble them randomly. The new genes are then cloned into *E. coli* to identify which of them produce usable or potentially interesting products, for example by screening clones for the ability to degrade a novel substrate. It is possible to extend the concept by using synthetic DNA fragments with random sequence to replace selected parts of the gene.

In part therefore this is a return to the older concept of selecting randomly generated strains that have the required characteristics, rather than the more specific methods that are usually associated with gene cloning technology, but using the technology to speed up and extend the generation of new variants.

8.6.3 Other bacterial hosts

Although *E. coli* is the most versatile host for gene cloning, it has its limitations. One of these is the difficulty of obtaining product secretion. Secretion of proteins is advantageous for commercial production since it facilitates purification of the product. In addition, expression of proteins at a very high level intracellularly can cause the solubility limit to be exceeded, so that the expressed proteins form insoluble aggregates within the cell. These can be damaging to the cell and are also difficult to re-solubilize. Since the extracellular volume of a culture is much larger, this problem is unlikely to arise with a secreted protein. The use of other bacteria, such as *Bacillus subtilis*, can facilitate secretion of the product.

Furthermore, genetic modification of bacteria is not limited to the production of foreign proteins. Specific alterations of the genetic composition of a bacterium can provide us with invaluable information about biochemical and physiological processes and other important attributes – notably pathogenicity (see Chapter 9). This has involved the development of vectors and other techniques for modifying a wide range of bacteria.

Shuttle vectors

A major problem in applying gene cloning and associated techniques to a wider range of bacteria is that the techniques available for these organisms are limited – for example, transformation (or electroporation) frequencies may be low and the methods for screening gene libraries may be difficult. It is therefore advantageous to carry out the initial cloning or manipulation of a gene using *E. coli* as the host organism. Then, when the correct piece of DNA has been obtained, it can be recovered from that recombinant and inserted into the target bacterium. This procedure is made much easier by the use of a vector that can replicate in either organism. Such a vector can be constructed by inserting, into an *E. coli* plasmid, an origin of replication that will function in the second host. This plasmid is then capable of replicating in either bacterium. Since such a vector can be used to carry pieces of DNA between two organisms, it is known as a *shuttle vector* (see Figure 8.15). Broad host range plasmids (see Chapter 5) can be used as an alternative to dual-origin shuttle vectors.

Protoplast transformation

The transformation system for *E. coli* as described earlier in this chapter, is not easily adapted for transformation of other bacterial hosts. Other transformation

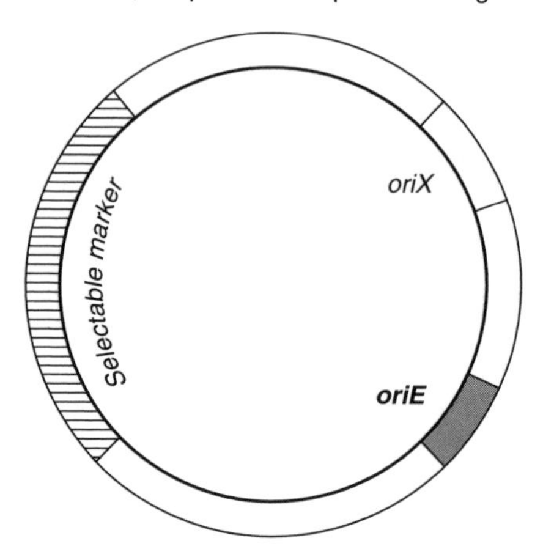

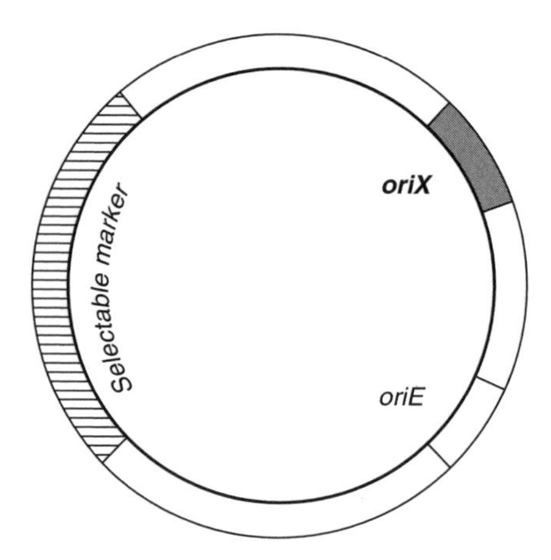

Figure 8.15 Shuttle plasmids

systems are needed. One of these is known as *protoplast transformation*. Enzymatic removal of the cell wall in the presence of an osmotic stabilizer such as sucrose, generates protoplasts in which large areas of the cytoplasmic membrane are exposed. Addition of the DNA solution together with polyethylene glycol (PEG) causes the cells to take up the DNA. The PEG is then removed and the cells allowed to regenerate on an osmotically-stabilized medium. If the correct conditions are used, a very high proportion of the resultant colonies will be transformed. This procedure can be used for transformation of many different bacteria, including *Streptomyces* and *Streptococcus* species.

Electroporation

Although protoplast transformation can be achieved with many bacterial species, the efficiency of the process is variable when applied to different species and in some cases may be too low to be of much practical use. It can also be difficult to optimize the conditions for formation and regeneration of protoplasts. Electroporation provides a more versatile alternative. In this procedure, the bacterial cells are mixed with plasmid DNA and are subjected to a brief pulse of high voltage electricity. This probably causes transient tiny holes in the cell membrane through which the DNA enters the cell. The transformants can then be selected on the basis of the antibiotic resistance conferred by the plasmid.

Although a certain amount of work is required to optimize the electroporation conditions, in principle electroporation can be used to introduce DNA into any type of cell.

Agrobacterium tumefaciens: *nature's genetic engineer*

As is often the case with new scientific developments, the exuberant scientist is often chastened to find that nature got there first. This is the case with a simple soil bacterium which essentially has been producing transgenic plants for millennia and certainly long before the advent of molecular biology. The soil-borne bacterium *Agrobacterium tumefaciens* is the causative agent of crown gall disease in plants. In this disease, the gall is actually a plant tumour. One of the most exciting discoveries associated with this bacterium was that formation of the tumour occurs because a specific fragment of bacterial DNA is transferred to the plant cells and incorporated into the plant chromosome. A key component in this mechanism is the tumour-inducing (Ti) plasmid (see Chapter 5). This DNA transfer process has been exploited to create transgenic plants and has become a cornerstone of plant molecular genetics.

8.6.4 Novel vaccines

Until recently, all the vaccines in use were of one of three types: (a) live attenuated vaccines (e.g. oral polio vaccine), which consist of a strain of the pathogen that has been mutated so as to lose its pathogenicity; (b) killed vaccines (e.g. influenza); and (c) toxoids (diphtheria, tetanus), which consist of a purified toxin that has been inactivated. More recent developments have added additional possibilities, such as subunit vaccines including the acellular pertussis (whooping cough) vaccine which consists of several components purified from whole cells of the pathogen *Bordetella pertussis* and the *Haemophilus influenzae* type B (Hib) vaccine which contains the polysaccharide capsule from *H. influenzae* conjugated to a protein carrier.

Genetic technology contributes in several ways to the development of novel vaccines. Firstly, for a subunit vaccine, it is possible to clone a gene and express it at a high level in a more convenient host. Not only does this make it easier to produce large amounts of purified protein, but it also avoids the hazard of growing large amounts of a pathogenic microorganism. The best example is the preparation of a vaccine against the hepatitis B virus. Cloning and expressing the gene that codes for the hepatitis B virus surface antigen enabled the production of a safe and effective vaccine that is now in routine use. (In this case, production is obtained using the yeast *Saccharomyces cerevisiae*, as the antigen is not produced in the correct form in *E. coli*).

For some diseases, a live vaccine is required. One way of achieving this is to insert the appropriate gene into the genome of an existing vaccine strain. This produces a recombinant vaccine which expresses the foreign gene and can therefore be administered as a live vaccine. The principal carrier that has been exploited for this purpose is the smallpox vaccine virus, vaccinia. It is in fact possible to include genes from several different infectious agents in the one virus, thus raising the possibility of using a single vaccination to confer protection against a number of diseases simultaneously. In the same vein, bacteria such as attenuated strains of *Salmonella typhi* and the tuberculosis vaccine BCG are being used for the development of live recombinant vaccines.

Another way of using gene technology for the development of live vaccines is by rational attenuation of the pathogen. If the genes required for pathogenicity but not needed for growth in the laboratory can be identified, then it is possible to use gene knock-out technology (as described in Chapter 9) to specifically inactivate one or more of these genes, thus producing a strain which will be unable to cause disease. This has the advantage over conventional mutation in that it can be ensured that the relevant gene is completely deleted and thus there is no possibility of reversion to virulence. The determination of the complete genome sequence of pathogenic bacteria (see Chapter 9) is playing an important part in identifying potential targets for rational attenuation by gene replacement.

Gene cloning provides yet another, even more novel, route for vaccination. If the gene coding for the key antigen is inserted into a plasmid vector under the control of a promoter which functions in an animal cell, then this recombinant plasmid is *by itself* able to elicit an immune response. These DNA vaccines are still at an experimental stage and much remains to be determined regarding their safety and efficacy.

8.7 Other uses of gene technology

In this chapter the focus has been deliberately concentrated on the applications of gene technology to bacteria. Some of the most publicized uses of this technology –

notably molecular diagnosis of inherited diseases, human DNA fingerprinting, gene therapy for the treatment of human diseases and the production of transgenic plants and animals – are beyond the scope of this book. These topics are covered in the further reading suggestions in Appendix A.

9

Genetic Methods for Investigating Bacteria

Genetics has been central to the development of our understanding of bacterial metabolism and physiology and some examples will be reviewed in this chapter. In Chapter 10 the discussion will be extended to include the lessons that can be learned from genome sequencing, amongst other things. Although genomics receives most of the publicity, the techniques in this chapter are important for understanding the function of the genes concerned in relation to the overall behaviour of the cell.

9.1 Metabolic pathways

The initial impetus for the use of genetic analysis to investigate metabolic pathways came from the work of Beadle and Tatum in the 1940s. Although their work was with the fungus *Neurospora*, many of the principles were applied subsequently to bacteria. The concepts will be illustrated in relation to an hypothetical pathway leading to the formation of an essential product P (see Figure 9.1). Auxotrophic mutants that are unable to grow unless P is provided should be of three types corresponding to the three steps involved and can be assigned to three groups by complementation analysis.

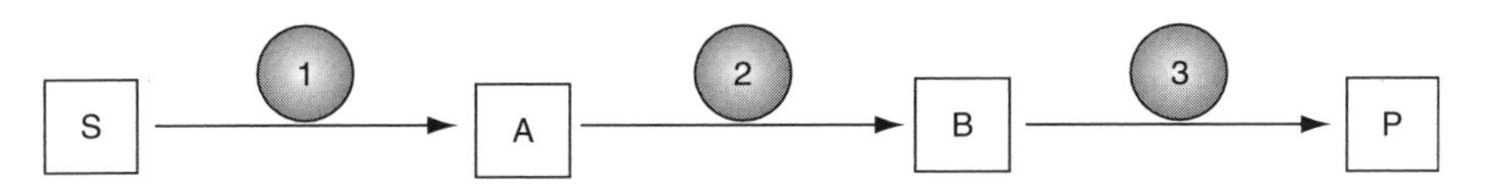

Figure 9.1 An unbranched metabolic pathway

Molecular Genetics of Bacteria, 4th Edition by Jeremy Dale and Simon F. Park

ISBN 0 470 85084 1 (cased) ISBN 0 470 85085 X (pbk)

9.1.1 Complementation

In Chapter 4, it was shown that a mixed infection with two mutant bacteriophages could result in *complementation* between the two mutants if the two phage strains carry mutations in different genes, so that between them they have a full set of functional genes. If complementation analysis is applied to mutants of a pathway such as that in Figure 9.1, the mutants could be assigned to three groups corresponding to the step in which they are deficient, assuming each enzyme contains a single polypeptide. For example, mutants defective in gene 1 would not be able to complement one another, but they would be able to complement mutants defective in either of the other two genes. Complementation analysis of this sort is easier to undertake in *Neurospora* than in bacterial systems (apart from phage).

In bacteria complementation is encountered more commonly in connection with plasmids. Introducing a plasmid can create a partial diploid with one version of a gene on the chromosome and another version on the plasmid. If the chromosomal gene is defective, it can be complemented by the plasmid. Partial diploids and F′ plasmids were discussed in Chapter 6 and in Chapter 8 plasmid complementation was described as a method of screening a gene library.

9.1.2 Cross-feeding

Complementation analysis does not identify the steps involved nor the order in which they occur. Additional information is obtained by a procedure known as *cross-feeding*, as illustrated in Figure 9.2. None of these mutants is able to grow on a medium that lacks the essential nutrient P. However if the plate is inoculated with each mutant in the pattern shown, then intermediates accumulated by a mutant may stimulate the growth of the neighbouring mutant. For example mutant I is unable to produce the intermediate A, but this is produced by mutant II. So the growth of mutant A is stimulated where it is near to mutant II. Similarly, growth of mutant II is stimulated by mutant III. A cross-feeding experiment can therefore categorize mutants in a way that corresponds to the steps in the biochemical pathway.

Neither of these procedures identifies the steps involved or the nature of the intermediates. However, an hypothetical pathway can be tested by examining the ability of the predicted intermediates to stimulate the growth of each mutant. Such an analysis was used to confirm the role of ornithine and citrulline as intermediates in the pathway of arginine biosynthesis. One class of mutants responded only to arginine, a second class responded to either arginine or citrulline, while the third group responded to arginine, ornithine or citrulline. This established that the synthetic sequence went from ornithine to citrulline to arginine.

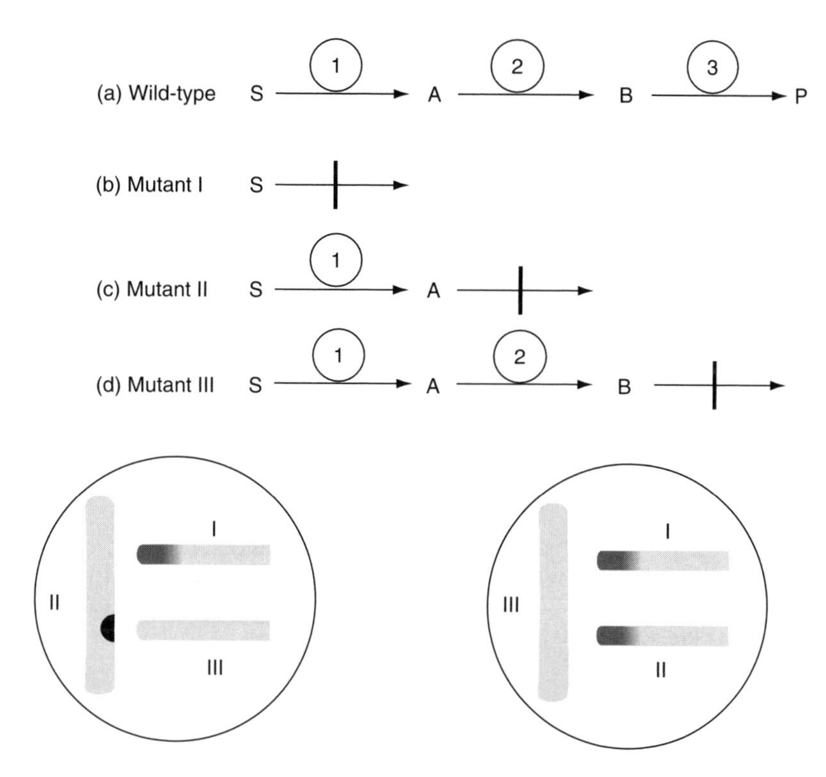

- Mutant I is unable to make A so cannot cross-feed either II or III
- Mutant II cannot make B so cannot cross-feed mutant III, but can cross-feed mutant I
- Mutant III cannot convert B to P but can make B, so cross-feeds mutants I and II

Figure 9.2 Cross-feeding

9.2 Microbial physiology

In the above examples of simple metabolic pathways, the genetic approach is essentially complementary to biochemical investigations. Genetics is useful, but is not absolutely essential to the investigation. However, a bacterial cell consists of much more than a series of straightforward biochemical pathways. There are complex structures, such as flagella, ribosomes and bacterial cell envelopes, as well as sophisticated systems such as the control of replication and cell division or the control of lysogeny in temperate bacteriophages. Although in some cases a reductionist approach can be applied – ribosomes for example can be disassembled and reassembled *in vitro* – very often this is extremely difficult or even impossible. Genetics therefore plays a central role in the investigation of such systems.

Some of the ways of analysing the genetic basis of physiological characteristics are summarized in Figure 9.3. The starting point in this case is the isolation of a

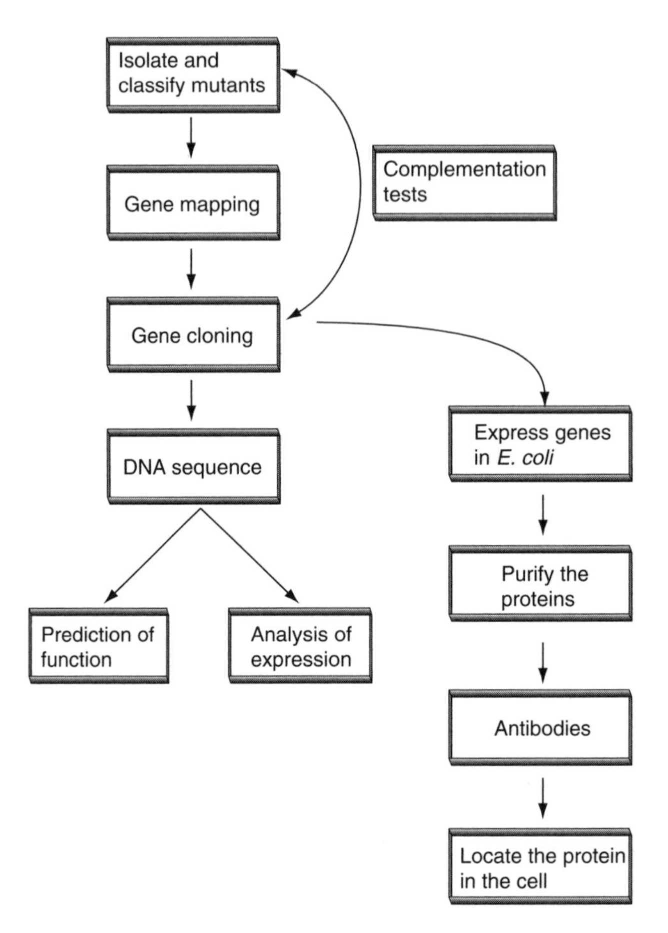

Figure 9.3 Techniques for genetic investigation of phenotypic characteristics. The flowchart illustrates some of the routes that start with specific mutants. There are many other methods of cloning and identifying genes

series of mutants that are altered in a specific characteristic (such as sporulation for example). These mutants can be classified according to the precise nature of their phenotype, as well as by complementation analysis. The genes involved can then be mapped (i.e. their positions on the chromosome can be determined) using methods described in Chapter 10. This information can be used to identify the genes in a library. (Note that there are a variety of other methods, as described in Chapter 8, for cloning genes). The identity of the clone can be confirmed by its ability to complement the original mutation, as described earlier in this chapter. The gene can then be sequenced and its function predicted from analysis of the sequence (Chapter 10). Probes can also be made, using the sequence data, to analyse the expression of the gene under different conditions, which may yield information as to its role. Furthermore the cloned gene can be expressed at a high level (Chapter 8) and thus obtained in pure form and antibodies to the purified protein can be produced to locate the position of the protein in the cell.

9.2.1 Reporter genes

Before considering examples where elements of the genetic approach have been invaluable in developing our understanding of bacterial physiology, another technique should be added to the toolbox. Instead of producing mutants, the expression of various genes under different conditions can be examined using the assumption that genes that are specifically needed for those conditions will be selectively expressed at that time. A convenient and widely used method of doing this is to employ *reporter genes*. This involves attaching the regulatory region of the gene concerned to another gene that is more easily detected so that the regulation by proxy can be followed by observing the expression of the reporter (Figure 9.4). For example, if a β-galactosidase reporter and a medium containing the chromogenic substrate X-gal is used, the colonies will only turn blue when the promoter in question becomes activated and the reporter gene starts to be expressed.

One use of reporter genes is to identify unknown genes whose expression is activated in response to a given stimulus. In this case, random fragments of DNA, some of which will contain promoter regions, are fused to the reporter gene to generate a library of promoter fusions. The library is then plated onto a medium containing X-gal (assuming that β-galactosidase is the reporter). The colonies of interest are those which are white initially but turn blue when the conditions are changed – for example if the plate is transferred to an anaerobic incubator. This indicates that the promoter is responsive to the new environment which in this case is growth under anaerobic conditions. From this we can infer that the gene which is normally expressed from that promoter is one that is needed for

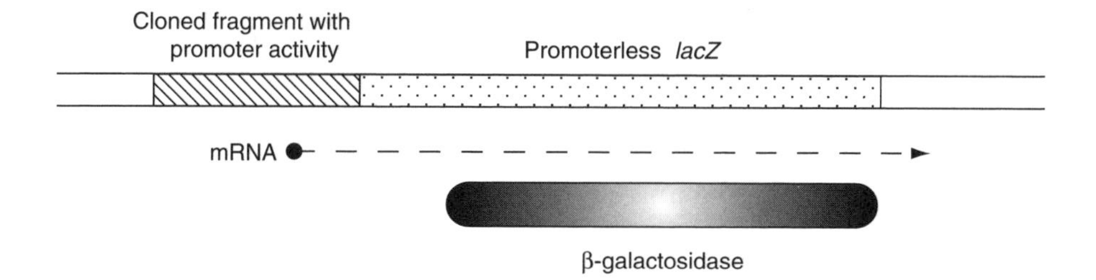

Figure 9.4 Use of reporter genes. The diagram shows a *transcriptional* fusion, in which a promoterless *lacZ* gene is fused to a promoter, thus enabling the activity of that promoter to be characterized. For some purposes, *translational* fusions are used, where the promoter fragment also provides translational signals and the 5′ end of a protein-coding sequence, leading to a fusion protein containing the reporter

anaerobic growth in the original host. Whilst this approach has been largely replaced by the advent of microarrays which enable global gene expression to be analysed more easily (see Chapter 10), it has been widely used to identify genes which are expressed in response to heat, starvation, osmotic shock and during sporulation for example.

Apart from β-galactosidase, two other reporter genes are worth a specific mention. The expression of luciferase results in the production of blue-green light and this allows the expression of a gene to be monitored simply by measuring light production. It also makes the host organism bioluminescent and by using sensitive imaging systems it is possible to trace the luciferase-labelled bacterium in complex environments and ecosystems, including tracking a pathogen in a living animal. Another reporter gene that has additional utility is Green Fluorescent Protein (GFP), a protein (originating from the jellyfish *Aequorea victoria*) that is intrinsically fluorescent and emits a green light when exposed to ultraviolet irradiation. This has the advantage of being readily detected *in situ*, without the need for an enzymic substrate. In addition to its use as a reporter of gene expression, it can also be expressed as a *translational fusion*, so that the target protein is labelled with GFP. This enables the location of the target protein within the cell to be determined.

In the next chapter other ways of studying changes in gene expression will be reviewed, especially the use of microarrays.

9.2.2 Lysogeny

In Chapter 4, aspects of the behaviour of bacteriophages were discussed. Much of this knowledge comes from the study of mutant bacteriophages. One example lies in the control of lysogeny of bacteriophage lambda (λ).

Lambda is a temperate bacteriophage, which means that when it infects a sensitive strain of *E. coli* it can establish a more or less stable relationship with the host cell (lysogeny) and in this state the λ DNA is inherited by the daughter cells at cell division. Its continued presence is indicated by the occasional breakdown of the lysogenic state which results in the production of bacteriophage particles.

Lysogens are resistant to infection by other lambda phage particles (they show *superinfection immunity*), so the plaques produced by infection of *E. coli* with λ are normally turbid, rather than the clear plaques produced by a virulent phage (such as T4). Within the plaque, lysogenic bacteria will continue to grow, which causes the turbidity of the plaque. However, if enough plaques are examined, occasionally one will be found that is clear rather than turbid. This is due to a mutant bacteriophage that is unable to establish lysogeny. These are known as clear plaque mutants, designated *c* (the designation of bacteriophage genes does not follow the normal three letter system that applies to bacterial genes).

Complementation analysis shows that there are three genes involved, *cI*, *cII* and *cIII*. These mutants behave rather differently, in that *cI* mutants always kill the infected cell, while *cII* and *cIII* mutants can (albeit very rarely) give rise to stable lysogens.

Referring back to Chapter 4, it is apparent that *cI* codes for the repressor protein which is needed for the establishment and maintenance of lysogeny, while *cII* and *cIII* code for proteins which are needed for the expression of *cI* during the establishment of lysogeny but are not needed for maintenance of the lysogenic state.

Although these mutants are defective in establishing lysogeny, they are not true virulent mutants. They are still susceptible to repression as can be demonstrated by their inability to infect a lysogenic *E. coli* (containing a wild type λ). However a different class of phage mutants (*vir*, for virulent) can be isolated which *are* able to infect a lysogen. These contain mutations in both the operator sites (O_L and O_R) to which the *cI* repressor normally binds. So they will be unaffected by the repressor present in the lysogenic cell. Analysis of these lambda mutants contributed extensively to the development of the model of lysogeny that was described in Chapter 4.

9.2.3 Cell division

The process of cell division is so central to bacterial multiplication that mutants would be expected to be non-viable. Surprisingly however, some mutants are still able to grow. Amongst these are the *min* mutants of *E. coli*, so-called because at cell division some of the daughter cells are small *minicells* that do not contain any chromosomal DNA (Figure 9.5). These minicells are of course non-viable, but the mutant is still able to multiply as some cell divisions occur as normal.

These mutants have been invaluable in attempting to answer a key question relating to cell division: Why does cell division normally occur only at the central point of the cell and not elsewhere? (A related question is how does the DNA partition between the two daughter cells so that each acquires a copy of the DNA?). The full answer is rather complex, but essentially the role of the products of the *min* genes is to inhibit cell division at sites other than the midpoint of the cell.

Of course many mutations affecting cell division *are* lethal, therefore conditional mutants are used to study these genes – especially temperature-sensitive mutants. One of the most likely consequences of a failure of cell division in a rod-shaped bacterium such as *E. coli*, is the formation of long filaments. Mutants that form filaments when grown at a higher temperature are known as *fts* mutants. Genetic analysis showed that mutation in any of a number of genes could give rise to this phenotype. One of the most important of these is the *ftsZ* gene. The techniques summarized earlier have established that the FtsZ protein initiates the formation of the septum that will divide the two cells by polymerizing into a ring-like structure at the site of cell division.

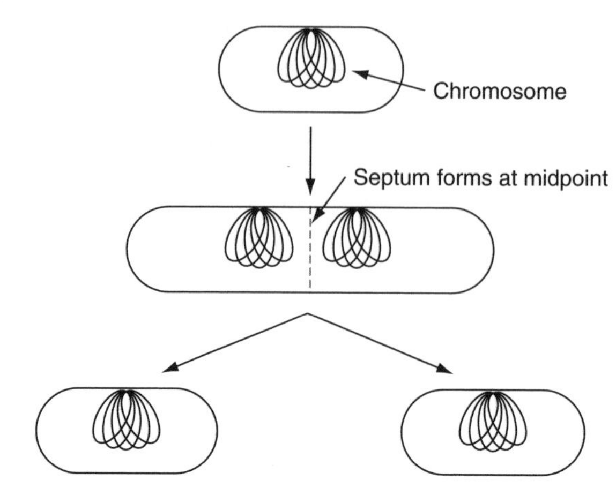

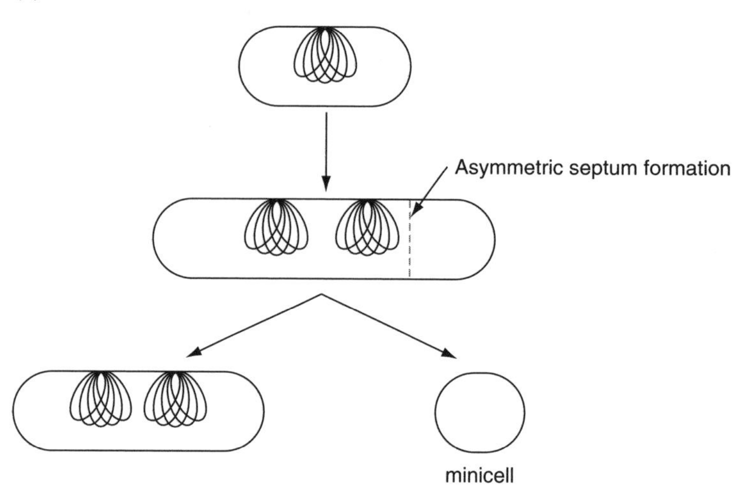

Figure 9.5 Cell division in *E. coli*: minicell mutants. (a) Cell division normally occurs at the midpoint of the cell. (b) With a *min* mutant, some cell divisions occur asymmetrically producing a small (non-replicating) cell that contains no chromosomal DNA

9.2.4 Motility and chemotaxis

In the absence of chemical or other stimuli, many bacteria including *E. coli*, swim about in an apparently random manner which is composed of periods of smooth motion in one direction (the run) interspersed by abrupt tumbling. This is related to the structure and control of the flagella which impart motion by rotating. When this rotation is counterclockwise, the filaments coalesce into a bundle which drives the cell evenly in a particular direction. Tumbling results from a reversal in the direction of rotation of the flagella. As soon as they start to rotate in a clockwise direction, the bundles of flagella fly apart, the cell tumbles briefly and then the flagella resume their normal counterclockwise rotation. The cell therefore starts to swim smoothly again, but now in a different direction.

The link with chemotaxis (the ability to swim towards, or away from, specific chemical stimuli) is in the control of the frequency of tumbling. When the cell is moving towards an attractant, the length of the run is increased (or the frequency of tumbling is decreased, which is the same thing). So those bacteria that happen to be swimming in the right direction will swim further before changing direction; those that are going the wrong way will change direction sooner.

Our understanding of this complex and fascinating system has been greatly helped by genetic analysis. Over 50 genes have been identified by the isolation of specific mutants. These include (1) defects in the production of flagellin (the protein subunit of which the flagella are composed) or in the assembly of the flagella; both lead to non-flagellated cells which are non-motile; (2) other flagellar defects that result in flagella that are unable to rotate (these cells are also non-motile); (3) defects in the control of rotation, leading to cells that tumble excessively or rarely, or general defects in chemotaxis, so that the cells have normal motility but cannot respond to any stimulus; and (4) specific chemotaxis defects, so that the cells can respond to some chemicals but not to others.

The genetic techniques summarized earlier, coupled with analysis of the phenotypes involved, enables the establishment of the identity of the genes involved. For example, analysis of specific chemotaxis mutants leads to the identification of the receptors that sense the presence of a specific stimulus.

9.2.5 Cell differentiation

Sporulation in Bacillus subtilis

When some bacteria are starved, they are able to respond by producing a resistant *endospore*. Sporulation has been most extensively studied in *Bacillus subtilis* and the regulation of the process was described in Chapter 3.

Sporulation in *Bacillus* takes about 7 h, and can be divided into a number of stages (see Figure 3.7) on the basis of the morphological changes that can be observed microscopically. Mutant strains of *B. subtilis* that are unable to form spores are divided into categories according to the stage at which development of the spores is arrested. Thus any gene in which mutation results in sporulation being blocked at stage II (and unable to progress to stage III) is referred to as *spoII*, while those in which sporulation proceeds to stage III before stopping are known as *spoIII* mutants and so on. At each stage there are a number of genes that are essential if the process is to continue to the next stage which can be distinguished by complementation and recombination analysis; the different genes are therefore denoted by the addition of further letters and numbers.

For example, one class of *spoII* mutants was designated *spoIIA* on the basis of genetic analysis. The gene concerned was isolated by cloning a DNA fragment that could complement a *spoIIA* mutant. The sequence of this fragment showed that it contained an operon of three genes, which were designated *spoIIAA*, *spoIIAB* and *spoIIAC*. The sequence of SpoIIAC is similar to that of known sigma (σ) factors, which determine the promoter-specificity of RNA polymerase. This suggested that SpoIIAC (now known as σ^F) is responsible for activating expression of the forespore genes required in stage III (see Chapter 3) and this was subsequently verified by *in vitro* tests of purified SpoIIAC. The other two *spoIIA* genes regulate the activity of SpoIIAC so that it only becomes active at the appropriate stage; SpoIIAB is an anti-sigma factor that inhibits SpoIIAC (σ^F), while SpoIIAA is an anti-anti-sigma factor that antagonizes SpoIIAB.

A large number of sporulation genes have been identified; the annotated genome sequence lists well over 100. In some cases, the function of the genes is known and they can be given more descriptive names. For example, *spoIIAC* is listed as *sigF*. Analysis of the genome sequence by itself would only have enabled a provisional identification of the function of a handful of these genes. The isolation and characterization of a large number of mutants provided the foundation for our understanding of the process of sporulation and in particular the regulation of the process, and the cross-talking between the forespore and the mother cell that was described in Chapter 3. However, there are still many aspects that are not understood, including the precise function of many of the genes that are known to be essential for sporulation.

Sporulation in Streptomyces

Streptomyces (especially *S. coelicolor*) produce a different sort of spore that is a dispersal mechanism rather than a survival strategy. These bacteria grow initially as filaments but after a few days they differentiate into aerial mycelia with chains of spores at their ends. Some mutants (designated *bld*, or bald) are unable to produce aerial mycelia, while another class (*whi*, or white – since the colonies lack the colour associated with the spores) produce aerial mycelia but no spores. Analysis of these mutants showed that the *bld* genes included functions needed for the secretion of, and response to, a series of extracellular signals, showing that the production of an aerial mycelium requires intercellular communication similar to the quorum sensing systems described in Chapter 3.

Amongst the *whi* genes, *whiG* codes for a σ factor that is needed for the switch in gene expression to activate spore formation and is highly similar to alternative σ factors in both *B. subtilis* and *E. coli* (see Chapter 3). Genome sequence data has shown that many of the genes needed for the control of sporulation in

S. coelicolor are related to genes that control different developmental processes in other, non-sporulating, bacteria.

Communication and differentiation

Some bacteria have more complex life cycles, involving distinct morphological stages (and in some cases multicellular behaviour). These form a useful bridge between the simpler bacterial systems and the complex developmental systems in multicellular organisms.

One example is the aquatic bacterium *Caulobacter crescentus*, which exists in two forms, stalked and swarmer cells. The former is attached to a substrate by means of its stalk (Figure 9.6). As it grows, a flagellum is formed at the end of the cell opposite to the stalk so that at cell division a new, motile, swarmer cell is liberated – the other cell remaining attached to the substrate. The swarmer cell does not replicate its DNA nor undergo cell division but eventually settles down at a new site, sheds its flagellum and forms a stalk to attach to the new substrate. It has now become another stalked cell which carries out DNA replication and a new cell division cycle. This poses some interesting questions about the mechanisms controlling this behaviour, including both the regulation of gene expression in the two types of cell and the reasons for the developmental asymmetry of the replicating cell.

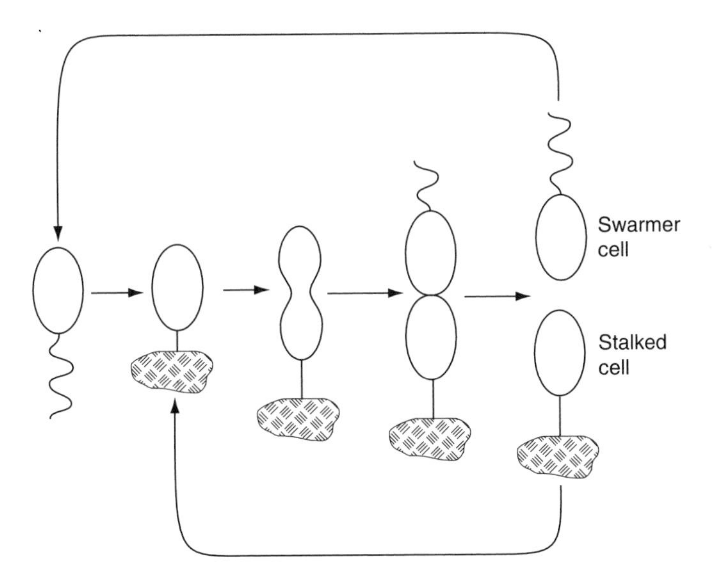

Figure 9.6 Division cycle of *Caulobacter crescentus*. The stalked cell attached to a surface, carries out DNA replication and cell division, producing a motile, non-replicating, swarmer cell. This eventually sheds its flagellum, attaches to a surface and becomes a replicating stalked cell

Reporter gene technology and microarray analysis (see Chapter 10) has shown differences in gene expression in the stalked and swarmer cells. One of the regulatory factors involved is a transcriptional activator called FlbD. The active form of FlbD is unequally distributed in the dividing cell, being only found in the swarmer pole. Amongst other things, this activates the genes needed for flagellum production in the swarmer cell.

Another bacterium with an interesting life cycle, that includes elements of multicellular behaviour, is the soil bacterium *Myxococcus xanthus* (see also Chapter 3). Under favourable conditions, individual cells swarm over surfaces in a coordinated manner. When food becomes scarce, the individual cells aggregate into a mound (containing some 10^5 cells) which develops into a fruiting body (Figure 9.7) containing a large number of spores (myxospores). The myxospores

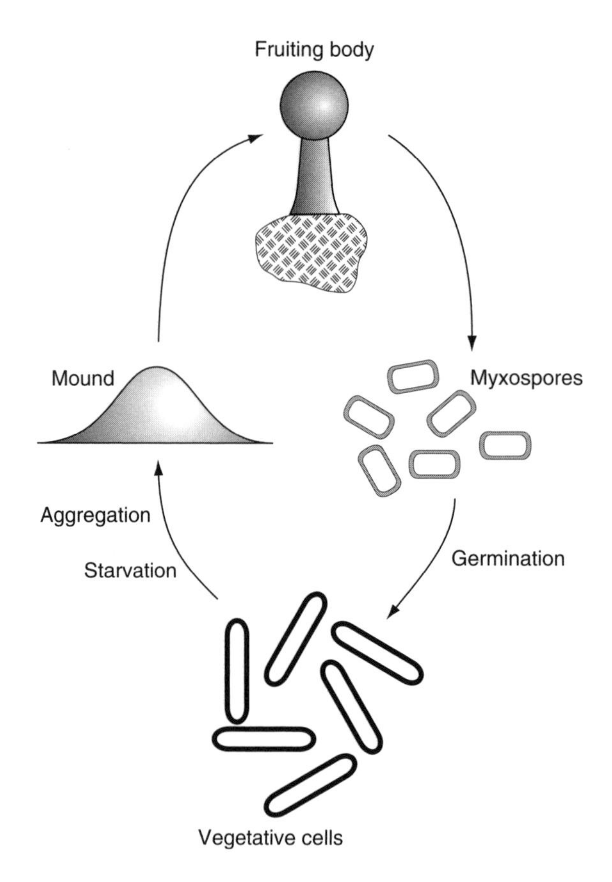

Figure 9.7 Schematic illustration of the life cycle of myxobacteria. This is a generalized and simplified diagram. In different species, the structure of the fruiting body varies substantially, from simpler forms without a stalk to more complex branched structures with multiple sporangia

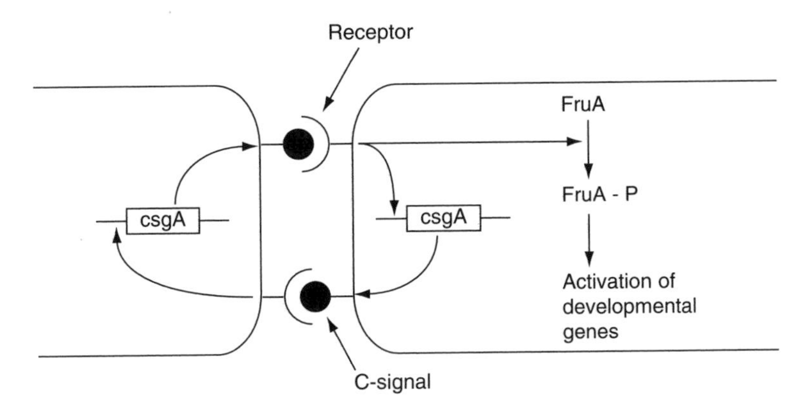

Figure 9.8 Simplified model of some elements of the C-signal pathway in *Myxococcus xanthus*

are resistant to drying and enable the bacteria to survive until conditions improve, when they germinate and form individual cells again.

Development from a swarm of individual cells to an aggregate and then on to sporulation is controlled by a series of signals. One of these, the A-signal is a mixture of amino acids that operates as a quorum-sensing mechanism (as described in Chapter 3) at an early stage of the process. Aggregation only occurs if the A-signal reaches a threshold level, thus ensuring that there are enough cells to form a functional fruiting body. Further development is controlled by another gene (*csgA*) coding for the C-signal, so that *csgA* mutants are also defective in sporulation. Unlike the A-signal, which is secreted, the C-signal requires cell contact, since it is a cell-surface protein which interacts with a surface receptor on another cell (Figure 9.8). This interaction has two main consequences. Firstly it stimulates transcription of *csgA*, so more of the signal is made, thus amplifying the effect. Secondly, it alters the regulation of more than 50 genes, including those needed for sporulation; this is, in part, mediated by phosphorylation of a regulatory protein called FruA. The overall effect is that sporulation will only occur in those cells within the mound which are in contact with other cells, thus contributing to a spatial difference in gene expression within the fruiting body.

9.3 Bacterial virulence

9.3.1 Wide range mechanisms of bacterial pathogenesis

Bacteria exploit a number of common molecular mechanisms to achieve a range of different objectives during infection. This section describes how bacterial molecular genetics has been used to uncover these key mechanisms of microbial pathogenicity.

From genome sequence information, it is clear that virulence genes often occur in clusters and that these regions are absent from closely related non-pathogenic bacteria. Furthermore, based on the %GC content of these regions compared to the rest of the genome, it became apparent that these large sections (e.g. 30–50 kb) of DNA had been acquired from other organisms. Together, these observations gave rise to the concept of *pathogenicity islands* which suggests that bacteria can acquire, *en masse*, complete systems that expand their ability to exploit different host environments. A previous example is *Vibrio cholerae* (Chapter 4), where the toxin genes are located on a pathogenicity island which was subsequently shown to be an integrated bacteriophage.

Although not expected at the time, a number of related studies on the plant pathogen *Pseudomonas syringae* were to have dramatic implications for our understanding of pathogenicity. In the mid-1980s transposon mutagenesis (see Chapter 10) was used to identify a cluster of genes which were essential for the generation of plant disease symptoms. The importance of this region was confirmed when the gene cluster was cloned into non-pathogenic bacteria using shuttle vectors and was shown to confer upon these avirulent cells the ability to generate specific disease symptoms. In the early 1990s it became clear that similar gene clusters existed in many other bacterial pathogens and that some of the proteins encoded by these gene clusters resembled parts of the machinery for exporting the bacterial flagellum. This information gave rise to the concept that a common secretion pathway existed in these bacterial pathogens.

It is now known that these regions encode a specialized secretory apparatus called a *type III secretion system* (for information on other types of secretion system see Chapter 1). These comprise at least 20 proteins and are the most complex transport systems known in bacteria. They are often referred to as 'molecular syringes' as they enable Gram-negative bacteria to secrete and inject effector proteins directly into the cytosol of eukaryotic host cells. The injected proteins can redirect the normal cellular signal transduction pathways and can result in disarmament of host immune responses or in cytoskeletal reorganization to establish pathways for bacterial colonization.

Much of the type III secretory apparatus is conserved between distantly related pathogens (although the effector proteins differ). Thus the same general bacterial pathogenicity mechanism is involved in a multitude of diseases from bubonic plague in humans to southern wilt in tomato plants. *Yersinia* spp., for example, inject at least three different effector molecules to destroy key functions of immune cells. The genes encoding type III secretion systems, including effector proteins and structural proteins, are clustered together as another example of pathogenicity islands that have been transferred between bacterial species.

9.3.2 Detection of virulence genes

Many bacterial pathogens have separate free-living and pathogenic life cycles, so they encounter very different environments and require very different functions for survival. As a consequence, pathogens must be able to recognize signals in the host that convey the need to express virulence genes. The expression of certain virulence gene functions in *Shigella*, for example, is triggered at body temperature (37 °C) but not at environmental temperatures (< 30 °C). Therefore, if the genes which are only expressed during infection can be identified, then the key virulence traits can also be identified. Reporter genes have been especially useful for this purpose.

In Vivo *Expression Technology (IVET)*

One of the most widely used methods of using reporters to identify virulence genes is known as *In Vivo* Expression Technology (IVET). This is a technique that selects bacterial promoters which are only expressed in the host and which thus drive the expression of virulence traits. In most examples of IVET, random fragments of DNA from the bacterial host are inserted adjacent to a promoterless reporter gene whose product confers a phenotype that can be positively selected for in the host. For example, a *purA* mutant of *Salmonella typhimurium* is unable to survive in an animal model because purine biosynthesis is essential for *Salmonella* in this environment. This defect can be complemented by a plasmid carrying a functional *purA* gene, but the *purA* gene on the plasmid will only be expressed if a promoter is inserted which is functional *in vivo* (Figure 9.9). (The construct shown also contains a *lacZ* gene whose expression is also controlled by the inserted promoter so that constitutively expressed promoters can be identified and excluded). When animals are infected with a pool of clones each carrying a different DNA fragment, only those clones containing an active promoter are able to survive and to be recovered from the animals, thus providing direct selection for those promoters which are active during infection. Identification of these promoter fragments then leads to identification of the genes that are normally regulated by them, which are likely to include genes which are essential for the virulence of *Salmonella*.

Signature tagged mutagenesis

As shown previously, the classical approach to identifying the genes responsible for a given phenotype would be to generate mutants that are defective in that phenotype. For virulence genes, the phenotype of the mutant would be the

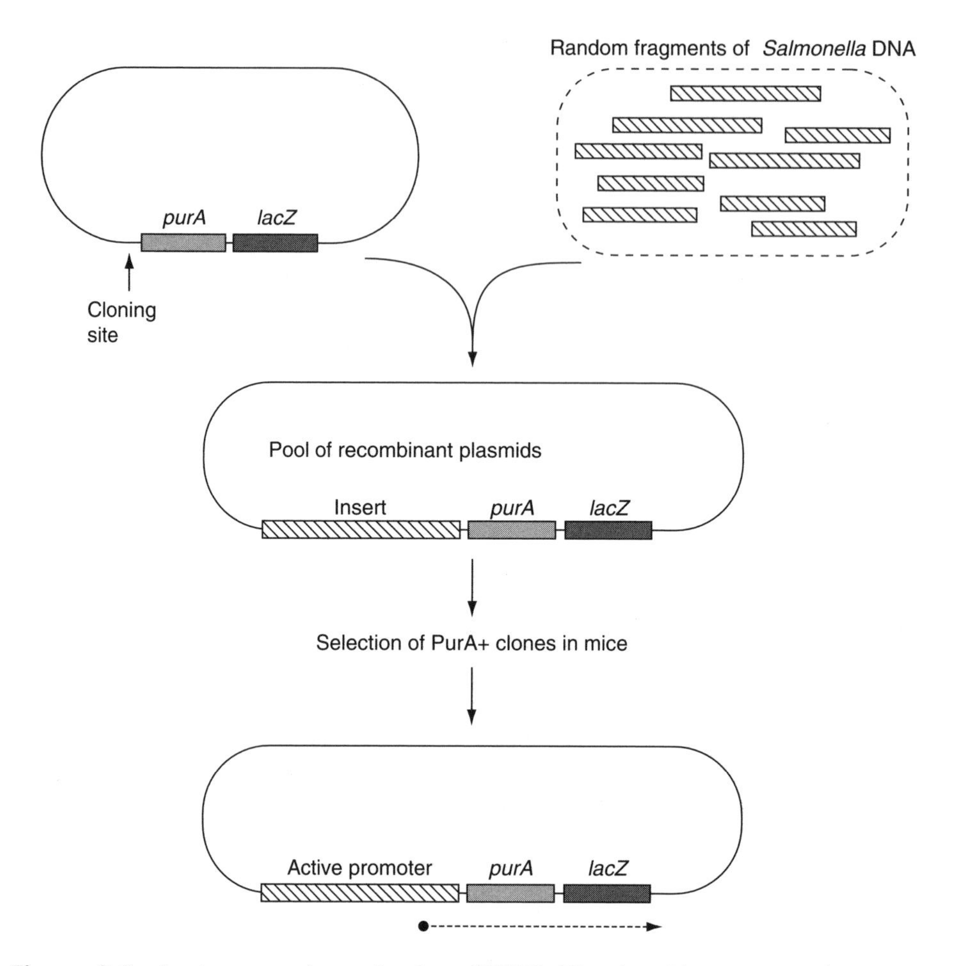

Figure 9.9 *In vivo* expression technology (IVET). The plasmid vector contains promoterless *purA* and *lacZ* genes; expression of these genes is dependent on insertion of a DNA fragment with promoter activity. Infection of mice with a pool of *Salmonella* containing recombinant plasmids is selective for $PurA^+$ clones which contain a promoter that is active under these conditions

inability to survive in an *in vivo* model for the disease. Isolation of the mutants cells will prove to be difficult as they are the very cells which do *not* survive. To circumvent this obstacle, a procedure called *signature tagged mutagenesis* (STM) has been invented. This is essentially a negative selection technique derived from transposon mutagenesis (see Chapter 10) in which each individual mutant is labelled with a unique DNA signature. By comparing the mutants present in the initial inoculum with those that are recovered after infection of the model, it is possible to identify those that did not survive and thus identify the genes that are required for virulence.

An overview of STM is given in Figure 9.10. A library of mutants is made for the bacterium of interest and each mutated gene tagged with a section of DNA containing a unique central region and two flanking arms which share their sequence with all of the other tags. The key to STM is that each individual mutant

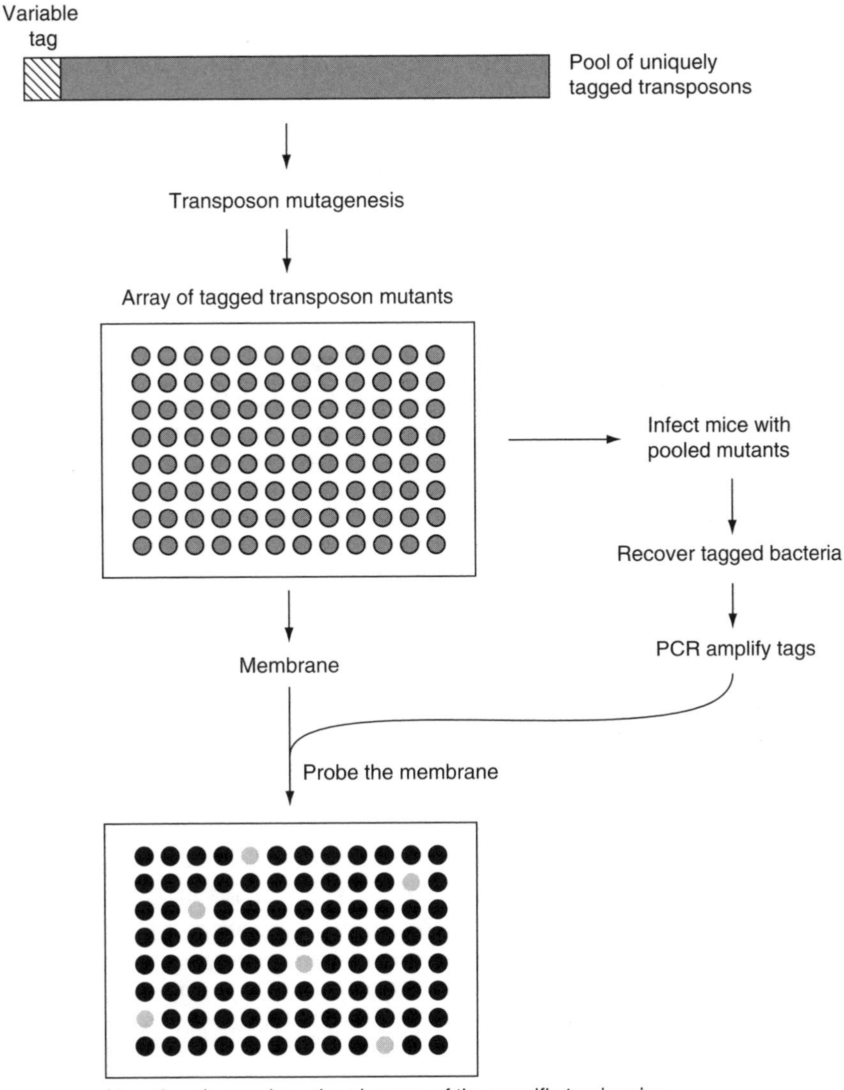

Figure 9.10 Signature tagged mutagenesis. This procedure uses transposon mutagenesis to inactivate genes that are needed for infection of mice and provides a method of identifying those genes

can be distinguished from every other mutant based on the possession of its own unique tag. The mutants are then stored individually in ordered arrays and DNA from each one spotted onto a membrane in an grid-like manner. The mutants are then pooled and inoculated into a relevant animal model and the bacteria that are able to survive and establish infection are recovered. PCR is then used to amplify all the tags present in the recovered bacteria and the mixed product is used to probe the gridded membrane. Mutants that fail to survive in the infected animal (those which are defective in virulence) can be identified since their tags will not be present in the output pool. They can be recovered from the original stored arrays for further study. This system has been widely used to identify novel virulence factors that are involved in colonization, immune system evasion and attachment to human cells in a number of bacterial pathogens.

9.4 Specific mutagenesis

So far in this chapter, conventional mutational techniques have been considered – producing mutants with an altered phenotype and then identifying the genes affected and determining their functions. These techniques can be complemented by the use of recombinant DNA technology. In contrast to conventional genetics, the recombinant DNA approach starts with an hypothesis that a specific gene is involved. This gene is then modified or deleted and the resultant phenotype characterized. So, while conventional genetics starts with the phenotype and works towards identifying the nature and function of the genes involved, the molecular approach starts by altering the gene and works towards an analysis of the phenotype.

9.4.1 Gene replacement

A key technique for determining the function of a specific gene is to inactivate it by a process known variously as gene replacement, allelic replacement or gene knock-out. Essentially, this uses homologous recombination to remove all or part of a specific gene or to replace it with an altered or inactivated gene. An example of how this can be done is shown in Figure 9.11. In this case, a plasmid has been constructed in which the central part of the cloned gene has been removed and replaced by an antibiotic resistance gene (*aph*, aminoglycoside phosphotransferase which confers resistance to kanamycin). The kanamycin resistance gene is flanked by regions of DNA that are the same as those in the host strain and it is within those regions that homologous recombination will occur. Note that two recombination events (a double crossover) are needed for gene replacement. Recombination at a single site (a single crossover) will merely integrate the plasmid into the chromosome. Additional techniques are needed to ensure that a double crossover is achieved.

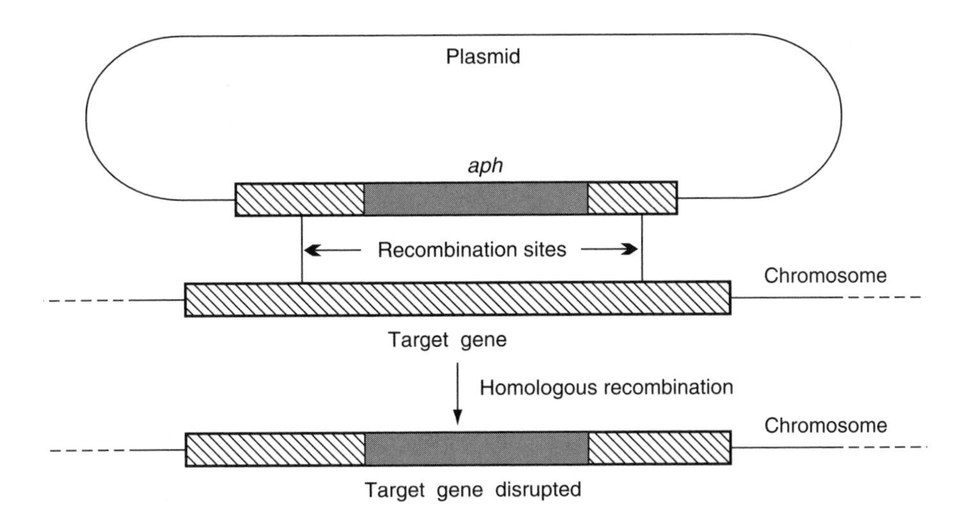

Figure 9.11 Gene disruption by allelic replacement. Homologous recombination between the disrupted cloned gene and the target gene on the chromosome leads to inactivation of the target gene. *aph*, aminoglycoside phosphotransferase (kanamycin resistance)

The plasmid that is used is one that cannot replicate in the chosen host cell. Thus after transformation, selection for kanamycin resistance will isolate cells in which the *aph* gene has been inserted into the chromosome by homologous recombination with the target gene, thereby inactivating that gene. Tests can be carried out to ascertain whether the expected phenotype is produced. For example, if it has been assumed that the target gene is necessary for survival and growth within macrophages, then testing the mutant for a deficiency in this respect will confirm or deny the assumption.

The results do not necessarily prove the case absolutely. Genes interact in many ways and knocking out one gene may have indirect effects on others. In particular, if the target gene is part of an operon, the gene knockout may affect the expression of other genes within the operon (in other words, the mutation may be *polar*). This possibility can be partly eliminated by the use of complementation. A plasmid carrying the wild-type gene (in this case using a plasmid that can replicate in this host) can be introduced into the mutant and if the original gene knock-out was responsible for the observed effect (e.g. loss of ability to grow within macrophages), then the introduced plasmid will restore the original, wild-type phenotype. Successful complementation therefore supports the contention that the product of this gene is necessary for survival in macrophages. But it needs to be said that it still does not finally establish a *direct* role for the gene product – for example, it could function by altering the expression of other genes.

Instead of knocking out a gene completely, it may be desirable to replace it with a specifically altered gene. For this gene replacement should be combined with the techniques of site-directed mutagenesis, as covered earlier in this chapter. In this way, the significance of a specific amino acid in the activity of an enzyme for example, can be tested by replacing it with a series of different residues and determining the activity of the product.

9.4.2 Antisense RNA

One problem with gene knock-outs is that if the gene is essential for the growth of the cell in the laboratory, then the complete loss of that gene would be lethal and therefore no recombinants would be obtained. It may therefore be useful to partially reduce the expression of the gene or the functionality of its product – thus leaving enough activity to cope with the comfortable conditions of normal laboratory growth but not enough to deal with the stress conditions that may be imposed on it subsequently. One alternative strategy is to use *antisense RNA*.

For this purpose, part of the gene would be cloned in an expression vector, so that it is transcribed from a promoter on the vector, but the insert would be deliberately put in the wrong orientation. The insert will therefore be transcribed in the opposite direction from normal – or in other words, the opposite strand will be transcribed. The RNA produced (the antisense RNA) will be complementary to the normal mRNA and will pair with it to produce a double-stranded RNA molecule. This may interfere with translation of the mRNA and thus reduce the level of the protein that is made. Using a stronger or a weaker promoter (or even better, using a promoter that can be switched on and off) will lead to different amounts of the antisense RNA being made and hence will alter the extent of the reduction in the amount of the protein product formed.

9.5 Taxonomy, evolution and epidemiology

9.5.1 Molecular taxonomy

It is possible to distinguish between even closely related organisms using a range of biochemical tests, but this approach does not necessarily give an accurate picture of the true taxonomic or evolutionary relationship between different organisms. Conventional taxonomy therefore resulted in some quite different organisms being erroneously grouped together in the same genus or family. Molecular approaches have played an important role in resolving these issues.

One simple molecular characteristic that is used to classify bacteria is the base composition of the DNA, defined as the number of guanine and cytosine residues as a percentage of the total number of bases (%GC). It is not necessary to sequence the genome to determine this value (although of course that gives the most accurate and precise value). The %GC can be determined using physical techniques. The base composition of bacterial DNA varies widely from one species to another – over a range of 20–80 per cent – but closely related organisms tend to have similar DNA base compositions. There are good reasons for this. Replication, transcription and translation are all, in different ways and to varying extents, sensitive to the base composition of the nucleic acids and so have evolved together. Earlier in this chapter it became evident that this can be used as evidence that a portion of the genome has been acquired more recently in the evolution of the organism, as its base composition is different from the rest of the genome.

If this concept is applied to bacterial taxonomy, it is sometimes found that organisms which are otherwise quite similar have quite different DNA base composition. For example the genera *Staphylococcus* and *Micrococcus* are morphologically similar Gram-positive cocci (although they can be distinguished biochemically). However *Micrococcus* has a high GC content (about 70 per cent GC), while *Staphylococcus* DNA has a low proportion of G+C (30–39 per cent).

Ribosomal RNA sequencing is a much more powerful technique. The ribosomal RNA genes are very highly conserved, being remarkably similar in all bacteria, and yet there are small variations in the sequence from one species to another. These variations (most commonly in the 16S rRNA) not only distinguish between species but also indicate the degree of difference. In other words, by counting the number of bases that are different in two species a measure of the evolutionary distance that separates them can be calculated. If a number of such sequences is compared, a *phylogenetic tree* can be constructed which will show a possible route by which these species have diverged from a common ancestor. A simple example (with a much shorter sequence than would be used in practice) is shown in Figure 9.12. The sequence of organism A is more similar to B than it is to C or D for example and this is reflected in the arrangement of the tree. A word of caution: construction of a phylogenetic tree is much more complicated than this simple description and many trees can be drawn from a single set of data. The computer produces the best fit, but it is only a model and does not necessarily reflect the true evolution of the organisms involved.

Cloning the ribosomal RNA genes to do this is not necessary. The variation in the 16S (or 23S) rRNA gene is not evenly spread across the gene. Some regions are particularly highly conserved, so a pair of PCR primers can be used which recognize conserved sequences on either side of a variable region and amplify the region which contains differences. This amplified product can then be sequenced (see Chapter 10). The degree of conservation is such that the same pair of primers can be used for any organism, without knowing anything about it. The sequence obtained can then be compared with sequences of rRNA from known organisms and thus the identity of the unknown bacterium and its relationship to known species can be determined, at least provisionally.

Organism	Sequence
A	C A U A G A C C U G A C G C C
B	C C U A G A G C U G G C G C G
C	C G A A G A G G U G G C G C C
D	G C U A G A A G U G G C G G C

Only a short sequence is shown as an example
In practice much longer sequences would be used

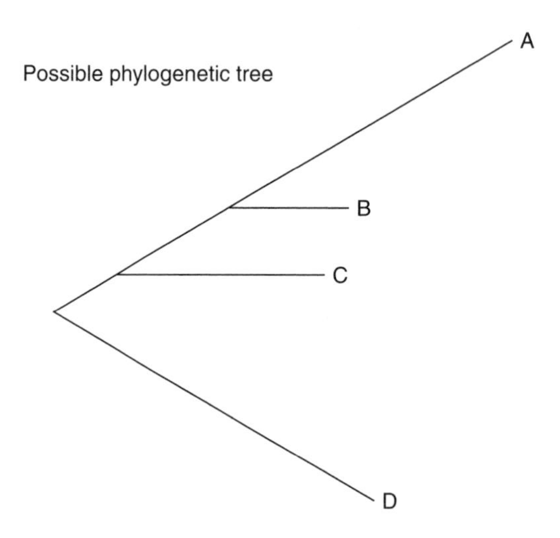

Figure 9.12 Construction of a phylogenetic tree from sequence data. This illustrates the principle only. In practice a much longer sequence would be used and extensive computer analysis is required to test the many possible trees that can be drawn

The power of PCR to amplify minute amounts of DNA means that the bacterium in question does not have to be cultured. This is significant as standard bacteriological techniques are designed to culture certain bacteria, especially medically important pathogens. The range of bacteria that can be isolated can be extended by using different media and different growth conditions. But however wide the range of conditions used, there will still be some bacteria – often a substantial majority – that are unable to grow. Applying PCR to such a sample, using primers directed at the 16S rRNA gene, will produce a very wide range of amplified products. Cloning this mixture of products, rather like constructing a gene library (Chapter 8), enables each one to be isolated and sequenced so that the bacteria present in the sample can be identified (within the limitations of the known sequences in the database). For example, the bacterial flora of the human colon has been extensively investigated using cultural techniques and its constituent bacteria were thought to have been thoroughly characterized. However, the genotypic approach described above showed that 76 per cent of the 16S rRNA genes generated did not correspond to known organisms and were clearly derived from hitherto unknown and uncultured bacterial species.

9.5.2 Diagnostic use of PCR

Traditional methods for the detection and identification of bacteria rely on growing the organism in pure culture and identifying it by a combination of staining methods, biochemical reactions and other tests. This applies equally to detection of environmental organisms (in soil or water), bacteria in food (including milk and drinking water) or pathogens in samples from patients with an infectious disease. However these methods are slow, requiring at least 24 h or several weeks for slow-growing organisms such as *Mycobacterium tuberculosis*. In addition, there are some bacteria, such as *Mycobacterium leprae* (the causative agent of leprosy) that still cannot be grown in the laboratory.

In principle, gene probes could be used to provide quicker results by directly detecting the presence of specific DNA in the specimen. However, this only works if the bacteria present are plentiful. Gene probes are not sensitive enough to detect the small numbers of organisms that may be present, and significant, in such specimens. The technique that is needed is the polymerase chain reaction (PCR) as described in Chapter 2. This provides greatly enhanced sensitivity, being capable (in theory) of detecting a single organism. In order to apply this to the detection of a specific species, it is necessary to know the sequence of a gene that is characteristic of that species – that is, it is always present (and the sequence is conserved) in that species, but is absent or significantly different in other bacteria. A pair of PCR primers can then be designed which will anneal to this target sequence so that PCR will amplify a DNA fragment that can be easily detected. Other bacteria, lacking the specific binding sites for those primers, will not give an amplified product. In a research laboratory the amplified product (*amplicon*) would commonly be detected by gel electrophoresis, sometimes combined with Southern blotting and hybridization with specific gene probes (see Chapter 2) to increase the sensitivity and specificity of the procedure. The commercial kits that are now available for detection of some bacterial pathogens (some using forms of gene amplification that are distinct from PCR) use other, quicker, ways of detecting product amplification. A technique known as *real-time PCR* which produces results more rapidly than gel electrophoresis and has the additional advantage of quantifying the target present in the sample, will be discussed in Chapter 10.

9.5.3 Molecular epidemiology

Epidemiology is the study of the occurrence and distribution of diseases. By identifying the source of infection, the measures necessary to control an outbreak can be determined. The microbiology laboratory contributes to this effort by identifying the pathogen and determining the strain involved (*strain typing*). Patients who have caught the disease from the same source will be infected with the same strain; if the strains are different, the cases do not belong to the same outbreak.

There are a wide variety of methods available for typing different bacterial species. Phage typing for *Staph. aureus* has already been discussed in Chapter 4. Serotyping, using variable antigens, is also widely used. In recent years, molecular typing methods have become increasingly popular. One of the most widely applicable of such methods is *restriction fragment length polymorphism* (RFLP). When a gene probe is hybridized to a Southern blot (see Chapter 2) of restriction enzyme-digested DNA from different strains, the size of the fragment(s) detected may vary from one strain to another. This effect can arise from point mutations that remove (or create) restriction sites or from the insertion or deletion of DNA fragments in the region detected by the probe (see Figure 9.13).

Insertion sequences can be extremely useful for this purpose, since there are often multiple copies of the element in a strain (giving rise to a number of bands on the Southern blot) and also because the site of insertion in the chromosome is often highly variable (giving rise to extensive polymorphism). One element that is widely used for epidemiological purposes is the insertion sequence IS*6110* in *M. tuberculosis*. With this probe, similar patterns (such as the arrowed tracks in Figure 9.14) are obtained only with strains from the same outbreak.

Another approach makes use of a different type of repetitive sequence. Bacterial genomes often contain *tandem repeats*, i.e. short (e.g. 50–100 bp) sequences that are present as several copies, in the same orientation and repeated without any intervening sequence. The number of copies of one of these sequences at a specific point can vary from one strain to another, probably due to slipped-strand mispairing (see Chapter 7). If PCR is used to amplify the DNA region containing a tandemly repeated sequence, the number of copies of the repeat can be determined from the size of the PCR product (Figure 9.15). This constitutes the typing method known as VNTR (variable number tandem repeats). The extent of variation at any one locus will be quite limited, but there are usually several such loci in the genome and the data from each of them can be combined to produce a test with a high degree of discrimination.

RFLP typing and VNTR methods have the disadvantage that they only examine relatively small regions of the genome. Methods that are available for analysing overall genomic structure, including pulsed-field gel electrophoresis (PFGE) and genomic microarrays will be described in Chapter 10.

The application of these typing methods can be extended to studying the evolution of bacterial strains. For this purpose it is necessary to decide whether the typing method successfully identifies a coherent strain. For example, if a particular RFLP type is identified, does this predict other properties of the

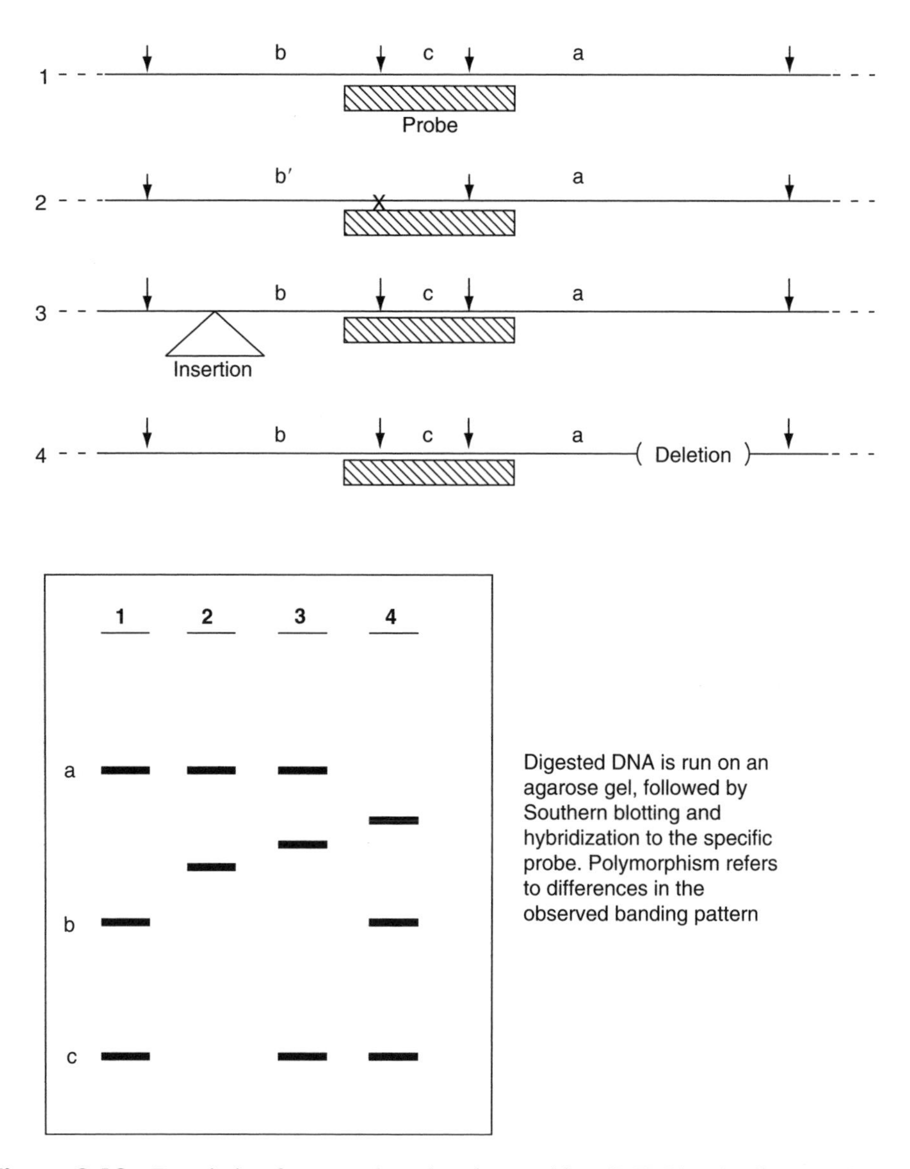

Figure 9.13 Restriction fragment length polymorphism (RFLP). The diagram shows a region of the chromosome that hybridizes to a specific probe and some of the possible reasons for polymorphism in the Southern blot pattern: (1) The 'original' sequence, in which the probe detects three DNA fragments labelled a,b and c. (2) Loss of a restriction site by mutation results in fragments b and c being replaced by a single larger fragment. (3) An insertion within fragment b changes its size, without altering the number of fragments. (4) A deletion within fragment a reduces its size

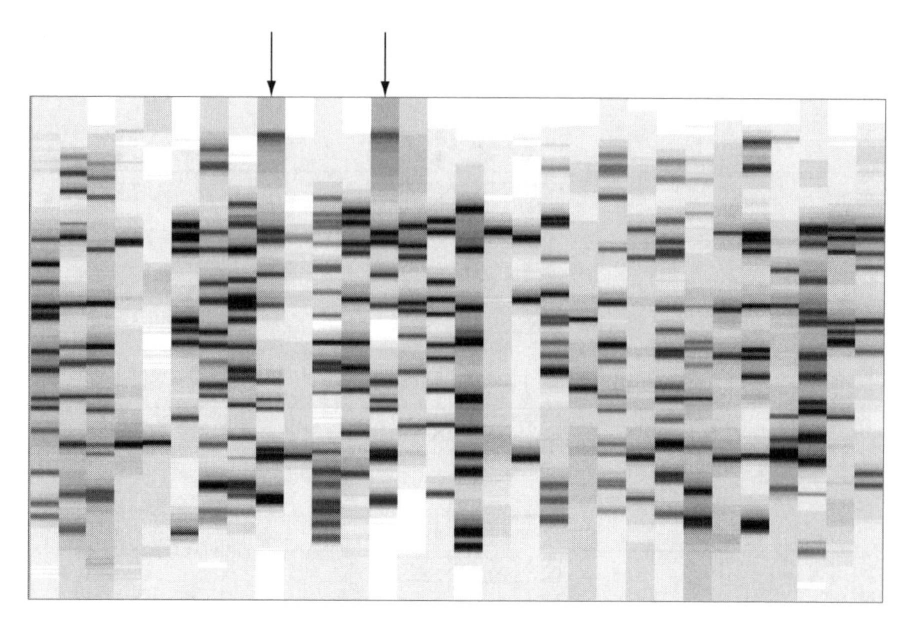

Figure 9.14 Fingerprinting of *Mycobacterium tuberculosis* using IS*6110*. Isolates from cases with a common source of infection show identical patterns (arrowed)

'strain'? If a bacterial species is truly clonal, i.e. all members of a strain (or clone) are descended from a single individual with no horizontal gene transfer, then it would be expected that members of that strain will be more like one another than like other strains. The typing method might therefore be expected to provide some prediction of other properties of the organism. This does sometimes occur: for example, the *E. coli* O157 serotype is associated with a high degree of virulence. Indeed much of diagnostic microbiology is based on such associations, for example the use of the coagulase test to differentiate pathogenic *Staphylococcus aureus* from the less pathogenic staphylococci such as *Staphylococcus epidermidis*. However, it has to be said that virulence and other significant characteristics, are often too complex for such simple analysis, especially as they are affected by host responses as well as the characteristics of the organism in question.

On the other hand, if horizontal gene transfer has occurred, the species will not be clonal. In that case, these associations will not necessarily hold and indeed the results of different typing methods may not agree with one another. So the apparent evolutionary relationships traced by one typing method may prove to be quite different when examined by an independent method.

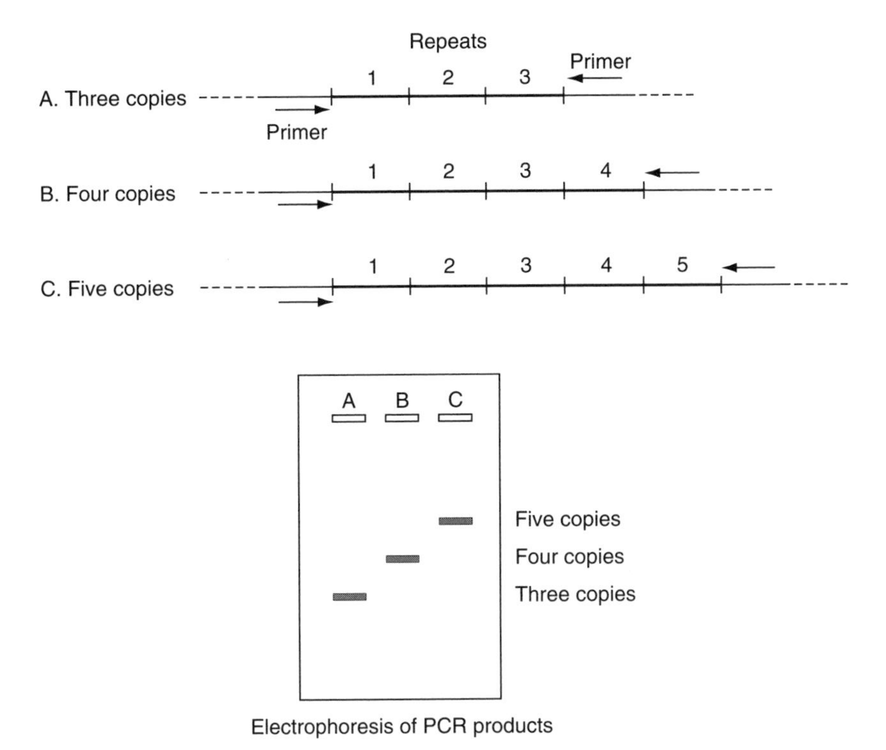

Figure 9.15 Variable Number Tandem Repeats (VNTR). PCR amplification of a region containing tandem repeats can be used for bacterial typing by determining the number of copies of the repeated sequence

10 Gene Mapping to Genomics

In this chapter, the methods available for studying the location, structure and expression of genes will be examined. This includes *in vivo* methods of gene mapping as well as genome sequencing and methods for studying the expression of genes individually and genome-wide. At the level of this book it is possible to provide only an introduction to some of the most exciting aspects of modern molecular biology and genetics.

10.1 Gene mapping

One of the main objectives in genetic analysis is the determination of the position of genes on the chromosome. In isolation, this may seem a rather arcane occupation, but knowledge of the organization of the chromosome does play a major role in understanding gene function and has contributed extensively to the advances described in Chapter 9.

In bacteria, the classical methods of gene mapping depend on the production of recombinants by gene transfer using conjugation, transformation and transduction. These methods have now been supplemented, although not entirely supplanted, by methods based on *in vitro* gene technology. Nevertheless, a basic understanding of these methods is valuable for an appreciation of the development of our knowledge of bacterial genetics.

10.1.1 Conjugational analysis

In an earlier chapter it was shown that integration of the F plasmid into the *E. coli* chromosome produces an Hfr strain which is capable of transferring a copy of the chromosome to a suitable recipient. Transfer of the whole chromosome would

Molecular Genetics of Bacteria, 4th Edition by Jeremy Dale and Simon F. Park
 ISBN 0 470 85084 1 (cased) ISBN 0 470 85085 X (pbk)

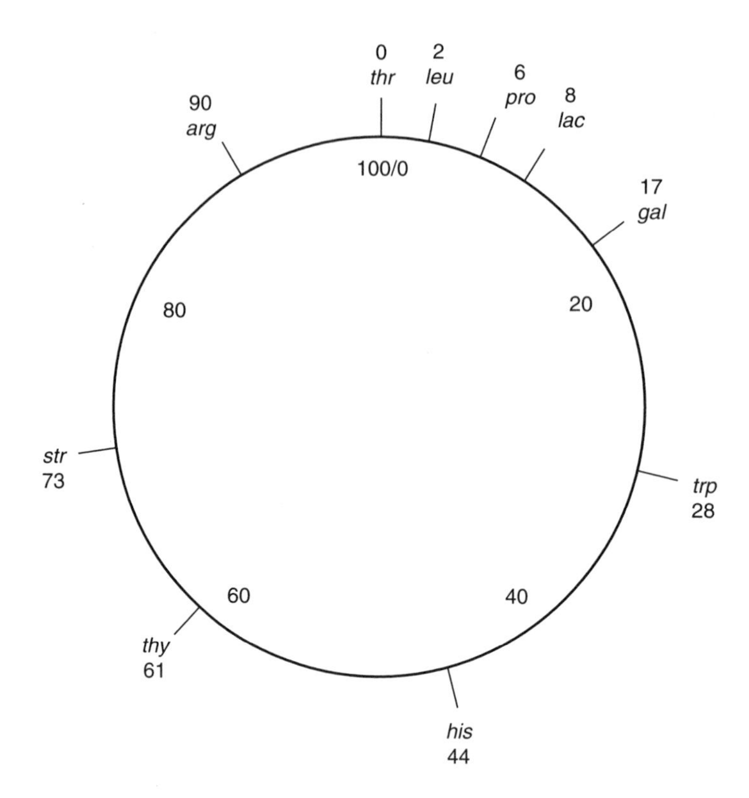

Figure 10.1 *E. coli* genetic map. The genes shown are those for synthesis of threonine, leucine, proline, tryptophan, histidine, thymine and arginine; for utilization of lactose and galactose; and for resistance to streptomycin. Note that the positions shown represent, in many cases, groups of genes rather than a single gene. The *arg* regulon also includes genes at other positions. The units shown (map times) represent minutes taken for conjugal transfer, starting with the threonine locus

take about 100 min. For this reason, the *E. coli* genetic map (Figure 10.1) is calibrated from 0 to 100 min, with each gene being assigned a position that corresponds to the time at which it is transferred from an arbitrary origin at the threonine locus (*thr*, 0 min) with transfer proceeding in a clockwise direction. The actual time at which transfer of a specific gene occurs and the direction of transfer will depend on the Hfr strain used, since the F plasmid can be integrated at different points and/or in a different orientation.

However, it is quite rare for the complete chromosome to be transferred. The mating pairs will tend to become separated at randomly distributed times. The longer it takes for transfer of a gene, the more chance there is that the mating pair will have separated before that gene is transferred. There will therefore be a *gradient of transfer* corresponding to the position of the genes with respect to the point at which transfer starts.

This provided a convenient way of determining the relative position of genes on the *E. coli* chromosome (Figure 10.1). If a prototrophic Hfr strain is mated with a multiply auxotrophic recipient (e.g. *thr leu trp his arg*), the number of recipients that have received each of the markers can be determined by plating aliquots of the mixture on a minimal medium supplemented with four out of the five amino acids. For example, the number of *thr* recombinants is measured using a medium that contains leucine, tryptophan, histidine and arginine, but not threonine. It is of course necessary to prevent growth of the prototrophic donor, for example by using a streptomycin-resistant recipient and including streptomycin in the medium. Streptomycin in this instance is used as a *counter-selecting* agent. On this medium, the donor will be unable to grow (because of the streptomycin) and the parental recipient will not grow (because of the absence of threonine). The only cells that can grow will be the recombinant recipients that have received the *thr* gene.

The result is illustrated by Figure 10.2. In this instance, the HfrH donor has been used, from which the genes are transferred in a clockwise direction starting very close to the *thr* locus. There is a linear relationship between the logarithm of the number of recombinants and the map position of the genes concerned. If it is assumed that the position of the *trp* gene is not known, determining the number of Trp$^+$ recombinants will allow the gene to be mapped as shown in Figure 10.2.

An alternative method for more accurate mapping of genes that are transferred relatively early in mating involves deliberately separating the mating pairs (by violent agitation) in samples of the mixture at different times after the start of mating (*interrupted mating*). Recombinants that have received a specific gene start

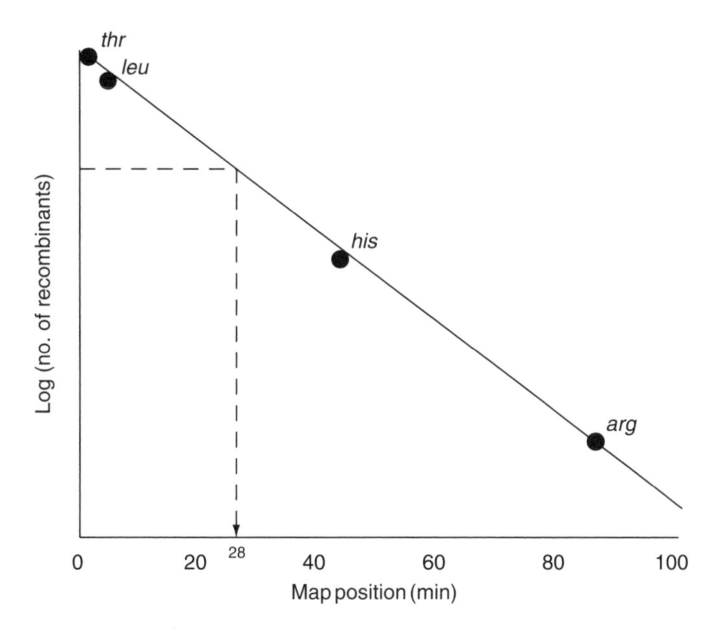

Figure 10.2 Gene mapping using the gradient of transfer by conjugation. Determination of the position of the *trp* gene

to appear at a certain time after the start of mating (the *time of entry*), which is a measure of the distance of that gene from the origin of transfer.

10.1.2 Co-transformation and co-transduction

Transformation can be used for mapping the relative positions of genes, by selecting recombinants in which one marker has been transferred and then determining the frequency of recombination for a second marker. If the two are close together, they will tend to be inherited together.

If the two markers are labelled A and B: the donor is wild type for both genes (A^+B^+), while the recipient is the double mutant (A^-B^-). After transformation of the recipient with chromosomal DNA extracted from the donor, the cells are plated on a medium that only allows A^+ cells to grow. The colonies that result from this (transformants) can then be tested for the presence of the B^+ gene (the unselected marker). Two recombination events (crossovers) are needed to incorporate a piece of linear DNA into the chromosome. Since transformants that have received gene A have been selected, one crossover must be to the left of A in Figure 10.3 and the second will be to the right of A – either between A and B or

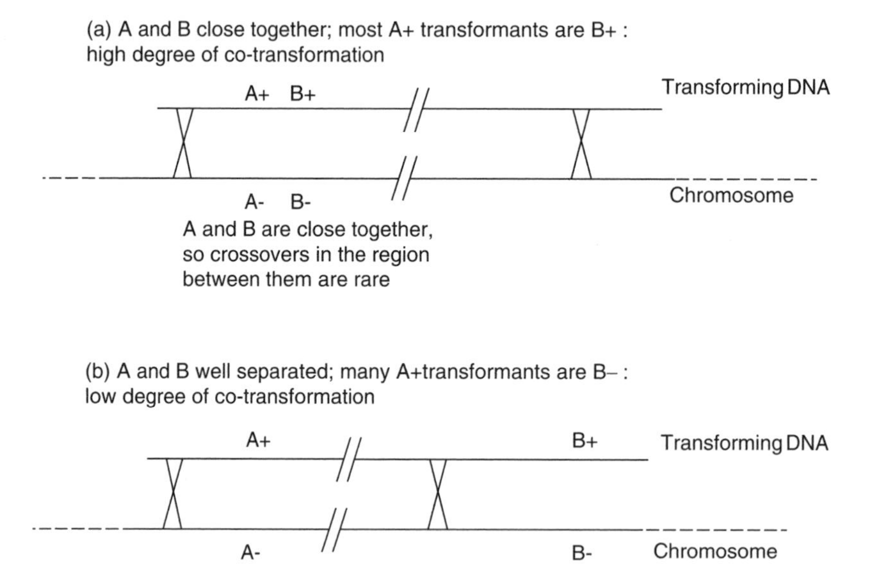

Figure 10.3 Determination of the relative position of two genes by co-transformation frequency

to the right of B. If A and B are close together, it is unlikely that the second crossover would occur in the short region between them, so both markers will be incorporated together (co-transformation). The further apart the two markers are, the more likely it is that the second crossover will happen in the intervening region and hence the frequency of co-transformation will be lower.

Generalized transduction (see Chapter 6) can be used in a similar way and provides a powerful tool for the short range mapping of the position of genes on the bacterial chromosome. As an example, if the phage P1 is grown on a prototrophic *E. coli* strain, the phage preparation can be used to infect cells of a recipient strain which is auxotrophic for threonine and proline. Plating these cells on a medium that contains threonine (but not proline) will detect transductants that have received the *pro* gene from the donor by transduction. These cells can then be tested for their ability to grow in the absence of threonine, i.e. whether they have also received the *thr* gene. A high degree of co-transduction indicates closely linked genes.

10.1.3 Molecular techniques for gene mapping

Gene libraries

Gene cloning techniques provide a powerful alternative to the classical gene mapping techniques described above. Since each clone in a genomic library (Chapter 8) carries a limited amount of DNA, it can be used to identify two genes that are close together on the chromosome, for example by testing the ability to complement two mutations simultaneously – essentially by a co-transformation test.

More commonly, the relative position of genes in a library would be determined by hybridization. If a clone that contains the first gene is selected using a specific probe (probe 1 in Figure 10.4), the insert from that clone can be used as a probe to identify a second clone that contains an overlapping portion of the genome. This second clone can in turn be used as a probe to identify a further overlapping clone and so on until a clone containing the second gene of interest is selected. This procedure, known as *chromosome walking*, can either be used to map the position of two known genes or conversely to clone an unknown gene that is known to be close to an identified marker.

An extension of this procedure can be used to obtain a complete set of overlapping clones that are arranged in the order they appear in the chromosome. This is known as an *ordered library*. Constructing an ordered library can involve a significant amount of work, but once established it is a valuable resource. Some of the genome sequencing projects (see below) have made good use of ordered gene libraries.

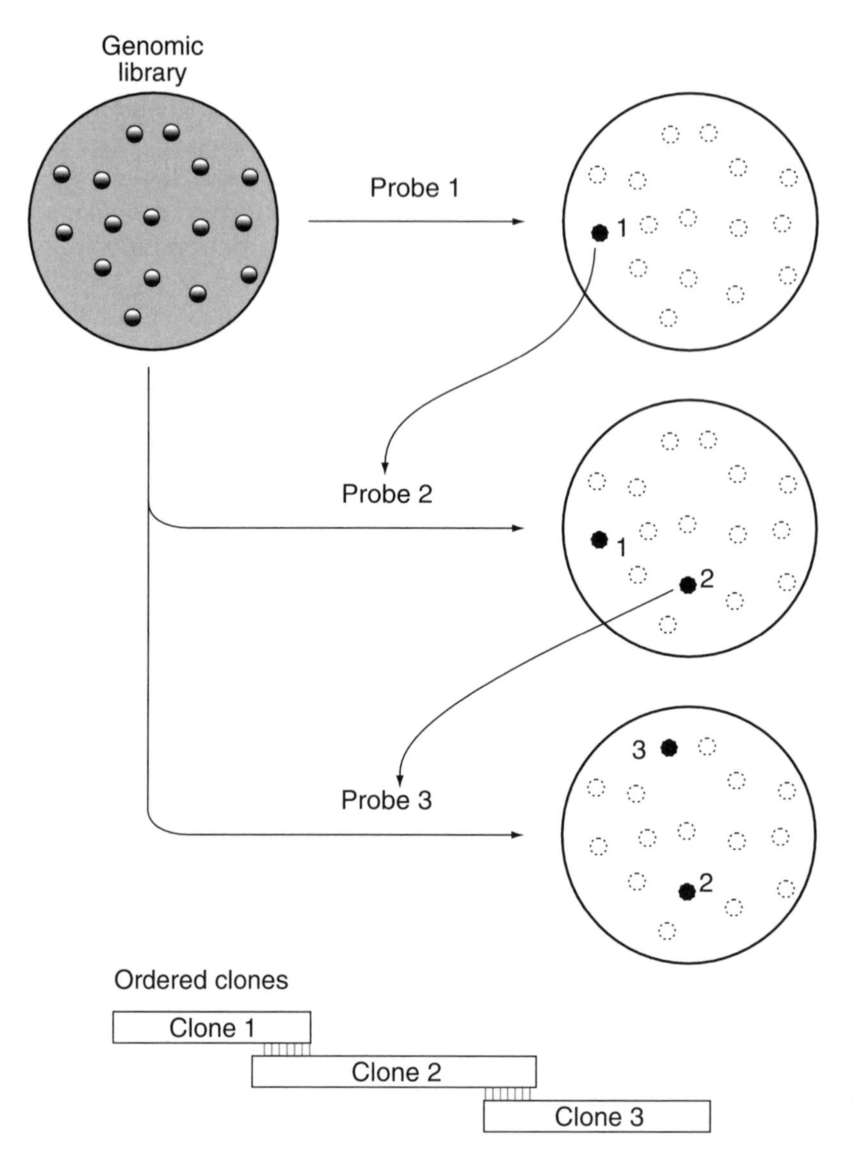

Figure 10.4 Chromosome walking and construction of an ordered library. Clone 1, identified by probing a genomic library with probe 1, is used to generate probe 2 for re-screening the library. This identifies an overlapping clone (clone 2), which is in turn used to identify clone 3 and so on

Restriction mapping and pulsed-field gel electrophoresis

It is possible to determine the position of restriction sites on a DNA fragment by digesting it with various restriction enzymes, singly and in combination, and analysing the fragment sizes obtained.

Restriction mapping used to be a key procedure for characterizing a cloned fragment of DNA, but this is now more easily achieved with DNA sequencing. Nevertheless, analysis of restriction sites (or fragment sizes) is still useful, for example in comparing the chromosomal organization of different strains. However, cutting the total bacterial DNA with most restriction enzymes results in such a large number of fragments that they cannot easily be resolved by conventional gel electrophoresis and usually appear as a smear. In Chapter 9 the use of Southern blotting with specific probes for RFLP strain typing was discussed. One alternative is to use enzymes that cut less often, such as *Not*I or *Pac*I (which both have 8 bp recognition sites, rather than the 6- or 4-bp sites required by more commonly used restriction enzymes) and are therefore likely to yield a relatively small number of very large fragments.

However, linear DNA fragments above a certain size (commonly about 20 kb, but varying according to the conditions) are not separated on a conventional agarose gel. The fragments do move through the gel but in a size-independent manner. This can be visualized as the large DNA molecules migrating end on; once they are lined up, their length is immaterial. This is in contrast to smaller fragments where the random configuration is small enough to pass through the pores of the gel.

These larger fragments can however be separated by periodically changing the direction of the electric field (Figure 10.5). All of the fragments actually move at the same speed through the gel, but when the direction of the field changes they have to re-orientate themselves in the new direction. The larger the molecule, the longer this takes. The overall effect therefore is that the larger molecules are found nearer to the origin of the gel.

This technique is generally referred to as Pulsed Field Gel Electrophoresis (PFGE). There are several variants of this technique. The simplest, but rather limited, method is Field Inversion Gel Electrophoresis (FIGE) which uses a conventional electrophoresis tank with short pulses of reversed polarity of the electrodes. More powerful techniques rely on complex arrangements of electrodes; one of the most widely used systems is known as Contour-clamped Homogeneous Electric Field Gel Electrophoresis (CHEF) which uses an hexagonal array of electrodes with the field intensity and direction being controlled so as to generate an homogeneous electric field, thus ensuring that the DNA molecules migrate in a straight line.

10.2 Gene sequencing

One of the key aims of molecular biology is to understand the structure of DNA and how the information held within it is translated into the complex workings of a cell. Obviously, this task is only feasible if the exact sequence of the bases in DNA can be determined. In 1977, Fred Sanger developed an enzymatic chain

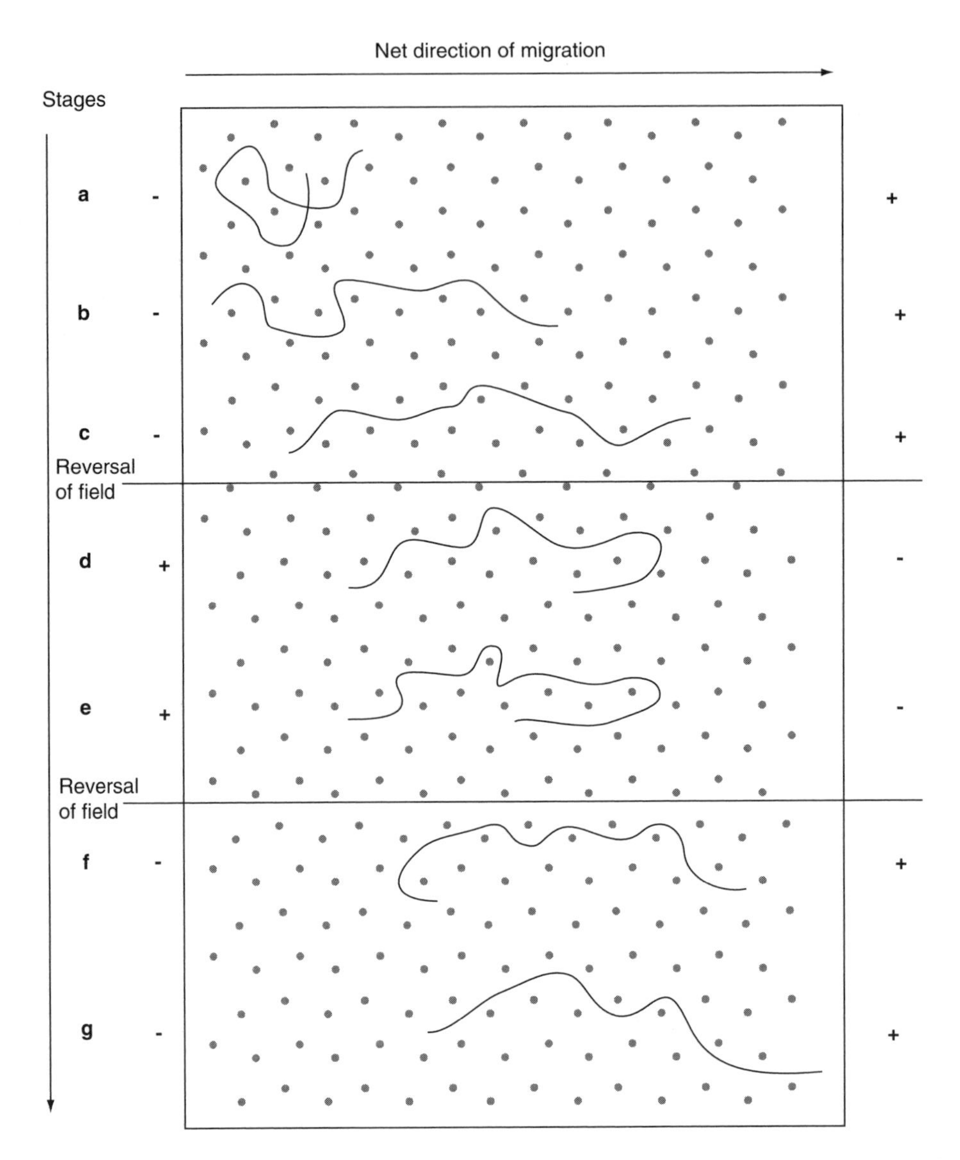

Figure 10.5 The principle of pulsed field gel electrophoresis (PFGE). The DNA molecule is too large to migrate through the gel in its initial state (a). Under the influence of an electric field, it gradually orientates itself (b, c) so that it can move, end on. When the direction of the field changes briefly (d), it becomes tangled again (e). Before it has completed re-orientation, the field changes again (f) and the molecule has to resume its linear form before it can start moving from left to right again (g). Once the molecule is orientated, its mobility is essentially independent of size; the main determinant of mobility (at given pulse conditions) will be how quickly it can re-orientate itself, which is a function of its length

termination procedure that made this possible and for the first time the sequence of strands of DNA could be deciphered.

10.2.1 DNA sequence determination

The method that Sanger developed involves *in vitro* DNA synthesis with DNA polymerase and the use of 2′,3′-dideoxynucleotides. These lack the 3′ OH group, so their incorporation into a growing DNA strand will stop the synthesis of the strand at that point, since there is nothing to attach the next residue to.

Synthesis of the new strand is initiated from a specific position by using a primer that hybridizes to a position adjacent to the 3′-end of the region to be sequenced (Figure 10.6). If the reaction is supplied with each of the four deoxynucleotide triphosphates plus, say, dideoxy ATP (ddATP), then at each point when an A residue should be added there is a chance of incorporating ddA instead. This will lead to a mixture of DNA chains each of which terminates at an A residue. Since these will be different in size, they can be separated on an acrylamide gel, thus indicating the position of each of the A residues. Similar reactions, carried out in parallel with each of the four dideoxy derivatives, lead to a gel with a pattern of bands as shown in Figure 10.6, from which the sequence can be read. Several hundred bases can be read from a single gel.

For obtaining longer sequences, a common strategy is to cut the DNA into random, overlapping, fragments and to clone and sequence each fragment (Figure 10.7). As each new fragment is sequenced, a computer is used to compare the sequence to all the previous ones. When an overlap is found, the computer joins the two sequences together to form a *contig*. As more sequences are determined, the contigs become larger and will overlap other contigs; the overlapping contigs are joined to form an even larger contig. Eventually, the process forms a single contig, covering the whole gene.

The development of gene sequencing also provided a route for deriving the sequence of a protein by using a computer to produce the sequence of the translated product. Before this, determining protein sequences was an expensive and arduous task. This technique almost put the protein biochemists out of work, as it was now much easier to sequence the gene than to sequence the protein. However, with manual DNA sequencing, as described above, it could still take several years to clone and sequence an individual gene. When Leroy Hood in 1986 developed a semi-automated device for DNA sequencing, commercial machines soon followed. This opened the way for larger scale sequencing projects, including sequencing complete genomes of independently living organisms such as bacteria and eventually humans.

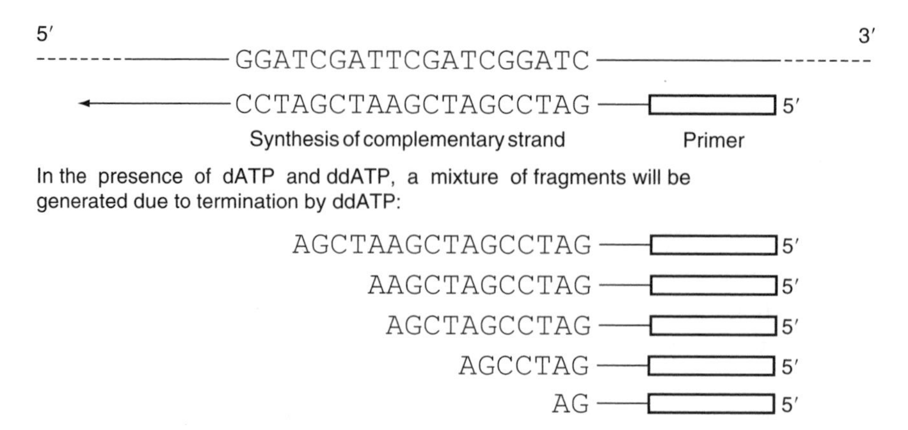

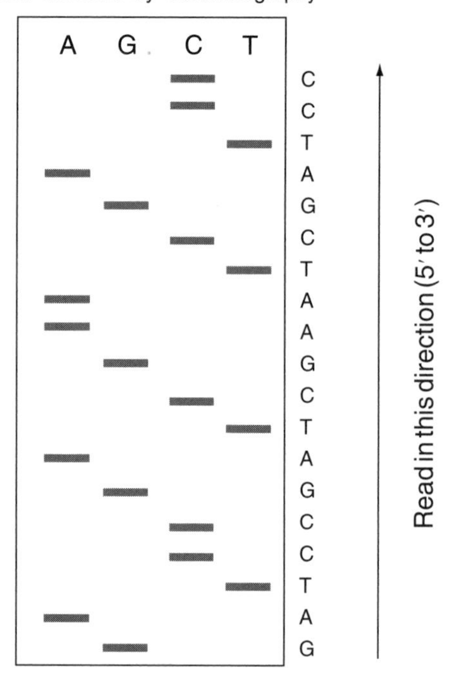

Figure 10.6 Determination of DNA sequence

10.2.2 Genome sequencing

The first DNA genome to be sequenced was that of the bacteriophage φX174 (5.4 kb). The sequences of other viruses, including bacteriophage λ (49 kb) followed gradually, but determining the complete sequence of a bacterial genome

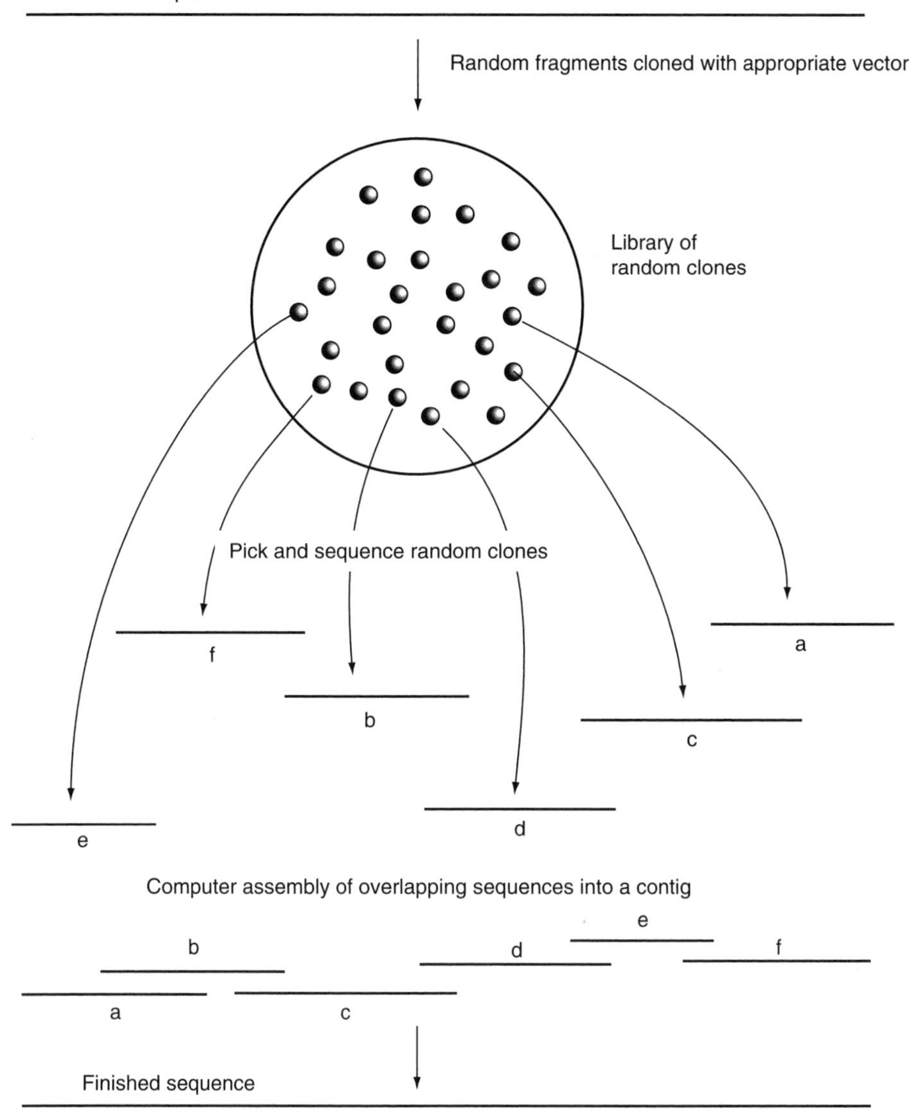

Figure 10.7 Shotgun cloning and sequencing

(about 100 times larger than that of λ) only became possible with the advent of automated sequencing machines and robotic methods for the associated procedures.

A key moment in the history of molecular biology was on 28 July 1995 when Venter, Smith and Fraser reported the first complete genome sequence for a free-living organism, *Haemophilus influenzae* (which has a genome of 1.83 Mb). The ability to sequence entire bacterial genomes has now revolutionized our understanding of bacteria and the entire field of bacteriology. To put this in perspective, by 2002 the sequences of nearly 100 bacterial genomes had been completed.

Analysis of these sequences provides tremendous biological insight into an organism, including information that cannot be obtained using classical approaches. For example, it has become clear that classical bacterial genetics has identified only a fraction of the genes within any genome and often half of the predicted coding regions had not been previously characterized and did not have any known function. By comparing entire genome sequences the evolutionary forces that have forged the physiology of a bacterium can be determined. Genome projects can also be the starting points for far-reaching projects which seek for example, to identify the function of all genes that an organism possesses or to characterize the expression of all genes in a bacterial cell.

Genome sequencing strategies

The aim of a bacterial genome sequencing project is to determine the sequence of several million bases (mega bases or Mb) with an error rate lower than 1 per 10 000 nucleotides. However, the available methods can produce, at the most, sequence data for 1000 bases for each reaction and consequently it is necessary to split up the total DNA of an organism, sequence the parts and then reassemble the information. The most widely used strategy for this is whole genome shotgun sequencing. The principle of this is essentially the same as that illustrated in Figure 10.7, but using thousands or millions of clones rather than a handful. In essence, the DNA of a chosen bacterium is shattered into millions of small fragments which are cloned using a plasmid vector (or sometimes M13), generating a library that represents the entire genome of the target organism. The inserts in each of these plasmids are then sequenced and assembled into contigs.

Since these are random fragments, many of them will overlap. This may seem wasteful, as sequences that have already been determined will be repeated. Actually, it is critical to the success of the procedure. Not only does the sequencing of overlapping fragments enhance the accuracy of the sequence, but the overlaps are crucial for contig assembly.

After this process the genome sequence is nearly complete but often small gaps are left in the sequence because certain regions cannot be cloned. Repetitive elements such as insertion sequences also cause difficulties in contig assembly. Closing these gaps is often the limiting factor which dictates the speed at which genomes can be sequenced. Vectors that are capable of accepting very large inserts (such as bacterial artificial chromosomes or BACs) are useful for bridging large gaps, while smaller gaps can be bridged using PCR.

An alternative strategy, as used for sequencing the largest bacterial genome determined so far (*Streptomyces coelicolor*), is to start by establishing an ordered gene library and then to determine the sequence of each of the clones separately before joining them together to obtain the complete sequence. This 'clone by clone' approach has the advantage that each clone sequenced is a coherent and usable piece of information. Contig assembly and gap closure are also easier, since this approach is less affected by problems with repetitive DNA. However, establishing an ordered library can involve extensive work. Many genome sequencing projects combine elements of both approaches.

10.2.3 Comparative genomics

With genome sequences being available for a number of bacteria, it is possible to compare them to investigate gene function and provide information on the evolutionary history of bacteria. If for example, RNA polymerases from two bacteria are compared, they will be seen to share a very similar sequence. They have descended, through divergence, from a common ancestral molecule. These two enzymes are called *homologues*. In this case it is apparent that the homologues have evolved from a common ancestral gene by speciation and thus have similar function, and consequently the two RNA polymerases can be described as *orthologues* (a specific type of homologue). Sometimes evolutionary events are not as straightforward. *Paralogues*, for example, are homologues that are produced by duplication of a gene within a genome followed by subsequent divergence of one of the copies. Although they share some amino acid sequences, they have different functions. In contrast, *analogues* are genes/proteins that have descended convergently from an unrelated ancestor but have similar functions although they are not related to each other – so they are non-homologous.

One of the most obvious differences in genome sequences is size. This can range from 450 and 580 kb for *Buchnera* and *Mycoplasma genitalium* respectively, to those of *Pseudomonas aeruginosa* (6.3 Mb), *Mesorhizobium loti* (7.04 Mb) and *Streptomyces coelicolor* (8.7 Mb). *Buchnera* is an intracellular symbiont of aphids and has through evolution, retained only the bare minimum of essential genes for survival in its own specialized niche. On the other hand, *P. aeruginosa* is a highly adaptable bacterium which can thrive in many different environments including the human host. *S. coelicolor* is a soil-dwelling bacterium which also shows nutritional versatility as well as a life cycle involving different morphological stages (see Chapter 9) and the ability to produce a range of secondary metabolites (*Streptomyces* species are the main source of naturally-occurring antibiotics). Not surprisingly, both of these organisms contain a large number of regulatory genes and the size and complexity of their genomes reflect an evolutionary adaptation that allows them to occupy diverse environments.

The study of *E. coli*, more than any other organism, has provided key biological insights. Whilst most strains of *E. coli* are regarded as commensals in the gut, a small number cause disease. In this context, the sequence of the non-pathogenic K-12 strain provides a baseline against which other *E. coli* sequences can be assessed. For example, *E. coli* O157:H7 is one of the most dangerous pathogens threatening our food supply. When the genome of this strain was sequenced it was found to possess over 1000 genes that the non-pathogenic K-12 strain lacked. Many of these 'new' genes appear to be associated with the ability of O157:H7 to cause disease. The nature and origin of this inserted DNA was discussed in more detail in Chapter 4. Important insights into other closely related pathogens have also been gained by comparing genomes. For example, *Mycobacterium leprae* (the causative agent of leprosy) has lost the function of around 1700 genes compared with its close relative *Mycobacterium tuberculosis* and only half of the genome actually encodes functional genes. The degree of this decay in genetic function represents a highly specialized adaptation, to such an extent that *M. leprae* still cannot be grown in laboratory culture.

Microarrays

The genome sequence of a bacterium can be used to study the variation that occurs within other strains of the same species. For example, the question could be asked, in what way does the tuberculosis vaccine strain BCG differ from the sequenced virulent strain of *Mycobacterium tuberculosis*? Rather than determining the complete sequence of BCG (although that will probably have been done by the time this book is published), *microarrays* can be used. A further application of microarrays – for comparing gene expression in two organisms – is dealt with later in this chapter.

From the complete genome sequence of a bacterium (in this example, *M. tuberculosis*), it is possible to generate a set of DNA fragments that correspond to part of each gene in the genome, either by PCR amplification or by using synthetic oligonucleotides. This might seem an impossible task – for example there would be about 4000 such DNA fragments for *M. tuberculosis* – and indeed it would be arduous if carried out manually. In fact, robotic devices are used. A tiny spot of each of these oligonucleotides is printed onto the surface of a glass slide in a grid-like arrangement (again using a robot to give very precise positioning of each spot).

To compare the genetic content of *M. tuberculosis* and BCG, DNA from each organism is extracted and labelled (as small fragments) with a different fluorescent dye. The two DNA samples are then mixed and applied to the glass slide so that they will hybridize to the array of spots (Figure 10.8). The binding of a labelled fragment to a spot on the microarray is read by a machine which is able to

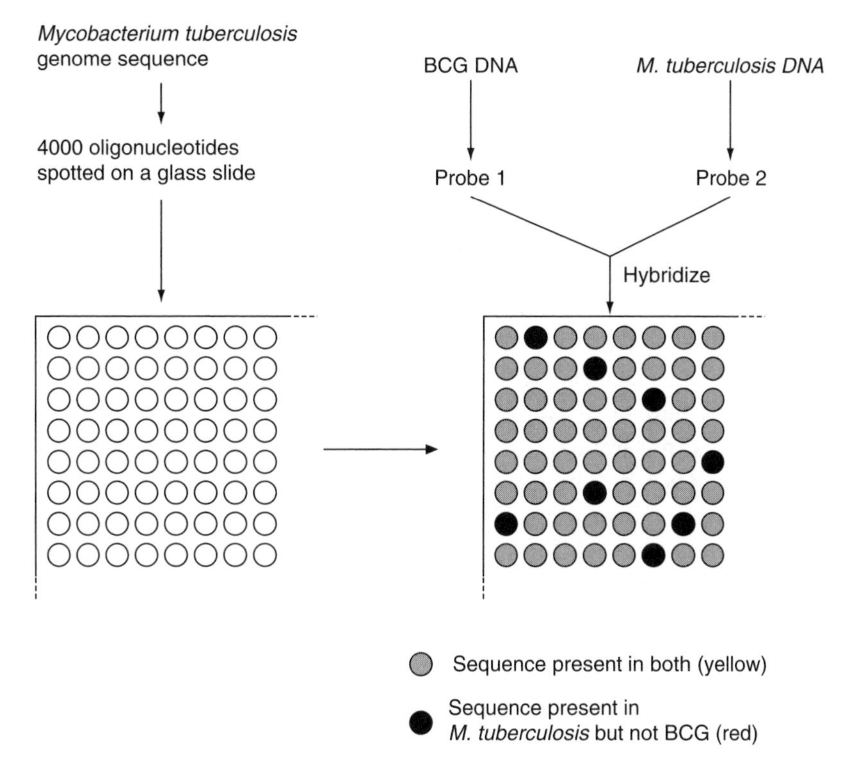

Figure 10.8 Genomic comparison using a microarray. The two probes are labelled with different fluorescent dyes which can be distinguished by the machine reading the microarray. Spots hybridizing only to probe 2 will be displayed as red, while spots hybridizing to both probes will be shown in yellow. Spots hybridizing only to probe 1 would be displayed as green, but are not expected in this example as probe 2 is from the same source as the array and is expected to hybridize to all spots (see also Figure 10.12)

distinguish the fluorescence of the two dyes and the results are displayed as a pattern of dots – red and green dots indicating binding of one or the other sample only, while if both labelled DNA samples bind, the result will be shown as a yellow dot. So in this example, all genes that are present in both *M. tuberculosis* and BCG will be shown in yellow, while those that are only present in *M. tuberculosis* (and missing in BCG) will appear in red. The microarray cannot show whether BCG has any *extra* DNA which is not present in *M. tuberculosis*, therefore green spots would not be expected.

This technique can therefore provide a rapid assessment of the differences between the genomes of two organisms – but it will only detect deletions. For other types of changes, in particular sequence polymorphisms, other techniques have to be used.

Sequence polymorphism

It is not (yet) possible to undertake a global analysis for all the sequence differences between two bacteria, short of complete genome sequencing. However, large numbers of strains can be rapidly surveyed for changes in specific genes. Primers derived from the sequence of a gene from one strain can be made and PCR used to amplify the corresponding gene from the other strains. The amplified product can be sequenced directly.

One example comes from studies of bacterial resistance to rifampicin which is (amongst other uses) a front-line drug in the treatment of tuberculosis. The target for rifampicin is the bacterial RNA polymerase and analysis of sequence polymorphism shows that most rifampicin-resistant strains of *M. tuberculosis* have base substitutions in a limited number of positions in the *rpoB* gene (coding for the β subunit of RNA polymerase).

Sequence polymorphisms can also be detected by specialized electrophoretic techniques which can detect differences as small as a single base change between two PCR products. Two such technologies are Single Strand Conformational Polymorphism (SSCP) in which denatured PCR products are run on acrylamide gels and Denaturing Gradient Gel Electrophoresis (DGGE) where PCR products are resolved on an acrylamide gel which contains a gradient of a denaturing agent. These techniques are often used to screen large numbers of PCR fragments to determine whether or not they contain base changes prior to nucleotide sequence determination.

10.2.4 Bioinformatics

The genome sequence consists only of millions of DNA bases in a defined order. Bioinformatics is the use of computers to obtain biologically interesting information from the sequence data. This starts with identifying genes, primarily by searching for open reading frames (ORFs) – although in practice it is more complex than that. The gene sequences are then automatically translated to determine the protein sequences they encode.

It is then necessary to determine the function of those genes and their products. One way of doing this is to compare each protein to all the proteins in the international databases in which all known DNA and protein sequences are stored. (There are three main databases, GenBank, EMBL and DDBJ, but all the information is exchanged between them so they are all equally informative). It is preferable to compare protein sequences rather than DNA, partly because the differences in codon usage between organisms means that the DNA can be quite different even though the proteins encoded are similar. These sequence comparisons are more complicated than simply lining up the two sequences and counting the matches, but fortunately there are programs available (one of the most widely used being known as BLAST) which will carry out this procedure. The results may show that a gene (or protein) is similar to one in the database with a known or predicted function, so it is therefore possible to assign a putative function to that gene from your organism of interest.

The second tool for assigning function looks for the presence of shorter sequences in the protein that are characteristic of proteins with specific functions. For example there may be a specific arrangement of hydrophobic regions which is characteristic of membrane proteins, or a predicted structure that indicates a DNA-binding protein (which would be likely to be a regulatory protein). There are computer-based libraries of these signatures and software that will compare proteins of interest to those libraries. This process, which essentially labels a gene with a possible function and a unique catalogue number, is known as *annotation*.

However, even with the wealth of information that is available and the sophisticated software that can be used for automated screening of complete genomes, a substantial proportion of the genes in sequenced bacterial genomes give no clues at all to their function, as they are neither similar to any genes in other bacteria nor do they have any structural features which indicate their function. Determining the function of these genes remains a major challenge. The next section in this chapter includes some of the ways in which this question can be addressed.

10.3 Physical and genetic maps

The techniques described so far can be divided broadly into two categories. Analysis of recombination between mutants (by conjugation for example) produces a *genetic map* in which the relative positions of genes are determined. On the other hand, restriction mapping and PFGE analysis give a *physical map* in which the position of restriction sites or other features such as deletions and insertions are located, without any reference to the location of specific genes. Even the genome sequence, in the absence of other information, is merely a physical description of the chromosome.

How are these maps correlated? In other words, if the function of a gene and its position on a genetic map is known, how can its position on a physical map of restriction fragments or on the genome sequence be found? The question can also be reversed, i.e. if an adequate physical map (ideally the complete genome sequence) is available, how can the genes that are associated with a specific function in the cell be identified?

Some of the approaches might be obvious from the previous discussion. For example, if a gene has been cloned, it is easy to determine its position on a restriction map or within an ordered gene library by testing its ability to hybridize to the relevant DNA fragments. If the sequence of the gene is known and the genome sequence has been determined, a simple computer analysis will locate it. There are however some additional techniques that enable these questions to be approached without such detailed information and also to refine the analysis of the relationship between gene structure and phenotype.

10.3.1 Deletions and insertions

If a mutant is selected on the basis of an altered phenotype, it may have arisen by a point mutation – a single base change in the gene concerned. Such mutations are not readily located on a physical map even if the complete genome sequence is known. It can however provide a route for cloning, and hence identifying the gene, for example by selecting a clone that complements the mutation (see Chapter 8).

On the other hand, if the loss of a gene function is due to the insertion or deletion of a larger piece of DNA, this can be visualized as a change in the size of a restriction fragment. It is then possible to clone and characterize this fragment to confirm its identity and its relationship to the observed phenotype. Alternatively, if the genome sequence has been determined, microarrays can be used as described earlier.

10.3.2 Transposon mutagenesis

Transposons (see Chapter 7) provide a more convenient way of doing the same job, since they are capable of inactivating genes by insertion and the inactivated genes are readily identified and recovered. A typical strategy, outlined in Figure 10.9, is to construct a plasmid carrying the transposon, using a temperature sensitive vector, i.e. a plasmid that is unable to replicate at an elevated temperature. When this plasmid is introduced into bacterial cells and the cells are grown at a temperature permissive for plasmid replication (say 30°C), all the cells will carry the plasmid. If the culture is subsequently shifted to a temperature at which the plasmid is unable to replicate (e.g. 42°C), the only cells to retain the transposon will be those in which the element has transposed onto the chromosome. If a transposon such as Tn*5*, is used, which carries a kanamycin resistance gene, these transpositions can be selected for by plating the culture at 42°C on agar containing kanamycin.

This generates a transposon mutant library, with the transposon inserted at a variety of chromosomal positions and therefore causing a range of mutations. Once the mutants of interest have been selected on the basis of a specific phenotypic change, the gene(s) into which the transposon has inserted can then be identified. For this, advantage is taken of the resistance gene on the transposon. If the total DNA is cut with a restriction enzyme, the collection of fragments can be cloned to produce a gene library. If this gene library is then plated on kanamycin agar, only those clones containing Tn*5* will be able to grow. These

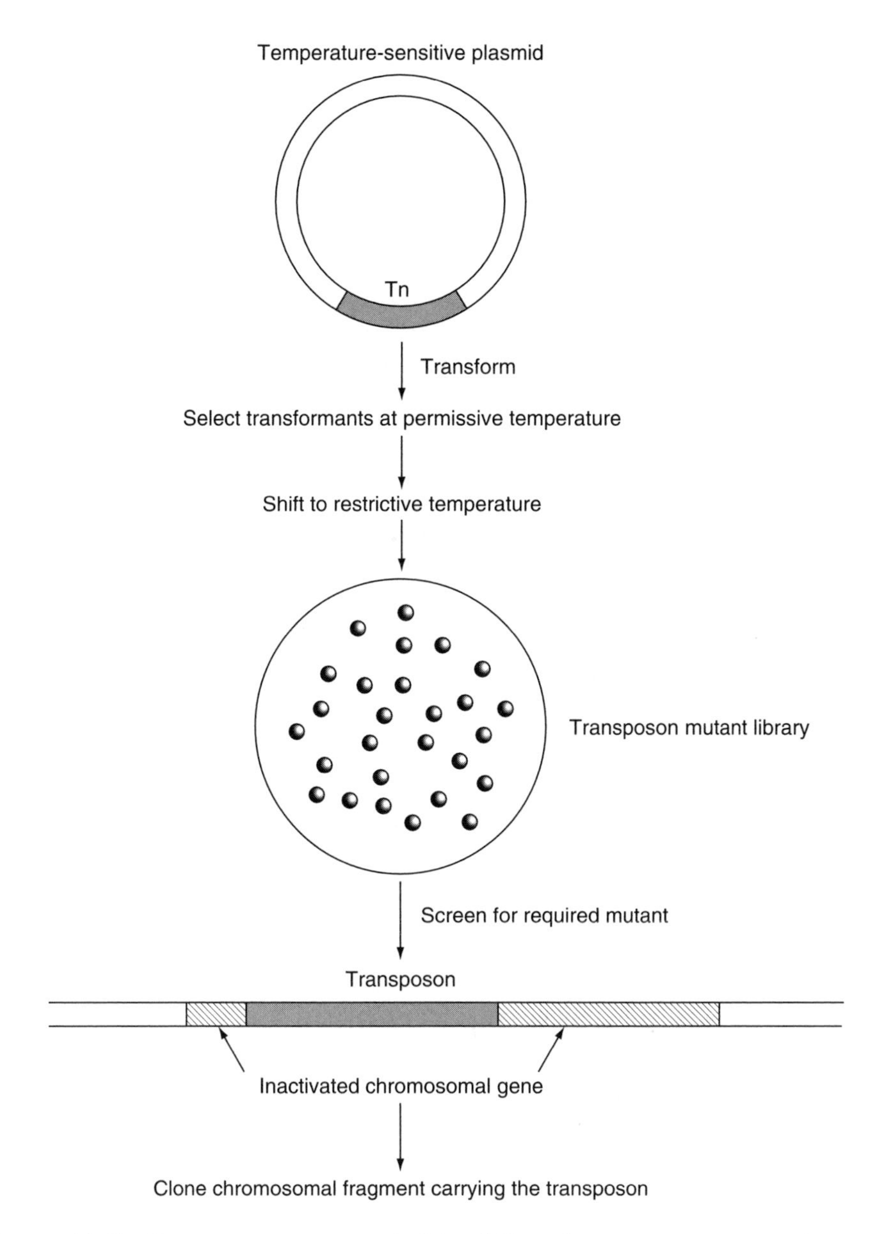

Figure 10.9 Transposon mutagenesis. The ability of some transposons to insert randomly into chromosomal DNA can be used to generate a library of mutants. One advantage of transposon mutagenesis is the ease of identifying the affected gene

clones will contain not only Tn*5* but also some of the sequence on either side of it; this is the gene that has been inactivated by Tn*5* insertion, which can now be characterized.

10.3.3 Gene replacement

Having identified a gene of potential interest, for example from the genome sequence, one of the best ways of determining its function and significance is to inactivate it. This may not be easy to achieve by the classical procedures of mutagenesis and selection, especially if the phenotype associated with that gene is not known or not readily selectable. However, a cloned gene that has been inactivated *in vitro* can be used to replace the wild-type chromosomal gene by homologous recombination, a process also known as gene knock-out. The procedure for doing this was described in Chapter 9. Once the gene has been knocked out, it will be possible to determine what the consequences are for the phenotype of the cell which will give an indication of the function of that gene.

10.3.4 Site-directed mutagenesis

Instead of knocking out a gene completely, it may be desirable to replace it with a specifically altered gene. For this gene replacement as described above would be combined with the techniques of site-directed mutagenesis, as covered in Chapter 8, either by direct replacement or by complementing a knocked-out gene with a plasmid carrying the mutated gene.

This enables the significance of a specific amino acid in the activity of an enzyme for example, to be tested by replacing it with a series of different residues and determining the activity of the product. Or the precise requirements for the function of a regulatory sequence on the DNA, such as a promoter site, could be investigated by introducing defined mutations into that region.

10.4 Analysis of gene expression

Bacteria are capable of recognizing many different environmental stimuli and changing their pattern of gene expression to optimize survival when faced with new environments (see Chapter 3). Much can be learned about the function of a gene and our understanding of these physiological responses can be enhanced by following the changes in expression that accompany alterations in the morphological or physiological state of the cell (e.g. during sporulation or transition to stationary phase) or in different external environments (such as bacterial growth within macrophages).

The most obvious way of monitoring gene expression, by assaying enzyme activity, is not always appropriate. For a start, not all gene products are enzymes; some of the most interesting are regulatory proteins. And even if it is an enzyme, the assay procedure may be too complex to use for this purpose.

10.4.1 Transcriptional analysis

In bacteria, gene expression is primarily regulated at the transcriptional level. It can therefore be reasonably assumed that measuring the amount of a specific mRNA in the cell is an appropriate measure of gene expression. Furthermore, in general, bacterial mRNA is short lived, so measuring the amount of a specific mRNA at a given time is likely to provide a good assessment of the current level of transcription of that gene. An *in vivo* method for doing just this, i.e. the use of reporter genes, has been previously encountered in Chapter 9. Here, we concentrate on the *in vitro* methods available for assessing mRNA levels.

Northern blots and RT-PCR

One way of looking at the production of a specific mRNA is to fractionate bacterial RNA by gel electrophoresis, and transfer the RNA to a filter by blotting, followed by hybridization of that filter with an appropriately labelled DNA probe. The intensity of the signal provides a measure of the amount of the specific mRNA in the preparation. Since this procedure is analogous to Southern blotting (see Chapter 2), it is referred to as *Northern* blotting.

Northern blotting is not ideal for the purpose being considered here. Firstly, many bacterial mRNA species are quite unstable, with a half-life of only 1 or 2 min. Secondly, the amount of RNA required may be more than can be readily obtained under the conditions being investigated. For example, we may want to investigate the levels of expression within different parts of a single colony or from bacteria growing within macrophages. A much more sensitive technique is provided by an adaptation of the polymerase chain reaction (PCR).

In standard PCR (see Chapter 2), a DNA polymerase is used to make copies of the template DNA. In order to obtain a sensitive method for detecting a specific mRNA, a DNA copy is made initially, using an RNA-directed DNA polymerase, reverse transcriptase (see the description of cDNA cloning in Chapter 8). Following the initial reverse transcription, a standard PCR can be used to amplify the DNA strand produced. This method, reverse transcript PCR or RT-PCR (Figure 10.10), provides a very sensitive method of detecting the presence of a specific transcript within small sections of a bacterial population.

Real-Time PCR

A serious limitation of RT-PCR, as described above, for the purpose being considered here, is that it is not easily quantifiable. It can be used to *detect* a transcript but not (or at least not easily) to determine how much of that transcript

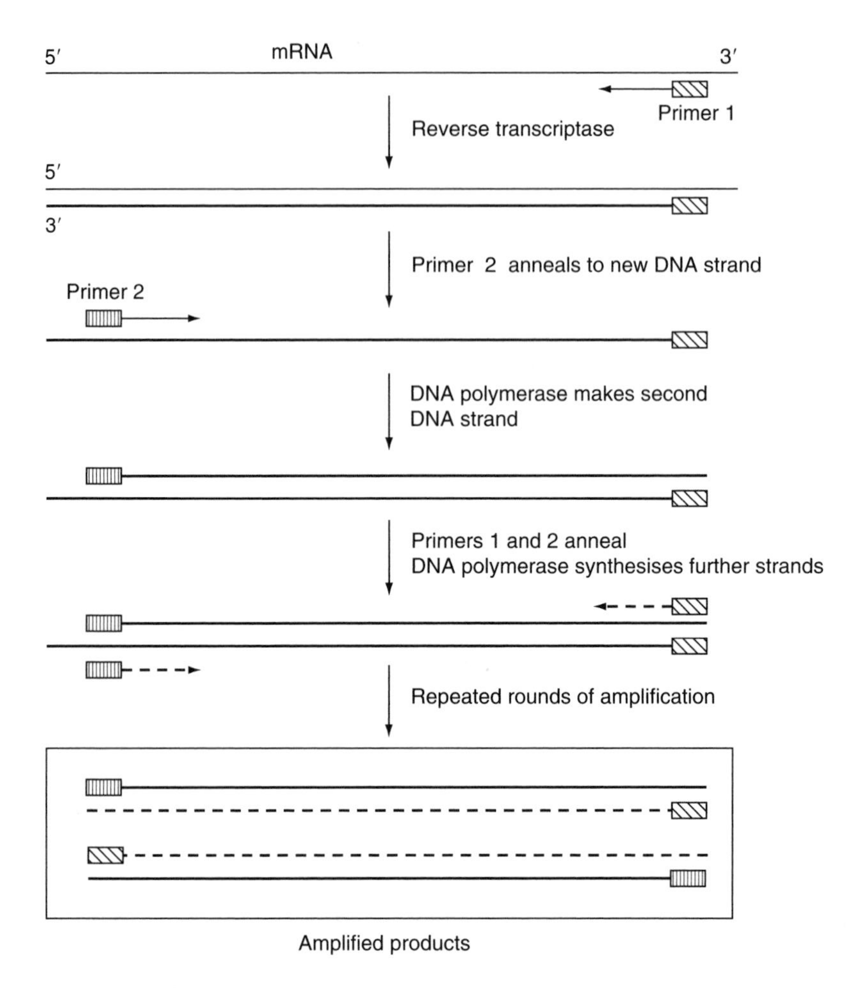

Figure 10.10 Reverse transcript PCR (RT-PCR)

is present. For that, another variant of PCR known as *real-time PCR* can be employed, since it enables continuous monitoring of the appearance of the amplified product during the reaction.

There are actually several quite different techniques for doing this, but the discussion here will be limited to the method which is conceptually the simplest. This involves including in the PCR mixture a dye which fluoresces when it binds to double-stranded DNA. This signal can be detected using a machine that not only carries out the temperature cycling but also measures the fluorescence of the sample. Initially, the DNA concentration is very low, so there will be no detectable signal. But as the reaction progresses, through one cycle after another, the amount of the amplified product increases exponentially, until eventually the machine registers a signal (Figure 10.11). The number of cycles needed to reach

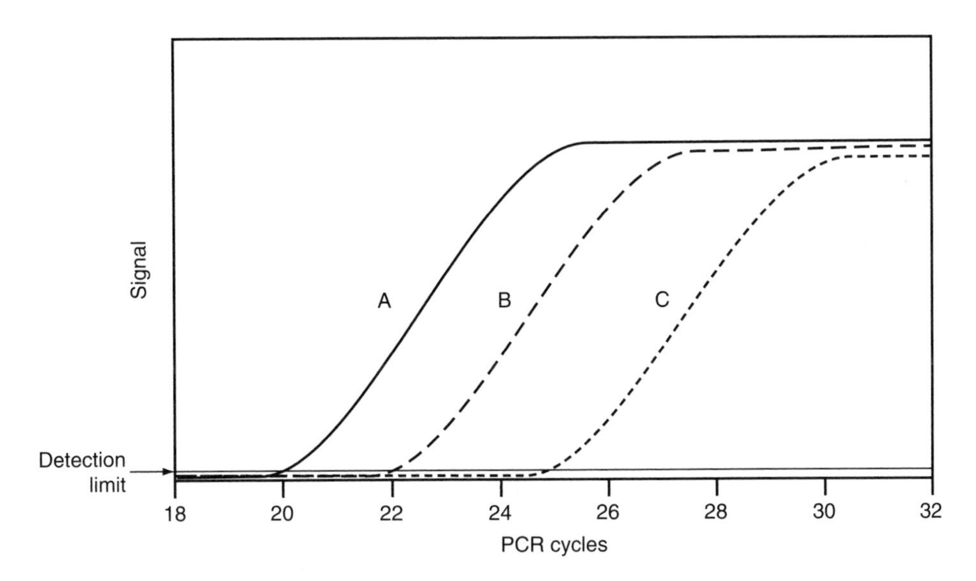

Figure 10.11 Real-time PCR. Several procedures are available for continuous monitoring of the progress of PCR. By determining the number of cycles required before a detectable signal is produced, the amount of initial template can be estimated

this point depends on the amount of target DNA in the original sample. So sample A in the figure, requiring only 20 cycles to reach a detectable level, contained more template than sample B (22 cycles) or sample C (25 cycles).

Of course, for measuring levels of a transcript, the process begins with RNA not DNA. So for this purpose it is necessary to include a reverse transcription step before carrying out the real-time PCR.

Apart from being quantifiable, the other advantage of real-time PCR is the ease and speed of detection of the product – it does not need gel electrophoresis. This is useful in diagnostic applications of PCR (Chapter 9).

Analysis of mRNA expression using microarrays

The methods for analysing gene expression described above and the use of reporter genes (Chapter 9), are designed for monitoring the expression of a small number of genes individually. However, with the availability of complete genome sequences, it is now possible to assess the expression of all genes in a bacterium simultaneously, using microarrays.

The concept of microarrays was described earlier in this chapter, using DNA samples. For assessing gene expression – or more specifically the amount of each mRNA – the labelled DNA would be replaced with labelled mRNA (or labelled cDNA produced by reverse transcription of the mRNA).

To assess the expression of the genes on the microarray, cells are grown under a control condition and in the presence of a stimulus of interest (Figure 10.12) – for example, a comparison of gene expression in actively growing cells (culture A) with those in stationary phase (culture B). The mRNA is isolated from both cultures and reverse transcribed to produce a spectrum of cDNA molecules. Each cDNA preparation is tagged separately with a different fluorescent dye as described previously. The cDNA samples are then mixed and incubated with the microarray to allow them to hybridize to the complementary spots on the array. The two differentially dyed cDNAs in the mixed sample compete for binding to each spot on the array, the outcome depending on the relative amounts of the specific cDNA within the two samples. Afterwards, the microarray is scanned using a specific reader as described previously. In this example, any gene that is more strongly expressed during active growth will be represented by more cDNA labelled with the first dye. The signal from that dye will therefore be stronger which will be displayed as a red spot (spot 1). Conversely, genes that are activated during stationary phase, with cDNA labelled with the second dye, will be shown as a green spot (spot 3). If a gene is expressed equally under both conditions, there will be no difference between the signal from the two dyes and this will show as a yellow spot (spot 2).

Thus under these conditions, the relative expression of all the genes in a genome can be assessed by simply examining the colour of the spots. The spectrum of genes in a cell which are transcribed under a specific set of conditions is known as the *transcriptome* by analogy with the *genome*, which is the complete genetic content of the cell. It is worth noting that while the genome for an individual strain is more or less fixed, the transcriptome is not – it will be different according to the conditions pertaining at the time the cells were harvested.

10.4.2 Translational analysis

As mentioned earlier, analysis of mRNA levels assumes that gene regulation occurs at the transcriptional level – but this may not be true. Consideration should be given to the possibility that the levels of specific proteins are affected by the efficiency of translation as well as, or instead of, transcription. Furthermore, if there are several similar genes coding for different proteins, it may be quite difficult to distinguish them by transcriptional analysis, as the products may cross-hybridize.

Western blots

If specific antibodies are available which recognize the protein under study, these can be used to monitor expression of the protein under different growth

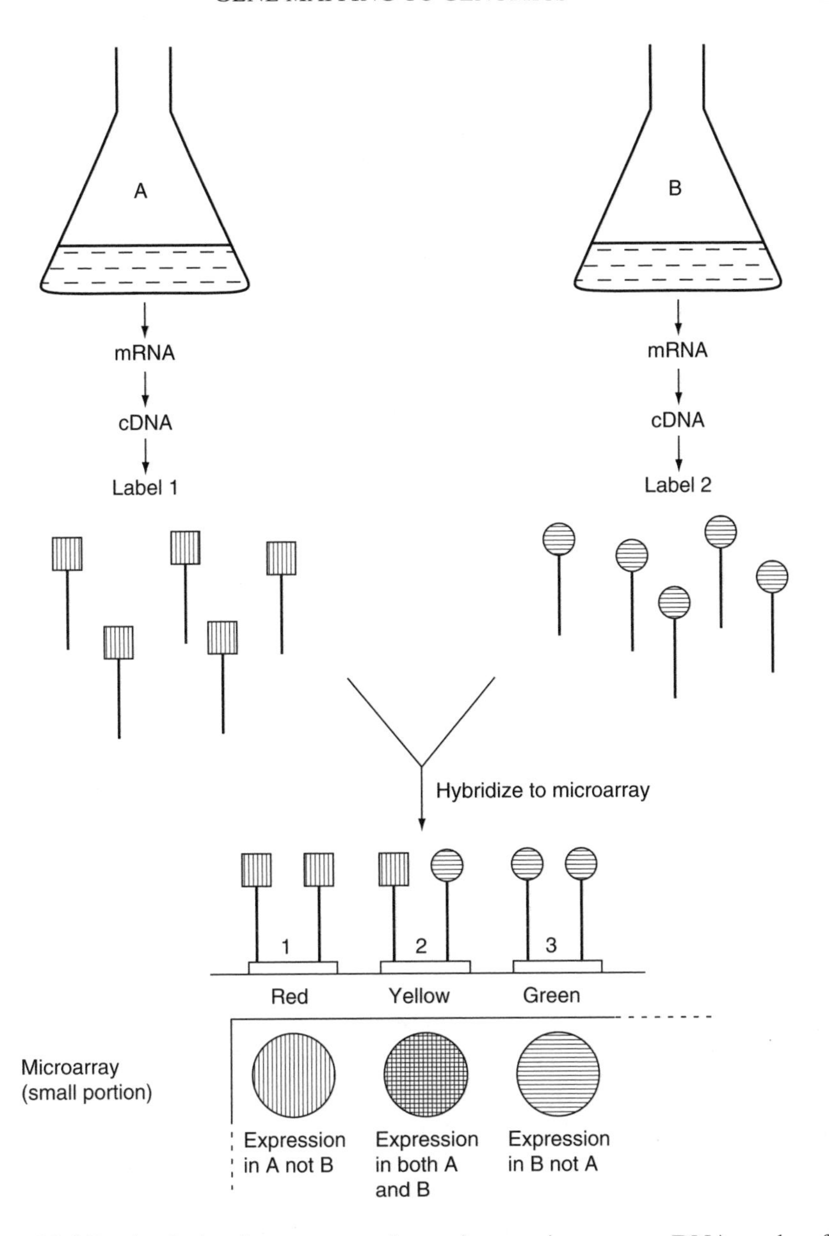

Figure 10.12 Analysis of gene expression using a microarray. cDNA probes from cultures A and B are labelled with different fluorescent dyes, as described in Figure 10.8. If a gene is expressed more in culture A, label 1 will bind more strongly to that spot on the array, which will appear red. A gene expressed more strongly in culture B will show as green, while if there is no difference the spot will be shown as yellow

conditions. For this purpose, a cell extract would be fractionated by acrylamide gel electrophoresis (SDS PAGE), the proteins transferred to a filter and the specific proteins detected with antibodies. These antibodies are either labelled directly or detected by using a second labelled antibody which reacts with the first one. The most common label is an enzyme (such as peroxidase) that can be visualized using a chromogenic substrate. This procedure is known as Western blotting, by analogy with the Southern blotting technique described in Chapter 2, although the process is rather different as it usually involves electrophoretic transfer. However, Western blotting is only suitable for monitoring the expression of specific proteins. For global analysis of protein expression (the *proteome*), two-dimensional gel electrophoresis may be employed.

Analysis of gene expression using proteomics

Whilst microarray technology allows the determination of the amount of all individual mRNAs in a cell, the scope of proteomics is to characterize the expression of every single protein in a cell at a specific time and under specific conditions. Bacterial cells may contain thousands of different proteins, so powerful methods are needed to separate them. The amino acid sequence of a protein determines both the mass and the overall charge of the molecule and these two characteristics can be used to separate the individual proteins of a cell. For example, many proteins have a similar mass and could not therefore be resolved by SDS-PAGE alone. However, most of those with similar mass have different compositions of acidic and basic amino acids which confer a different overall electrical charge to the protein. The overall charge of the protein depends on the pH. At high pH values they will be negatively charged, while at low pH the charge will be positive. In between is the isoelectric point (pI) at which the protein is neutral. The pI varies from one protein to another, depending on the amino acid composition.

Two-dimensional polyacrylamide electrophoresis (2D-PAGE) separates proteins in two sequential steps: first by their charge and secondly by their mass (Figure 10.13). The first step involves isoelectric focusing (IEF) during which the proteins present in the cell extract are resolved on a gel in which a stable pH gradient is generated using chemicals called ampholytes. When an electric field is applied to the gel, each protein will migrate through the gradient until it reaches a point (the pI) at which the net charge of the protein is zero and it becomes stationary. Because they have different pI values, many of the proteins will travel to different positions on the IEF gel. However the resolution is not complete, as some proteins will have similar isoelectric points. Further resolution is achieved by applying the IEF gel to an SDS-PAGE gel, so as to separate the proteins on the basis of their mass alone. Using this technology as many as 1000 proteins can be resolved simultaneously.

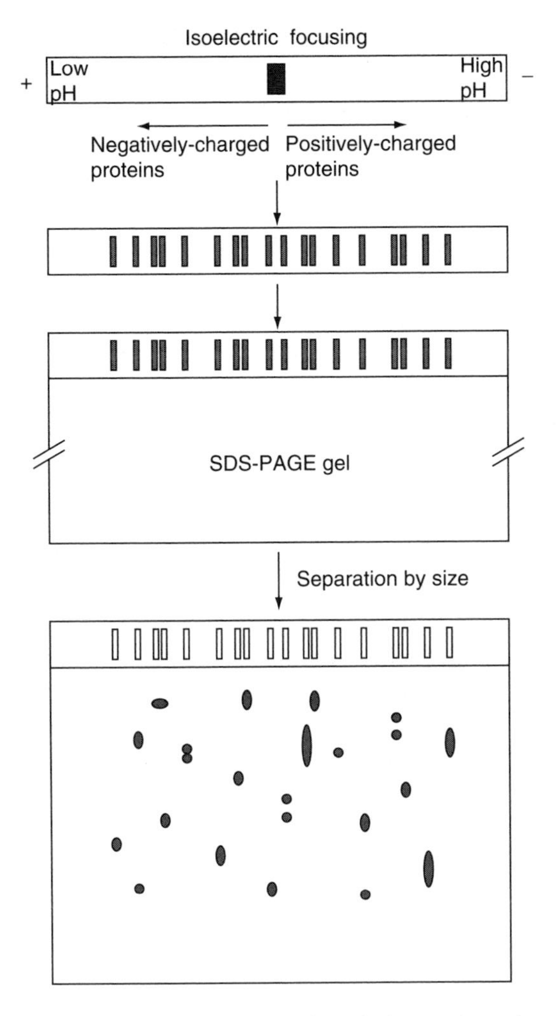

Figure 10.13 2D protein gel electrophoresis

After the gel has been run, a stain is applied to highlight the resolved proteins which appear as individual and separated spots. The location of a known protein in the final gel gives it a unique signature that is formed by its charge and mass and this can be used to identify it. Alternatively, the identity of an unknown protein can be determined by cutting the protein spot from the gel and subjecting it to micro-chemical analysis such as direct amino acid sequencing or mass spectrometry. These techniques will provide partial but unique amino acid sequences, which can then be linked back to the corresponding gene in the genome by reverse translation. Using 2D-PAGE it is now possible to study the total expression profile of nearly all the proteins in a bacterial cell under any chosen condition.

10.4.3 Systematic analysis of gene function

The genome sequence represents a catalogue of all the genes of a bacterium but many of the genes identified in these projects do not have a known function. Consequently, following the completed genome sequences for some bacteria, systematic programs to identify the function of all known reading frames have been initiated. The aim of these is to inactivate all the genes (where this is possible) of an individual species and by assessing the resulting phenotype, to determine the function of every individual protein. These ambitious and far-reaching projects are only now possible with the availability of complete bacterial genome sequences.

10.5 Conclusion

These final chapters can give no more than a brief overview of the range of techniques available and the extent to which molecular genetics has enhanced our understanding of the basic biology of bacteria (to say nothing of the revolution that has occurred with higher organisms). The expansion of the subject shows no signs of abating. The rapidity with which genome sequencing is advancing will by itself have a major impact on our understanding of all aspects of bacteriology. This includes not only the molecular biology of these organisms, but their physiology, biochemistry, pathogenicity, evolution – nothing escapes. However, in order to make the most of this information, the information gleaned from molecular studies has to be related back to the behaviour of the cell as a whole and its interaction with its environment. It is this synthesis of different fields that is the major challenge in the years ahead.

Appendix A
Further Reading

Comment

This is a highly selective list of the major texts that we find most useful – the absence of a book from this list does not necessarily mean that it has no merit.

In a rapidly moving field such as molecular genetics, information rapidly becomes out-of-date. It is not always wrong, but sometimes the language changes subtly or the emphasis alters or a tentative explanation becomes an accepted truth. So, a student relying too heavily on old books can convey a rather old-fashioned feel to the answers. The most recent edition should always be used and anything more than a few years old should not be referred to unless you are really sure of your ground. Wherever possible, this list should be supplemented with review articles and original literature. We have not included any such references as they date even more rapidly than do textbooks; however, if the reader wishes to search them out for themselves, *Trends in Genetics/ Microbiology* or *Microbiological Reviews* are recommended. *Scientific American* is also a valuable source of information. Some journals (such as *Molecular Microbiology*) also publish short review articles as well as research papers. For the more ambitious, reference to the *Annual Reviews* series (*Biochemistry, Microbiology* or *Genetics*) is suggested – but these articles do tend to be rather heavy going.

Symposium proceedings are often of limited use for student reading; the Society for General Microbiology symposia are a notable exception.

General texts and basic molecular biology

B. Lewin. *Genes VII.* (2000). Oxford University Press. First choice for further reference, but it may be found to be too big a jump from the level of this book. Wide ranging, but very useful for bacteria as well as eukaryotes.

C. R. Calladine and H. R. Drew. *Understanding DNA: The Molecule and How It Works.* 2nd edn (1997). Academic Press. (3rd edn due in 2004). An unusual book. Aimed partly at non-scientists, it starts in a deceptively simple manner, but progresses to dealing with many difficult concepts relating to the structure and function of DNA. Especially useful on supercoiling. Strongly recommended as an antidote to some of the rather superficial accounts of DNA structure in most molecular biology textbooks.

Molecular Genetics of Bacteria, 4th Edition by Jeremy Dale and Simon F. Park
 ISBN 0 470 85084 1 (cased) ISBN 0 470 85085 X (pbk)

B. Alberts *et al. Molecular Biology of the Cell*, 4th edn (2002). Garland. Remarkable value for money, but difficult to carry around. Although relating primarily to eukaryotes, a valuable reference source for some of the molecular aspects.

H. Lodish *et al. Molecular Cell Biology*, 4th edn (1995). W. H. Freeman. (5th edn due in 2003–2004). Another text aimed primarily at eukaryotic cell biology, but containing enough information of general relevance, or specifically about bacteria, to be useful for reference on the broader molecular aspects.

More focus on bacteria

L. Snyder and W. Champness (2003). *Molecular Genetics of Bacteria*, 2nd edn. American Society for Microbiology. First choice for further reference on bacterial genetics.

S. Baumberg (ed.) (1999). *Prokaryotic Gene Expression*. A multi-author book, which makes for a very detailed account of selected topics rather than a fully integrated storyline. A big jump from the level of this book.

M. T. Madigan, J. M. Martinko and J. Parker (2000). *Biology of Microorganisms* (better known as 'Brock'), 9th edn. Prentice Hall International. Included in case the reader needs to revise the biological background to some of the systems described.

Gene cloning

J. W. Dale and M. von Schantz (2002). *From Genes to Genomes*. John Wiley & Sons. Companion to this book, taking recombinant DNA technology much further and into the world of eukaryotes.

T. A. Brown (2001). *Gene Cloning – An Introduction*, 4th edn. Blackwell Science. An excellent introductory book.

S. B. Primrose, R. Twyman and R. W. Old (2001). *Principles of Gene Manipulation*, 6th edn. Blackwell Science. Much heavier going, but mandatory for the more advanced student.

B. R. Glick (2003). *Molecular Biotechnology: Principles and Applications of Recombinant DNA*, 3rd edn. American Society for Microbiology.

Websites which give more information on bacterial genome sequences and access to genomic data bases

http://www.sanger.ac.uk/ The website for the The WellcomeTrust Sanger Institute is one of the leading genomic centres in the world.

http://www.tigr.org/ The website for the Institute for Genomic Research (TIGR) which is involved in the comparative analysis of genomes and gene products from a wide variety of organisms including bacteria.

http://www.ncbi.nlm.nih.gov/ A website that contains public databases and software tools for analysing genome data.

General genetics

Some suggestions for those who want to explore further the relationship between bacterial genetics and the broader field of genetics.

D. P. Snustad and M. J. Simmons (2000). *Principles of Genetics*, 2nd edn. John Wiley.

W. S. Klug and M. R. Cummings (2000). *Concepts of Genetics*, 6th edn. Prentice Hall.

L. H. Hartwell and others (2000). *Genetics*. McGraw-Hill.

P. J. Russell (2002). *Genetics*. Benjamin Cummings.

A. J. F. Griffiths, W. M. Gelbart, R. C. Lewontin and J. H. Miller (2002). *Modern Genetic Analysis*, 2nd edn. W. H. Freeman.

Special topics

R. W. Hendrix *et al* (1983). *Lambda II*. Although well over our usual time limit, this is still an invaluable source of reference about bacteriophage λ, but only for the more dedicated student at this level.

M. Wilson, R. McNab and B. Henderson (2002). *Bacterial Disease Mechanisms*. Cambridge University Press. Provides further information about many of the virulence mechanisms that provide important examples of the applications of bacterial genetics.

A special mention

W. Hayes (1968). *The Genetics of Bacteria and their Viruses*, 2nd edn. Blackwell Scientific Publications. The classic text. Recommended for those interested in the historical development of the subject – it still contains a surprising amount of useful material.

Appendix B Abbreviations

Amino acid notations

	Three-letter notation	One letter notation
Alanine	Ala	A
Arginine	Arg	R
Asparagine	Asn	N
Aspartate	Asp	D
Cysteine	Cys	C
Glutamate	Glu	E
Glutamine	Gln	Q
Glycine	Gly	G
Histidine	His	H
Isoleucine	Ile	I
Leucine	Leu	L
Lysine	Lys	K
Methionine	Met	M
Phenylalanine	Phe	F
Proline	Pro	P
Serine	Ser	S
Threonine	Thr	T
Tryptophan	Trp	W
Tyrosine	Tyr	Y
Valine	Val	V

Other abbreviations

5-BU	5-Bromouracil
A	Adenine
BAC	Bacterial artificial chromosome

Molecular Genetics of Bacteria, 4th Edition by Jeremy Dale and Simon F. Park
 ISBN 0 470 85084 1 (cased) ISBN 0 470 85085 X (pbk)

C	Cytosine
cAMP	Cyclic AMP
CAP	Catabolite activator protein (see CRP)
ccc	Covalently closed circular (plasmid DNA)
cDNA	Complementary DNA
CHEF	Contour-clamped homogeneous electrophoretic field electrophoresis
CRP	Cyclic AMP receptor protein
ddNTP	2′,3′-Dideoxynucleoside triphosphate; any of ddATP, ddCTP, ddGTP or ddTTP (or a mixture of all four)
DGGE	Denaturing gradient gel electrophoresis
dNTP	Any one of dATP, dCTP, dGTP or dTTP (or a mixture of all four)
DR	Direct repeat
ds	Double-stranded (RNA or DNA)
e.o.p.	Efficiency of plating (of a bacteriophage)
EMS	Ethyl methane sulphonate
EtBr	Ethidium bromide
FIGE	Field inversion gel electrophoresis
f-Met	Formyl-methionine
G	Guanine
GFP	Green fluorescent protein
GSP	General secretory pathway
Hfr	High frequency of recombination
HPK	Histidine protein kinase
IEF	Isoelectric focusing
IPTG	Iso-propylthiogalactoside.
IR	Inverted repeat
IS	Insertion sequence
IVET	*In vivo* expression technology
kb	Kilobase (=1000 bases)
Mb	Megabase (1 million bases)
MNNG	1-Methyl-3-nitro-1-nitroso-guanidine
mRNA	Messenger RNA
NTP	any one of ATP, GTP, CTP, or UTP (or a mixture of all four)
oc	Open circular (nicked form of plasmid DNA)
ORF	Open reading frame
PAGE	Polyacrylamide gel electrophoresis
PCR	Polymerase chain reaction
PEG	Polyethylene glycol
PFGE	Pulsed field gel electrophoresis
pfu	Plaque-forming units
RF	Replicative form
pI	Isoelectric point (pH at which a protein has no net charge)
RFLP	Restriction fragment length polymorphism
RR	Response regulator

rRNA	Ribosomal RNA
RT-PCR	Reverse transcript PCR
SDS	Sodium dodecyl sulphate
ss	Single-stranded (RNA or DNA)
ssb	Single-strand DNA binding protein
SSCP	Single-strand conformational polymorphism
STM	Signature tagged mutagenesis
T	Thymine
Ti	Tumour-inducing
Tn	Transposon
TRAP	Tryptophan-RNA-binding attenuator protein
tRNA	Transfer RNA
U	Uracil
UAS	Upstream activator sequence
uv	Ultraviolet radiation
VNTR	Variable number of tandem repeats
X-gal	5-Bromo-4-chloro-3-indolyl-β-D-galactoside

Appendix C
Glossary

See Appendices D and E for information about enzymes and genes referred to in this book.

Adaptive mutation: see *directed mutation*.

Allelic replacement: see *gene replacement*.

Allosteric effect: binding of a substance to a site on a protein can alter its conformation and hence alter the activity of a distinct site on the protein.

Amber mutation: the introduction of a stop codon (UAG) within the coding sequence of a gene which results in premature termination of translation.

Aminoglycosides: a group of antibiotics, including kanamycin and streptomycin, which interfere with protein synthesis.

Ampicillin: an antibiotic of the penicillin group.

Amplification: (a) increasing the copy number of a plasmid by inhibiting replication of the chromosome while allowing plasmid replication to continue; (b) an increase in the number of copies of a gene due to repeated duplication (normally referred to as 'gene amplification').

Analogues: unrelated genes with similar function (see *homologues*).

Annealing: formation of double-stranded DNA from single-stranded DNA (see *hybridization*).

Anti-terminator: a protein that allows RNA polymerase to read through a terminator.

Anticodon: the region of tRNA which pairs with the codon.

Antimetabolite: an analogue of the end-product of a pathway that causes feedback inhibition or repression, but cannot replace the genuine product; used for selecting feedback-deficient mutants.

Antisense RNA: RNA that is complementary to the mRNA which can interfere with translation.

Attenuation: (a) reduced expression of those genes in an operon that are situated beyond a certain point; (b) reduction in the virulence of a pathogenic microorganism (not the usual meaning in the context of this book, but more common elsewhere).

Autogenous control: regulation of the expression of a gene by its own product.

Autoradiography: detecting radioactively labelled material using X-ray film.

Auxotroph: a mutant that requires the addition of one or more special supplements to its growth medium (see *prototroph*).

Back mutation: exact reversal of a mutation to restore the wild-type sequence of the gene.

Bacterial artificial chromosome: use of a vector based on the F plasmid to clone large DNA fragments.

Molecular Genetics of Bacteria, 4th Edition by Jeremy Dale and Simon F. Park
 ISBN 0 470 85084 1 (cased) ISBN 0 470 85085 X (pbk)

Bacteriocin: a protein/polypeptide with antibiotic activity, usually against a narrow range of closely related bacteria. Usually plasmid mediated.

Bacteriophage: a virus that infects bacteria (often shortened to 'phage').

β-lactams: a group of antibiotics (including the penicillins) that affect bacterial cell wall synthesis.

Bioinformatics: computer-based analysis of biomolecular data, especially large-scale datasets derived from genome sequencing.

BLAST: Basic Local Alignment Search Tool; a computer program for searching sequence databases for related sequences.

Box: a short sequence of bases in DNA conforming (more or less) to a consensus for that particular type of box; usually has a regulatory function, e.g. by providing a binding site for a regulatory protein.

Catabolite repression: the expression of the gene is turned off by the presence of an easily metabolized substrate such as glucose.

cDNA library: see *gene library*.

Chaperones: proteins that affect the folding of other proteins or the assembly of complex structures.

Chloramphenicol: an antibiotic which interferes with protein synthesis.

Chromosome walking: a technique for identification of DNA regions adjacent to a known marker by sequential hybridization of clones.

Circular permutation: a population of linear DNA molecules carrying genes in the same order on a circular map, but starting at different positions.

Cis-acting: a control region that influences genes on the same DNA molecule only (usually adjacent genes) and has no effect on other DNA molecules (cf. *trans-acting*).

Cistron: a region of DNA that codes for a single polypeptide. No longer in common use except for emphasis (as in polycistronic mRNA).

Clonal: variation only by mutation, without horizontal gene transfer.

Clone: population of organisms descended asexually from a single individual.

Cloning: obtaining an homogeneous population of cells by repeated single colony isolation. Since this also yields in pure form a large number of copies of any recombinant DNA molecule carried by such a cell, the term is also used in this context (often as 'gene cloning') and by further extension (with less justification) for the reactions used in the construction of such recombinants.

Co-transduction: the simultaneous transfer of two genetic markers by transduction.

Co-transformation: the simultaneous transfer of two genetic markers by transformation.

Codon: A group of three bases in mRNA that codes for a single amino acid in a polypeptide chain.

Codon usage: a measure of the relative use of synonymous codons (i.e. different triplets coding for the same amino acid).

Cohesive ends: Single-stranded ends on DNA molecules that have complementary sequences.

Cointegrate: plasmid formed by the complete fusion of two smaller plasmids; usually an intermediate in transposition. Can also be used to refer to transposon-mediated integration of a whole plasmid into the chromosome. (see *resolution*).

Colicin: a bacteriocin produced by a strain of *E. coli*.

Competent: a bacterial cell that is able to take up added DNA.

Complementary DNA (cDNA): DNA synthesized using mRNA as a template, a reaction carried out by reverse transcriptase.

Complementary strand: a nucleic acid strand that will pair with a given single strand of DNA (or RNA).

Complementation: Restoration of the wild-type phenotype by the introduction of a second DNA molecule without recombination.

Composite transposon: mobile element formed by two copies of an insertion sequence flanking a region of DNA.

Conditional mutant: the effects of the mutation are only seen under certain conditions, such as elevated temperature. Under other conditions (permissive conditions), the cell/virus behaves normally.

Conjugation: transfer of genetic material from one cell to another by means of cell to cell contact (see *transformation*, *transduction*)

Conjugative transposon: mobile element that promotes conjugation and transposes from one cell to the other without having any stable independent existence.

Consensus: for some functions, such as promoters, ribosome binding sites, etc., all the known sequences with that role are related to a common sequence, the consensus sequence for that function.

Constitutive: the gene product is formed irrespective of the presence of inducers or repressors (see *induction*, *repression*).

Contig: DNA sequence built up from a number of smaller overlapping sequences, during a sequencing project.

Copy number: the number of copies of a plasmid per cell; also used to refer to the number of copies of a specific gene (gene copy number).

***cos* site:** the sequence of bases of bacteriophage λ which is cut asymmetrically during packaging, generating an unpaired sequence of 12 bases at each end of the phage DNA (cohesive ends).

Cosmid: a plasmid that contains the *cos* site of bacteriophage λ. After introducing a large insert, the recombinant cosmid forms a substrate for *in vitro* packaging.

Cross-feeding: stimulation of growth of a mutant by material liberated from another cell.

Cross-talking: regulation of gene expression by signalling between cells or between compartments within a cell.

Curing: elimination of a plasmid or prophage from a bacterial culture.

Denaturation: (a) separation of the two strands of DNA by disruption of the hydrogen bonds (usually by heat or high pH); (b) disruption of the secondary and tertiary structure of proteins.

Diauxic growth: growth in the presence of two substrates (e.g. glucose and lactose) where one substrate is not utilized until the other has been exhausted (see *catabolite repression*)

Direct repeat: two identical or very similar DNA sequences reading in the same direction (see also *inverted repeat*).

Directed mutation: mutation that occurs in response to external conditions that favour the mutant; contrast *random mutation*. Also referred to as adaptive mutation, but that term is also used in other contexts.

Distal: sequences beyond a given point (see *proximal*).

Domain: a region of a protein with a (partly) independent structure, connected to other domains by flexible loops.

Dyad symmetry: a short DNA sequence immediately repeated in the opposite direction; often involved in binding regulatory proteins.

Early genes: bacteriophage genes that are expressed soon after infection (see also *late genes*).

Electrophoresis: separation of DNA molecules (or other molecules, including RNA or protein) by application of an electric field, usually across an agarose or acrylamide gel.

Electroporation: inducing cells to take up DNA by subjecting them to brief electric pulses.

Enhancers: DNA regions that increase the level of transcription; often remote from the transcriptional start, upstream or downstream.

Epigenetic: a change to the DNA that affects expression of a gene but is not inherited.

Episome: originally defined as a genetic element (plasmid or bacteriophage) that can exist either autonomously or integrated into the chromosome. Now often used to refer only to the extrachromosomal state. Mainly avoided in this book.

Error-prone repair: an inducible repair mechanism, part of the SOS response, which functions by reducing the specificity of replication to allow a damaged region to be copied. The cause of mutations following ultraviolet irradiation.

Eukaryote: a cell that has a discrete nucleus bounded by a membrane, e.g. fungi, protozoa, plant and animal cells. (cf. *prokaryote*).

Excision repair: repair of damaged DNA by removal of that part of the DNA strand containing the affected region, followed by filling in the gap by repair synthesis.

Exon: coding sequence, part of a single gene but separated by an *intron*.

Expression vector: a cloning vector designed for expression of the cloned insert using regulatory sequences present on the vector.

F′ (F-prime) plasmid: an F plasmid carrying some chromosomal genes. Formed by inaccurate excision of F from the chromosome.

Feedback: many metabolic pathways (especially biosynthetic ones) are regulated by their end-products. This can be via repression or by inhibition of specific enzymes.

Fimbriae: filamentous appendages on the surface of a bacterial cell. Similar to *pili* and the use of the two terms overlaps.

Fluctuation test: a procedure for distinguishing between *random* and *directed mutations*.

Frameshift: insertion or deletion of bases, other than in multiples of three, changes the reading frame of protein synthesis beyond that point.

Fusion protein: a protein made from all or part of two (or more) different proteins, e.g. by gene cloning or naturally by ribosomal frameshifting.

Gene knockout: see *gene replacement*.

Gene library: a collection of recombinant clones which together represent the entire genetic material of an organism. More specifically, a genomic library represents the entire genome, while a cDNA library represents the mRNA present in the cells at the time of extraction.

Gene replacement: inactivation of a chromosomal gene by recombination with an homologous sequence, inactivated *in vitro*.

Genome sequencing: determination of the complete sequence of the DNA of an organism.

Genome: the entire genetic material of an organism; in bacteria, usually confined to the chromosome; in eukaryotes, to the nuclear DNA.

Genotype: the genetic make-up of an organism (cf. *phenotype*).

Global regulation: a change in external conditions may affect the regulation of a large number of otherwise unrelated genes.

Gratuitous inducer: an inducer that is not a substrate for the enzymes induced (e.g. IPTG as an inducer of the *lac* operon).

Hairpin: a region of DNA or RNA that contains a short inverted repeat can form a base-paired structure resembling a hairpin (similar to a *stem–loop structure*).

Head-full: packaging method used by bacteriophages such as T4.

Hemi-methylated: DNA in which only one strand is methylated.

Heteroduplex: a double-stranded DNA molecule formed by base pairing between two similar but not identical strands. Since the two strands are not identical, some regions will remain single stranded.

Hfr: high frequency of recombination: conjugation donor in which F plasmid is integrated into chromosome.

Holliday junction: an intermediate in *homologous recombination*.

Homologues: genes (or enzymes) with similar sequence and function, descended from a common ancestor (see *analogues*, *orthologues*, *paralogues*)

Homologous recombination: recombination between two DNA molecules which share an extensive region of homology; requires RecA (see also *illegitimate recombination* and *site-specific recombination*).

Homologous: DNA (or RNA) molecules with the same sequence or sufficiently similar for complementary strands to hybridize. (Note that this is a rather loose use of the word 'homology' – see *homologues*).

Horizontal gene transfer: transfer of DNA from one bacterium to another, to contrast with vertical transfer by normal inheritance.

Hybridization: The formation of double-stranded DNA, RNA or DNA/RNA molecules by the production of hydrogen bonds between wholly or partially complementary sequences.

Hydrophilic: 'water-loving'. Substances, or parts of a structure which interact with water and therefore tend to be exposed to water.

Hydrophobic: 'water-hating'. Substances, or parts of a structure, that do not interact with water and hence tend to remove themselves from an aqueous environment.

Illegitimate recombination: recombination that requires neither extensive homology nor specific sites (contrast *homologous recombination* and *site-specific recombination*).

***in vitro* packaging:** the assembly of mature bacteriophage particles *in vitro* by mixing suitable DNA with cell extracts which contain bacteriophage heads and tails and the enzymes needed for packaging.

Incompatibility: inability of two related plasmids to co-exist in the same cell; often used for plasmid typing.

Induction: (a) increasing the synthesis of a gene product by a cell through the addition of a specific substance to the growth medium or by some other change in growth conditions such as a temperature shift. Most commonly used to refer to the induction of an enzyme by the addition of its substrate; (b) applying a treatment to a lysogenic bacterium that causes lysogeny to break down, resulting in the bacteriophage entering the lytic cycle; (c) also used somewhat loosely in a variety of contexts (as in 'uv-induced mutations').

Insertion sequence: a DNA sequence that is able to insert itself or a copy of itself into another DNA molecule; carries no information other than that required for transposition (see also *transposon*).

Insertion vector: a λ cloning vector into which DNA can be inserted at a single site (cf. *replacement vector*).

Insertional inactivation: destruction of the function of a gene by insertion of a foreign DNA fragment, either by transposition or by gene cloning.

Integrity: maintenance of the structure of a plasmid without deletions or rearrangements.

Integron: DNA integration element; a region that is able to acquire additional genes by site-specific integration.

Intercalation: the action of certain mutagens such as the acridines that can stack between the bases in the centre of a double helix. Commonly results in *frameshift mutations*.

Intermolecular: interactions between two different molecules.

Intramolecular: interactions between different parts of the same molecule.

Intron: Interruption (intervening sequence) in the coding sequence of a gene; removed by splicing (see also *exon*).

Inversion: change in the orientation of a DNA fragment with respect to the sequences either side.

Inverted repeat: two identical or very similar DNA sequences reading in opposite directions (see also *direct repeat*).

Isoelectric focusing: separation of proteins by electrophoresis in a stable pH gradient so that each protein will move to its *isoelectric point*.

Isoelectric point: the pH at which a specific protein has no net overall charge.

Isogenic: strains that are identical in their genetic composition; normally used to mean identical in all genes except the one being studied.

Iteron: DNA region containing a number of short repeated sequences; involved in the replication and copy number control of some types of plasmid.

IVET: *in vivo* expression technology; a procedure for identifying bacterial genes that are expressed during infection rather than during growth in the laboratory.

Kanamycin: an antibiotic of the aminoglycoside group.

Kilobase: a nucleic acid region that is 1000 bases long (abbreviated to kb).

Knock-out: see *gene replacement*.

Lamarckism: inheritance of acquired characteristics.

Late genes: bacteriophage genes that are expressed later in the lytic cycle, requiring the product of one or more *early genes*.

Leader: (a) nucleotide sequence at the 5′ end of mRNA before the start point for translation of the first structural gene. Often involved in regulation of gene expression; (b) also used to refer to the *signal peptide* but preferable to limit it to the first meaning.

Leaky: a mutation where the loss of the relevant characteristic is not complete.

Ligation: joining two or more DNA molecules using DNA ligase.

Linking number: a measure of the overall interlinking of two DNA strands; composed of *twist* and *writhe*.

Lysogeny: a (more or less) stable relationship between a bacteriophage (*prophage*) and a host bacterium (lysogen).

Lytic cycle: multiplication of a bacteriophage within a host cell leading to lysis of the cell and re-infection of other sensitive bacteria.

Mapping: determination of the position of genes (genetic map) or of physical features such as restriction endonuclease sites (restriction map, physical map).

Melting: separation of double-stranded DNA into single strands (see *denaturation*).

Messenger RNA (mRNA): an RNA molecule that is used for translation into a protein.

Microarray: a set of oligonucleotides representing part of each gene of the bacterium, spotted onto a glass slide. Used for genome-wide comparison of gene expression or for identifying genomic deletions.

Minicells: small, non-replicating, cells that lack DNA; produced at cell division in mutants defective in the *min* genes.

Mismatch repair: a method for removing and replacing incorrect bases from a new DNA strand.

Mobilization: conjugative transfer of a non-conjugative plasmid in the presence of a conjugative plasmid.

Modification: alteration of the structure of DNA (usually by methylation of specific residues) so that it is no longer a substrate for the corresponding restriction endonuclease (see also *restriction*).

Mosaic genes: genes composed of regions from different sources.

Motif: a short sequence of amino acids or bases which is conserved in proteins or nucleic acid sequences with similar functions.

Multiple cloning site: a short region of a vector containing a number of unique restriction sites into which DNA can be introduced.

Mutagenesis: treatment of an organism with chemical or physical agents so as to induce alterations in the genetic material.

Mutant: a cell (or virus) with altered properties due to a change in its genetic material (cf. *mutation*).

Mutation: an alteration in the genetic material giving rise to a cell (or virus) with altered properties (cf. *mutant*).

Nalidixic acid: an antibiotic which interferes with the action of DNA gyrase.

Nick: A break in one strand of a double-stranded DNA molecule.

Nonsense mutation: base substitution creating a *stop codon* within the coding sequence causing premature termination of translation.

Northern blot: transfer of RNA molecules from an agarose or acrylamide gel to a filter for hybridization.

Okazaki fragment: short fragment of DNA produced during replication of the lagging strand.

Oligonucleotide: a short nucleic acid sequence (usually synthetic).

Open reading frame (ORF): a nucleic acid sequence with a reading frame that contains no stop codons; it can therefore potentially be translated into a polypeptide.

Operator: a region of DNA to which a repressor protein binds to switch off expression of the associated gene. Usually found adjacent to or overlapping with the promoter.

Operon: a group of contiguous genes that are transcribed into a single mRNA from a common promoter and hence are subject to coordinated induction/repression.

Ordered library: a collection of clones containing overlapping fragments in which the order of the fragments has been determined.

Orthologues: equivalent genes (or enzymes) with similar sequence and function which have evolved from a common ancestor (see *homologues*).

Packaging: the process of incorporating DNA into a bacteriophage particle.

Paralogues: genes (or enzymes) with similar sequence but different function (see *homologues*).

Partitioning: distribution of copies of a plasmid between daughter cells at cell division; sometimes referred to as *segregation* but this term is best avoided as it also has the opposite meaning, i.e. the *failure* of partitioning.

Pathogenicity island: a DNA region carrying virulence determinants; often with a different base composition from the remainder of the chromosome.

Penicillin: see β-*lactams*.

Penicillin enrichment: technique for increasing the proportion of auxotrophic mutants.

Permissive host/conditions: see *conditional mutant*.

Phage: see *bacteriophage*.

Phage conversion: infection of an non-pathogenic bacterial strain by a specific phage renders the bacterium pathogenic – usually because the phage carries a toxin gene.

Phage display: using gene cloning to produce a recombinant phage with a foreign protein incorporated into the phage coat. This enables identification of proteins that bind to specific ligands (e.g. hormones, drugs).

Phage therapy: using bacteriophages to treat bacterial infections.

Phage typing: differentiating between bacterial strains by their profile of sensitivity to a collection of bacteriophages.

Phase variation: a reversible but inherited switch in the expression of one ore more genes.

Phenotype: the observable characteristics of an organism (cf. *genotype*).

Pheromone: secreted chemical used for signalling between two or more individuals.

Pilus (pl. pili): filamentous appendage on the surface of a bacterial cell. Some types of pili are specified by plasmids and are involved in mating-pair formation in conjugation (see also *fimbriae*).

Plaque: a region of clearing or reduced growth in a bacterial lawn as a result of phage infection.

Plasmid: an extrachromosomal genetic element capable of autonomous replication.

Plasticity: the concept that the genetic composition of an organism is not constant but is subject to large-scale variation between strains.

Plus strand: with RNA sequences, the mRNA is defined as the plus (+) strand and the complementary sequence as the minus (−) strand. DNA sequences maintain the same convention, meaning that it is the (−) strand of DNA that is transcribed to yield the (+) strand of mRNA. This distinction is relevant to single-strand DNA phages and by extension to plasmids which replicate via single-stranded intermediates.

Point mutation: an alteration (or deletion/insertion) of a single base in the DNA.

Polar mutation: a mutation in one gene may affect the expression of others (e.g. genes downstream in an operon). The phenotypic effect may not be directly caused by the original mutation.

Polarity: differences in the level of translation of genes within an operon, especially the effect of a polar mutation in one gene in reducing or abolishing the expression of subsequent genes.

Polycistronic mRNA: messenger RNA coding for several proteins (see *operon*).

Polymerase chain reaction: enzymatic amplification of a specific DNA fragment using repeated cycles of denaturation, primer annealing and chain extension.

Polymorphism: the existence of different forms of a characteristic within a population. Originally applied to a phenotype but now often used in other contexts, as in *restriction fragment length polymorphism*, or *sequence polymorphism*

Post-replication repair: DNA repair process involving the exchange of DNA between damaged and undamaged strands.

Post-translational modification: modification of the structure of a polypeptide after synthesis, e.g. by phosphorylation, glycosylation, proteolytic cleavage.

Pribnow box: consensus sequence within a *promoter*, centred at the -10 position with respect to the start of transcription.

Primary structure: the sequence of a nucleic acid or a protein.

Primer: synthesis of a new DNA (but not RNA) strand can only occur by extension of a pre-existing partial DNA strand. If a specific oligonucleotide (a primer) is provided, complementary to a defined region of the template strand, all the new DNA strands made will start from that point.

Probe: a DNA or RNA molecule that will hybridize to a specific target sequence. Labelling the probe (using radioactive isotopes or non-radioactive markers) enables it to be used to detect the specific target DNA/RNA.

Prokaryote: a cell that does not have a discrete nucleus bounded by a membrane; in this book referring to bacteria, but also includes the *Archaea* (cf. *eukaryote*).

Promoter: region of DNA to which RNA polymerase binds in order to initiate transcription.

Proof-reading: the ability of DNA polymerase to check the accuracy of the newly made sequence.

Prophage: the repressed form of bacteriophage DNA in a lysogen; it may be integrated into the chromosome or exist as a plasmid.

Protein engineering: altering a gene so as to produce defined changes in the properties of the encoded protein, e.g. thermal stability, substrate profile.

Proteome: the complete content of different proteins in a cell (cf. *genome, transcriptome*).

Protoplast: formed by complete removal of the cell wall using osmotically-stabilized conditions. Used for transformation and for protoplast fusion.

Prototroph: a nutritionally wild-type organism that does not need any additional growth supplement (cf. *auxotroph*).

Proximal: sequence before a given point, usually referring to the direction of transcription or translation (see *distal*).

Pulsed-field gel electrophoresis: separation of large DNA molecules by application of an intermittently varying electric field; generic term for a number of ways of achieving this.

Purine: one of the two types of bases in nucleic acids (adenine, guanine; see *pyrimidine*).

Pyrimidine dimer: covalent linkage between adjacent pyrimidines on a DNA strand caused by UV irradiation. Commonly referred to as thymine dimers but not restricted to thymine.

Pyrimidine: one of the two types of bases in nucleic acids (cytosine and thymine in DNA; cytosine and uracil in RNA; see *purine*).

Quorum sensing: mechanism whereby bacteria respond to cell density.

Random mutation: mutation occurring irrespective of its benefit to the cell (contrast *directed mutation*).

Reading frame: a nucleic acid sequence is translated in groups of three bases (*codons*); there are three possible ways of reading the sequence (in one direction) depending on where it is started. These are the three reading frames.

Real-time PCR: a PCR technique which allows monitoring of the amplification of the product as it happens. Especially useful for quantitative applications of PCR.

Recombinant: product of recombination (q.v.) using either definition, which leads to recombinant bacteria resulting from some form of gene transfer or recombinant plasmids arising from *in vitro* manipulation

Recombination repair: see *post-replication repair*.

Recombination: (a) the production of new strains by mating two genetically distinct parents; (b) the generation of new DNA molecules by breaking and re-joining the original molecules; this may occur naturally within the cell (*in vivo*) or artificially *in vitro*. There is considerable overlap between these definitions but they are not always synonymous.

Regulon: a set of genes that are coordinately regulated without being contiguous (see *operon*).

Relaxation: conversion of supercoiled circular DNA to an open circular form.

Relaxed plasmid: (a) open circular structure after nicking one strand of a plasmid; (b) plasmid which replicates to high copy number without being tied to chromosomal replication.

Replacement vector: a λ cloning vector in which a piece of DNA (the *stuffer fragment*) can be removed and replaced by the cloned fragment (cf. *insertion vector*).

Replica plating: using a velvet pad or some equivalent apparatus to transfer a number of colonies to several different media in order to compare their growth requirements or other characteristics.

Replication: synthesis of a copy of a DNA molecule using the original as a template.

Replicon: (a) a DNA molecule (such as a plasmid) that contains an origin of replication and is capable of autonomous replication within a suitable host cell; (b) the replication control region of a plasmid.

Reporter gene: a gene which codes for a readily detected product (such as β-galactosidase); study of a regulatory region of DNA is facilitated by fusion with the reporter gene.

Repression: (a) reduction in transcription of a gene usually due to the action of a repressor protein; (b) also applied to the natural repression of conjugal transfer that occurs with many plasmids and to the establishment of lysogeny with temperate bacteriophages.

Resolution: production of two smaller plasmids from a *cointegrate*.

Restriction: reduction in the apparent titre of a phage (or transforming ability of DNA) when certain strains are used as a host. These strains produce restriction endonucleases which degrade foreign DNA when it enters the cell (see also *modification*).

Restriction fragment length polymorphism: variation between strains in the size of specific restriction fragments; used for strain typing and for locating particular genes.

Restriction mapping: determination of the position of restriction endonuclease recognition site on a DNA molecule.

Reverse genetics: starting with specific alterations to the DNA *in vitro* and then examining the phenotype; contrasts with classical genetics which relies on selecting mutants on the basis of their phenotype and then studying the nature of the mutation. (This term is also used in other ways and is hence avoided in this book).

Reversion: a mutation that reverses the effect of the original mutation.

Ribosomal frameshifting: a change in reading frame used by the ribosome resulting in a fusion protein or re-initiation from an adjacent start codon.

Ribosome binding site: the region on an mRNA molecule to which ribosomes initially attach.

Rifampicin: an antibiotic which interferes with RNA polymerase.

Rolling circle replication: (a) production of a multiple length linear molecule of λ DNA; (b) replication of some plasmids and phages via single-stranded intermediates or during conjugation.

RT-PCR: reverse transcript PCR. Technique for producing an amplified DNA product from an mRNA template.

Scaffolding: proteins used to assist in the assembly of a bacteriophage particle and removed during maturation.

SDS-PAGE: polyacrylamide gel electrophoresis in which proteins are separated according to molecular weight in the presence of sodium dodecyl sulphate.

Secondary structure: the spatial arrangement of amino acids in a protein or of bases in nucleic acid.

Segregation: in eukaryotic genetics, this refers to the distribution of genes among the progeny following meiosis. By extension, when bacteria contain two versions of a gene (e.g. following mutational alteration of one strand of DNA), two types of colonies will result. Similarly, a plasmid-containing strain may produce plasmid-free segregants

Selectable marker: a gene that causes a phenotype (usually antibiotic resistance) which can be readily selected.

Sequence polymorphism: variation in the sequence of a gene usually between otherwise closely related organisms such as members of the same species.

Shine–Dalgarno sequence: see *ribosome binding site.*

Shotgun cloning: insertion of random fragments of DNA into a vector.

Shuttle vector: a cloning vector that can replicate in two different species, one of which is usually *E. coli*. Facilitates cloning genes in *E. coli* initially and subsequently transferring them to an alternative host without needing to re-clone them.

Sigma (σ) factor: polypeptide that associates with RNA polymerase core enzyme to determine promoter specificity.

Signal peptide: amino acid sequence at the amino terminus of a secreted protein; involved in conducting the protein through the membrane.

Signal transduction: extracellular conditions alter the conformation of a transmembrane protein which in turn alters the regulation of metabolic pathways within the cell.

Signature tagged mutagenesis: a system for identifying virulence genes using mutagenesis with a transposon carrying a variety of unique tags.

Silent mutation: a change in the DNA structure that has no effect on the phenotype of the cell.

Site-directed mutagenesis: a technique for specifically altering (*in vitro*) the sequence of DNA at a defined point.

Site-specific recombination: recombination between two DNA molecules at a specific sequence; does not require extensive homology (see also *illegitimate recombination* and *homologous recombination*).

Slipped strand mispairing: replication error changing the number of copies of short repeated units of DNA.

SOS response: a number of genes involved with DNA repair and related functions are induced by the presence of unrepaired DNA.

Southern blot: the transfer of DNA fragments from agarose or acrylamide gels onto a membrane.

Specialized transduction: transfer of DNA by a bacteriophage that has incorporated a piece of chromosomal DNA.

Splicing: removal of introns from RNA and joining together of the exons.

Start codon: position at which protein synthesis starts; usually AUG or GUG, but occasionally other codons are used.

Stem–loop structure: a nucleic acid strand containing two complementary sequences can fold so that these sequences are paired (stem) with the region between them forming a loop of unpaired bases.

Stop codon: a codon with no corresponding tRNA which signals the end of a region to be translated.

Streptomycin: an antibiotic of the aminoglycoside group.

Stringency: conditions affecting the hybridization of single-stranded DNA molecules. Higher stringency (higher temperature and/or lower salt concentration) demands more accurate pairing between the two molecules.

Stringent response: amino acid starvation leads to a reduction in synthesis of ribosomal and transfer RNAs.

Structural genes: genes coding for enzymes (or sometimes other proteins) to distinguish them from regulatory genes.

Stuffer fragment: piece of DNA that is removed from a replacement λ vector and replaced by the cloned DNA fragment.

Supercoiling: coiling of a double-stranded DNA helix around itself.

Superinfection immunity: resistance of a lysogen to infection by the same (or related) bacteriophage.

Suppression: the occurrence of a second mutation which negates the effect of the first without actually reversing it.

Synonymous codons: two or more codons that code for the same amino acid.

Tag: a short sequence of amino acids added (by gene cloning) to one end (usually the N-terminus) of a protein to facilitate purification and/or antibody recognition.

Tandem repeat: occurrence of the same sequence two or more times directly following one another.

Tautomerism: representation of a chemical as an equilibrium between two alternative structures.

Temperate: a bacteriophage that is capable of establishing a lysogenic relationship with a susceptible host cell.

Template: the use of a nucleic acid strand to carry the information required for the synthesis of a new (complementary) strand.

Terminal redundancy: sequence of bases at one end of a linear molecule is repeated at the other end (as in T4).

Termination: (a) synthesis of a mRNA molecule will stop when it reaches a terminator site; (b) protein synthesis stops (usually) when it reaches a stop (termination) codon.

Terminator: site at which transcription stops.

Tertiary structure: folding of secondary structure components of a protein.

Tetracycline: antibiotic that affects protein synthesis.

Theta replication: replication of bacteriophage λ as a circular molecule (see *rolling circle replication*).

Thymine dimer: see *pyrimidine dimer*.

Ti plasmids: tumour-inducing plasmids of *Agrobacterium tumefaciens*. Used for introducing DNA into plant cells.

***Trans*-acting:** a control region that influences other DNA regions whether or not they are on the same molecule, e.g. by means of a diffusible repressor protein (cf. *cis*-acting).

Transconjugant: a recipient cell that has received DNA by means of conjugation.

Transcription: synthesis of RNA using a DNA template.

Transcriptome: the complete mRNA content of a cell (cf. *genome, proteome*).

Transduction: bacteriophage-mediated transfer of genes from one bacterium to another.

Transfection: introduction of bacteriophage DNA into competent bacteria.

Transformation: introduction of extraneous DNA into competent bacteria. Also used to mean the conversion of an animal cell into an immortalized tumour-like cell.

Translation: synthesis of proteins/polypeptides by ribosomes acting on a mRNA template.

Translocation: movement of a ribosome along a mRNA molecule.

Transposon mutagenesis: disruption of a gene by insertion of a transposon.

Transposon: a DNA element carrying recognizable genes (e.g. antibiotic resistance) that is capable of inserting itself into the chromosome or a plasmid independently of the normal host cell recombination machinery.

Twist: a measure of the turning of the DNA double helix (see also *linking number* and *writhe*).

Two-component regulation: a sensor in the cell envelope detects environmental changes and responds by phosphorylating a cytoplasmic protein (the response regulator); this activates or inhibits its regulatory function.

Two-dimensional gel electrophoresis: separation of a complex mixture of proteins by a combination of *isoelectric focusing* and *SDS-PAGE.*

Upstream activator sequence: a sequence upstream from the promoter which is required for efficient promoter activity.

Vector: a replicon (plasmid or phage) into which extraneous DNA fragments can be inserted, forming a recombinant molecule that can be replicated in the host cell.

Virulent: (a) able to cause disease; (b) a bacteriophage that does not establish lysogeny and hence results in lysis of the bacterial host.

Western blot: transfer of proteins from an acrylamide gel onto a nitrocellulose filter, usually for detection by means of antibodies.

Wobble: the ability of tRNA anticodons to pair with more than one synonymous codon.

Writhe: a measure of the degree of DNA supercoiling (see also *linking number* and *twist*).

X-gal: 5-bromo-4-chloro-3-indolyl-β-D-galactoside. Chromogenic substrate for β-galactosidase

Z DNA: alternative left-handed form of DNA helix.

Zygotic induction: conjugative transfer of DNA from a lysogen to a non-lysogen leads to a sudden drop in the number of recombinants when the prophage is transferred due to induction of the phage.

Appendix D Enzymes

This list contains a brief description of a selection of the enzymes (and some other proteins) mentioned in the text.

aminoacyl-tRNA synthetase: responsible for charging a specific tRNA molecule with the appropriate amino acid.

adenylate cyclase: produces cAMP.

alkaline phosphatase: (1) in gene cloning, used to remove phosphate groups from the 5′ end of DNA molecules; (2) used as a reporter gene for identification of secretion signals.

AraC: regulator of the *ara* operon.

aspartokinase: catalyses the initial step in the pathway for the biosynthesis of several amino acids – a key regulatory step.

β-galactosidase: splits lactose to glucose and galactose. The first step in lactose fermentation.

β-lactamase: hydrolyses the β-lactam bond in the nucleus of penicillins and cephalosporins. Responsible for penicillin resistance.

chloramphenicol acetyltransferase (CAT): causes chloramphenicol resistance, by inactivating the antibiotic.

cI repressor: protein responsible for repression of bacteriophage λ in lysogenic state.

Cro: bacteriophage λ protein with a key role in the lytic-lysogenic switch; essentially acting in opposition to cI (q.v.)

CRP (or CAP): cAMP receptor protein (or catabolite activator protein); in combination with cAMP, activates transcription of the *lac* operon. Responsible for some forms of catabolite repression.

Dam: deoxyadenosine methylase.

DnaA: required for initiation of chromosome replication.

DNA polymerase I: primarily known as a repair polymerase, which fills in single-stranded gaps; also involved in repair of the gaps formed on the lagging strand during replication. Also possesses both 5′–3′ and 3′–5′ exonuclease activity.

DNA polymerase III: the main replication polymerase.

endonuclease: an enzyme able to cut DNA at internal positions.

exonuclease: an enzyme that removes nucleotides from the ends of DNA fragments. A 5′–3′ exonuclease removes nucleotides from the 5′ end, while a 3′–5′ exonuclease removes nucleotides from the 3′ end.

FIS: Factor for Inversion Stimulation, a DNA-binding protein which influences local topology of DNA (and hence gene expression).

FlgM: anti-σ-factor, regulates flagella synthesis in *Salmonella*.

Molecular Genetics of Bacteria, 4th Edition by Jeremy Dale and Simon F. Park
 ISBN 0 470 85084 1 (cased) ISBN 0 470 85085 X (pbk)

glutamate dehydrogenase: produces glutamate from α-ketoglutarate and ammonia; important in ammonia assimilation.

green fluorescent protein (GFP): intrinsically fluorescent protein used as a reporter.

gyrase: a specific type of topoisomerase; introduces negative supercoils into DNA.

helicase: unwinds DNA, e.g. in conjugal plasmid transfer.

Hin: site-specific recombinase, responsible for inversion of the control element in *Salmonella* phase variation.

H-NS: histone-like nucleoid structuring protein, a DNA-binding protein which influences local topology of DNA (and hence gene expression).

HU: DNA-binding protein, influences local topology of DNA (and hence gene expression).

Int: integrase, needed for integration of λ DNA into the chromosome by site-specific recombination. Related proteins catalyse integration of other elements, such as conjugative transposons (see also Xis).

integration host factor (IHF): a host protein required for efficient integration of λ DNA into the chromosome.

kinase: phosphorylating enzyme.

LacI: repressor of the *lac* operon.

ligase: Seals single-stranded gaps (nicks) in double-stranded DNA. Also used for the formation of recombinant DNA molecules in gene cloning.

LexA: regulator of the SOS response.

luciferase: catalyses a light-emitting reaction; used as a reporter.

LuxI: acyl homoserine lactone synthase, responsible for synthesis of a messenger used in quorum sensing.

LuxR: transcriptional regulator, responding to level of homoserine lactone.

OxyR: global regulatory protein, responding to oxidative stress.

photolyase: responsible for light-activated removal of UV-induced pyrimidine dimers.

pilin: structural subunit of pili.

polynucleotide kinase: transfers a phosphate group from ATP to the 5′-OH end of DNA or RNA.

primase: a special RNA polymerase, which makes a short primer required for DNA synthesis.

RecA: key protein in homologous recombination and in DNA repair.

RecBCD: multifunctional enzyme involved in homologous recombination.

replicase: RNA-directed RNA polymerase, used in replication of some RNA viruses.

resolvase: catalyses site-specific recombination to resolve the cointegrate intermediate in transposition into two separate molecules.

restriction endonuclease: recognizes DNA at a specific site and degrades it by internal cleavage. Most commonly used are the type II restriction enzymes which cut within the recognition site.

reverse transcriptase: RNA-directed DNA polymerase; synthesizes DNA (complementary DNA) using mRNA template.

ribonuclease (RNase): degrades RNA molecules.

RNA polymerase: synthesizes RNA using a DNA template.

RNaseH: a specific RNase which cuts RNA–DNA hybrids; involved in replication of ColE1-like plasmids.

Ruv: several proteins involved with events at the Holliday junction during homologous recombination.

sigma (σ) factor: dissociable component of RNA polymerase, responsible for promoter specificity.

SSB: single-strand DNA binding protein; stabilizes single-stranded DNA, e.g. during replication.

topoisomerase: a class of enzymes that alters the conformation of DNA, e.g. by changing the degree of winding or supercoiling.

transposase: catalyses the initial steps in transposition.

TRAP: *trp* RNA-binding attenuator protein; responsible for attenuation of the *trp* operon in *Bacillus*.

TrpR: repressor of the *trp* operon.

XerC,D: resolution of plasmid dimers by recombination at *cer* sites.

Xis (pronounced 'excise'): interacts with Int (integrase) to promote excision of λ prophage from the chromosome. Related proteins have similar functions with other elements such as conjugative transposons.

Appendix E
Genes

A simplified selection of the genes/operons referred to in the text. Note that in many cases, a number of genes are needed for the indicated function and also that some genes with the same three letter designation may be concerned with other functions such as production of the specific tRNA.

Gene	Gene product/phenotypic trait
aac	aminoglycoside acetyltransferase (gentamicin resistance)
aad	aminoglycoside adenylyltransferase (streptomycin resistance)
amp	(=*bla*); ampicillin resistance
aph	aminoglycoside phosphotransferase, kanamycin resistance
ara	arabinose utilization
araC	regulator of arabinose operon
arg	arginine biosynthesis
bio	biotin synthesis
bla	ampicillin resistance (β-lactamase production)
bld	in *Streptomyces coelicolor*, mutants are unable to produce aerial mycelia
bom	nick site for mobilization of ColE1 (=*oriT*)
cat	chloramphenicol acetyltransferase (chloramphenicol resistance)
ccd	post-segregational killing of plasmid-free segregants
cer	recombination site for resolution of multiple length plasmid concatemers; deletion leads to instability
cml	see *cat*
col	colicin production
dhfr	dihydrofolate reductase (trimethoprim resistance)
dsrA	regulatory RNA, regulates *hns* and *rpoS*
fimA	structural subunit of fimbriae
ftsZ	initiation of cell division; conditional mutants produce filaments at the restrictive temperature
gal	galactose fermentation
hin	inversion of regulatory element of H2 operon (*Salmonella* flagellar antigens)
his	histidine synthesis
hok/sok	post-segregational killing of plasmid-free segregants

Molecular Genetics of Bacteria, 4th Edition by Jeremy Dale and Simon F. Park
 ISBN 0 470 85084 1 (cased) ISBN 0 470 85085 X (pbk)

ilv	synthesis of isoleucine and valine
inc	plasmid incompatibility
kan	kanamycin resistance (=*aph*)
lac	lactose fermentation; operon composed of *lacZ* (β-galactosidase), *lacY* (permease), and *lacA* (transacetylase).
lacI	repressor of the *lac* operon; constitutive mutations
leu	synthesis of leucine
lexA	LexA protein represses the genes of the SOS response
lys	lysine synthesis
mer	mercuric ion resistance
micF	antisense RNA, regulates *ompF*
min	regulate the position of septum formation at cell division; mutants produce minicells that lack DNA.
mob	nuclease required for initiation of plasmid transfer
nal	nalidixic acid resistance (*gyrA*)
nif	nitrogen fixation
nifA	positive regulator of *nif*
ompF	outer membrane protein (porin)
oriT	origin of (conjugal) transfer of a plasmid (see *bom*)
oriV	origin of replication
par	plasmid partitioning at cell division
pheA	phenylalanine synthesis
pil	pilin; subunits of pili of e.g. gonococci
pro	proline synthesis
recA	RecA protein mediates general recombination, repair processes and phage induction (amongst others)
recB,C	codes for a multifunctional enzyme with exo- and endonuclease activity, required for RecA-mediated recombination
rep	plasmid replication
res	resolution site in transposition
rpoB	RNA polymerase β-subunit; some mutations cause rifampicin resistance
rpoS	stationary phase σ-factor
rpsL	ribosomal protein S10. Streptomycin resistance (chromosomal mutation)
spoIIA	controls one stage in sporulation in *Bacillus subtilis*; *spoIIAC* (*sigF*) encodes a σ-factor, while *spoIIAB* and *spoIIAA* encode an anti-σ-factor and and antagonist of SpoIIAB respectively
str	streptomycin resistance (as a chromosomal marker, usually *rpsL*; on a plasmid or transposon, usually an aminoglycoside modifying enzyme such as AadA)
sul	sulphonamide resistance
tet	tetracycline resistance
thr	threonine synthesis

thyA thymidylate synthetase; thymine/thymidine production

tnpA transposase (e.g. in Tn*3*)

tnpR resolvase (e.g. in Tn*3*)

tra transfer; group of genes required for conjugal plasmid transfer

trp tryptophan biosynthesis.

trpA,B subunits of tryptophan synthase

trpR repressor of tryptophan operon

umuC, D ultraviolet mutation; defects in these genes cause increased sensitivity to the lethal effects of UV but the incidence of mutations is reduced

uvr ultraviolet repair. Several genes (*uvrA,B,C*) code for nuclease responsible for excision of thymine dimers

whiG in *Streptomyces coelicolor*, encodes a σ-factor needed for spore formation

A selection of lambda genes

att Site for integration of lambda DNA into the chromosome. The phage sequence is known as *attP* and that on the chromosome as *attB* or *att*λ. After integration the sites at the left and right ends of the prophage are known as *att*L and *att*R respectively

cI CI repressor, necessary for lysogeny; binds to O_L and O_R

cII, *cIII* ancillary genes required for the initial synthesis of repressor, by activation of P_E

cos cohesive end sites; position of asymmetric cuts during packaging

cro the Cro protein essentially functions as an anti-repressor, antagonizing the action of the CI repressor

int integration of lambda DNA into chromosome; together with Xis, excision of prophage

N anti-terminator; allows read-through the transcriptional termination sites, resulting in expression of delayed early genes

O_L, O_R operator sites to which CI and Cro bind, regulating expression from P_L and P_R

P_E promoter for the expression of CI during the establishment of lysogeny; activated by CII and CIII

P_L, P_R promoters for major early transcripts in left and right directions respectively

P_M maintenance promoter for expression of cI after lysogeny has been established

$P_{R'}$ promoter for the major late transcript

Q required for expression of late genes, from $P_{R'}$

S lysis

t_{R1}, t_{R2} terminator site for immediate early transcripts from P_R; anti-termination activity of N allows read-through

t_{R3} terminator site for delayed early transcript from P_R

xis with Int, excision of prophage when lytic cycle is induced

Appendix F
Standard Genetic Code

UUU	Phe	UCU	Ser	UAU	Tyr	UGU	Cys
UUC	Phe	UCC	Ser	UAC	Tyr	UGC	Cys
UUA	Leu	UCA	Ser	*UAA*	*stop*	*UGA*	*stop*
UUG	Leu	UCG	Ser	*UAG*	*stop*	UGG	Trp
CUU	Leu	CCU	Pro	CAU	His	CGU	Arg
CUC	Leu	CCC	Pro	CAC	His	CGC	Arg
CUA	Leu	CCA	Pro	CAA	Gln	CGA	Arg
CUG	Leu	CCG	Pro	CAG	Gln	CGG	Arg
AUU	Ile	ACU	Thr	AAU	Asn	AGU	Ser
AUC	Ile	ACC	Thr	AAC	Asn	AGC	Ser
AUA	Ile	ACA	Thr	AAA	Lys	AGA	Arg
AUG	Met	ACG	Thr	AAG	Lys	AGG	Arg
GUU	Val	GCU	Ala	GAU	Asp	GGU	Gly
GUC	Val	GCC	Ala	GAC	Asp	GGC	Gly
GUA	Val	GCA	Ala	GAA	Glu	GGA	Gly
GUG	Val	GCG	Ala	GAG	Glu	GGG	Gly

A list of amino acid abbreviations is provided in Appendix B.

Molecular Genetics of Bacteria, 4th Edition by Jeremy Dale and Simon F. Park
 ISBN 0 470 85084 1 (cased) ISBN 0 470 85085 X (pbk)

Appendix G
Bacterial Species

Note that throughout the text, *E. coli* is referred to unless otherwise specified.

Bacterial name	Abbreviation	Other names	Comment
Agrobacterium tumefaciens	*A. tumefaciens*		Plant pathogen, used for DNA transfer to plant cells
Bacillus anthracis	*B. anthracis*		Cause of anthrax
Bacillus subtilis	*B. subtilis*		Gram-positive spore former. Widely used in genetics research
Bacteroides			Gram-negative intestinal commensal
Bordetella pertussis	*B. pertussis*		Cause of whooping cough
Borrelia			Genus of spirochate bacteria associated with relapsing fevers and Lyme disease
Brevibacterium flavum	*B. flavum*		Now reclassified as *Corynebacterium glutamicum*
Buchnera			Genus of bacterial symbionts of aphids
Campylobacter fetus	*C. fetus*		Cause of abortion in sheep and cattle
Campylobacter jejuni	*C. jejuni*		A leading cause of food-borne illness
Caulobacter crescentus	*C. crescentus*		Aquatic bacterium, differentiating into stalked and swarmer cells
Clostridium botulinum	*C. botulinum*		Cause of botulism

Molecular Genetics of Bacteria, 4th Edition by Jeremy Dale and Simon F. Park
 ISBN 0 470 85084 1 (cased) ISBN 0 470 85085 X (pbk)

Bacterial name	Abbreviation	Other names	Comment
Clostridium tetani	*C. tetani*		Cause of tetanus
Corynebacterium diphtheriae	*C. diphtheriae*		Causes diphtheria, due to phage-encoded toxin
Corynebacterium glutamicum	*C. glutamicum*		Used in industrial processes, e.g. for amino acid production
Enterococcus faecalis	*E. faecalis*	Formerly *Streptococcus faecalis*	Gram-positive intestinal inhabitant. Opportunist pathogen
Erwinia carotovora	*E. carotovora*		Plant pathogen
Escherichia coli	*E. coli*		Standard organism for genetic studies. Gram-negative intestinal inhabitant, some strains cause serious disease
Haemophilus influenzae	*H. influenzae*		Respiratory tract inhabitant. Can cause serious disease, including meningitis
Helicobacter pylori	*H. pylori*		Stomach inhabitant, causing ulcers and other diseases
Mesorhizobium loti	*M. loti*		Important in agriculture due to its nitrogen fixation activity
Micrococcus spp.			Gram-positive coccus, common on skin
Mycobacterium leprae	*M. leprae*	Leprosy bacillus	Cause of leprosy
Mycobacterium paratuberculosis	*M. paratuberculosis*	*M. avium* var *paratuberculosis*	Animal pathogen
Mycobacterium tuberculosis	*M. tuberculosis*	TB bacillus	Cause of human tuberculosis
Mycoplasma genitalium	*M. genitalium*		Very small bacterium, lacking a cell wall
Myxococcus xanthus	*M. xanthus*		Soil bacterium, showing cooperative behaviour
Neisseria gonorrhoeae	*N. gonorrhoeae*	Gonococcus	Cause of gonorrhoea
Neisseria meningitidis	*N. meningitidis*	Meningococcus	Causes meningitis and septicaemia

Bacterial name	Abbreviation	Other names	Comment
Pseudomonas aeruginosa	*P. aeruginosa*		Gram-negative opportunist pathogen. Widely used in genetics research
Pseudomonas putida	*P. putida*		Free-living bacterium found in water and soil
Pseudomonas syringae	*P. syringae*		Plant pathogen
Rhizobium			Genus important in agriculture due to their nitrogen fixation activity
Salmonella typhi	*S. typhi*		Cause of typhoid fever
Salmonella typhimurium	*S. typhimurium*	*Salmonella enterica* serovar Typhimurium	Common cause of food poisoning. Widely used in genetics research. We have used the old name, which is still more familiar than the new, rather clumsy designation
Shigella spp.			Causes bacillary dysentery
Staphylococcus aureus	*Staph. aureus, S. aureus*		Major cause of hospital infections
Staphylococcus epidermidis	*Staph. epidermidis, S. epidermidis*		Common skin organism, less pathogenic than *S. aureus*
Streptococcus mitis	*S. mitis*		Oral commensal and opportunist pathogen
Streptococcus pneumoniae	*S. pneumoniae*	Pneumococcus	Gram-positive coccus, causes pneumonia, meningitis
Streptococcus pyogenes	*S. pyogenes*		Cause of a wide range of infections
Streptomyces			Gram-positive, filamentous soil organisms. The major source of currently used antibiotics
Vibrio cholerae	*V. cholerae*		Cause of cholera, due to toxin encoded by a prophage

Bacterial name	Abbreviation	Other names	Comment
Yersinia enterocolitica	*Y. enterocolitica*		Food-borne pathogen
Yersinia pestis	*Y. pestis*		Cause of plague

Index

Molecular Genetics of Bacteria, 4th Edition by Jeremy Dale and Simon F. Park
 ISBN 0 470 85084 1 (cased) ISBN 0 470 85085 X (pbk)